GENES, DETERMINISM AND GOD

Over the past centuries the pendulum has constantly swung between an emphasis on the role of either nature or nurture in shaping human destiny – a pendulum often energised by ideological considerations. In recent decades the flourishing of developmental biology, genomics, epigenetics and our increased understanding of neuronal plasticity have all helped subvert such dichotomous notions. Nevertheless the media still report the discovery of a gene 'for' this or that behaviour, and the field of behavioural genetics continues to extend its reach into the social sciences, reporting the heritability of such human traits as religiosity and political affiliation. There are many continuing challenges to notions of human freedom and moral responsibility with consequent implications for social flourishing, the legal system and religious beliefs. In this book, Denis Alexander critically examines these challenges, concluding that genuine free will, often influenced by genetic variation, emerges from an integrated view of human personhood derived from contemporary biology.

Denis Alexander is Founding Director (Emeritus) of The Faraday Institute for Science and Religion at St. Edmunds College, Cambridge University, where he is an Emeritus Fellow. He previously spent fifteen years in the Middle East where he helped establish the National Unit of Human Genetics at the American University Hospital in Beirut. More recently, he has been involved in immunology, genetics and cancer research in the United Kingdom, latterly at The Babraham Institute, Cambridge. Dr. Alexander was previously editor of the journal *Science & Christian Belief* and writes and broadcasts widely in the field of science and religion. He gave the Gifford Lectures at St. Andrews University in 2012.

Genes, Determinism and God

DENIS ALEXANDER

University of Cambridge

CAMBRIDGE
UNIVERSITY PRESS

CAMBRIDGE
UNIVERSITY PRESS

One Liberty Plaza, 20th Floor, New York, NY 10006, USA

Cambridge University Press is part of the University of Cambridge.

It furthers the University's mission by disseminating knowledge in the pursuit of education, learning, and research at the highest international levels of excellence.

www.cambridge.org
Information on this title: www.cambridge.org/9781316506387
DOI: 10.1017/9781316493366

First published 2017

Printed in the United States of America by Sheridan Books, Inc.

A catalogue record for this publication is available from the British Library.

Library of Congress Cataloging-in-Publication Data
Names: Alexander, Denis, author.
Title: Genes, determinism and God / Denis Alexander.
Description: New York, NY: Cambridge University Press, 2017. | "Amplified version of the Gifford lectures given at the University of St. Andrews in December 2012" – Preface. | Includes bibliographical references and index.
Identifiers: LCCN 2017004969 | ISBN 9781107141148 (hardback) | ISBN 9781316506387 (paperback)
Subjects: | MESH: Genetic Determinism | Genetic Variation | Religious Philosophies
Classification: LCC QH447 | NLM QU450 | DDC 572.8/6–dc23
LC record available at https://lccn.loc.gov/2017004969

ISBN 978-1-107-14114-8 Hardback
ISBN 978-1-316-50638-7 Paperback

Contents

v

Preface

This book is an amplified version of the Gifford Lectures given at the University of St. Andrews in December 2012. I am grateful to the then principal of St. Andrews, Professor Louise Richardson, now Vice-Chancellor of Oxford University, for her kind hospitality. My thanks also go to the members of the Gifford Committee for the invitation and to Professor Alan Torrance in particular for hosting and organising the occasion.

The Gifford Lectures were endowed by the will of Lord Adam Gifford, associate judge of the Court of Session, in a ceremony held in Edinburgh on 21 August 1885. A man of generous spirit and liberal values, Lord Gifford's only restraint as to content was that the lectures should address 'natural theology in the widest sense of the term'. Although theology per se does not appear in the present volume until the final chapter, it is the author's contention that a rich discourse of natural theology can only emerge when theology is woven together with many other disciplines including, in the present book, history, sociology, philosophy and, especially, science.

At the time of giving the Lectures I gave acknowledgement to two previous Gifford lecturers whose life and work have been of particular influence in my own thinking and writing. The first is the late Arthur Peacocke, my tutor in biochemistry whilst an undergraduate at St. Peter's College, Oxford, who gave the Gifford Lectures at St. Andrews University in 1992–3. The second is the late Donald MacCrimmon MacKay who gave the Gifford Lectures at Glasgow University in 1986. Donald was born in Caithness in the fishing village of Lybster in 1922 and graduated in physics from St. Andrews in 1943, later to become one of Britain's leading neuroscientists. Not many neuroscientists can claim to publish papers in *Nature* and in the philosophy journal *Mind* within a short space of time. Peacocke and MacKay were amongst the sharpest thinkers in the field of science and religion in the latter half of the twentieth century, representing two very different theological traditions, and yet their thinking was remarkably convergent on a number of key points. I wish to acknowledge my debt to them both, and now dedicate this book in their memory.

I am also hugely indebted to the many people who have helped with the research that provides the bedrock on which this book is based. My special thanks goes to Nell Whiteway who pursued literature searches with great energy and thoroughness during her three-year stint as a staff member of The Faraday Institute for Science and Religion based at St. Edmund's College, Cambridge. Preceding her for a shorter period was Nicole Maturen and, more recently, Lizzie Coyle, all likewise staff members of the Institute, who have both made valuable contributions.

I am also very grateful to those many kind friends and colleagues who have read and corrected draft chapters relating to their own particular areas of expertise, in many cases going the extra mile in fitting reading into tight schedules or in pointing to otherwise overlooked publications: Kathryn Asbury, Duncan Astle, John Coffey, Samuel Cohen, Tim Crane, Andrew Davison, Caroline Eade, John Evans, Jeff Hardin, Rodney Holder, David Lahti, Neil Levy, Hilary Marlow, Harvey McMahon, Alfred Mele, Michael Murray, Adam Nelson, Ronald Numbers, Stephen Oakley, Christopher Oldfield, Julian Rivers, Alan Torrance, Eric Turkheimer, Christopher Watkin and John Wyatt. It goes without saying that all the infelicities that remain are entirely the responsibility of the author. Indeed, in such a wide-ranging book, the danger of factual errors is ever-present and readers are welcome to e-mail any that they might find (to dra24@hermes.cam.ac.uk) with a view to correction in a further edition. Finally, my many thanks are due to Christopher Akhurst for his sub-editing and to Beatrice Rehl, Isabella Vitti, Laura Morris, Kaye Tengco and all the editorial staff at the Press for their support and help in seeing the manuscript published.

Introduction

The purpose of this book is to consider the relationship between genetic variation and human behaviour in the context of ideas about human freedom and determinism. Achieving this aim requires a prior examination of the dichotomous language that has tended to shape discussion on this topic: nature and nurture; hereditarian and behaviourist; innate and learned; genes and environment. It will be suggested that all forms of dichotomous thinking have been thoroughly subverted by recent biological findings, generating a richer and more nuanced picture of human identity, impinging on our understanding of the human as a freely choosing biological organism. A survey of genetic variation amongst organisms possessing nervous systems rather less complex than our own informs the discussion, but certainly does not resolve it.

This is not primarily a book about philosophy, although philosophy is brought into conversation with biology in Chapter 11. The framework here is provided more by biology than by philosophy. Nevertheless, the word 'determinism' appears in the title and requires some definition. For the purposes of this book 'hard determinism' is defined not in the way generally used by philosophers, but rather as the thesis that 'given our particular genomes our lives are not really up to us and are constrained to follow one particular future', where the word 'genome' refers to the sum total of information contained in our DNA. There is also a softer form of genetic determinism which states that 'given our particular genomes our lives are more likely to follow one particular future', which arguably is not really determinism at all. In reality the various positions adopted are located somewhere on a spectrum lying between the two poles provided by these two definitions. We will leave these definitions as 'place holders' until they receive greater attention in later chapters, and in Chapter 11 in particular.

As a biologist, I see 'free will' as a Darwinian trait which all adult humans in good health display in the same kind of way that they are typically characterised by having two arms and two legs. It refers to the universal feeling of up-to-usness that all humans experience during the process of making a decision, unless, that is, they are under the influence of drugs, suffering from a debilitating illness or psychiatrically impaired. Of course, having an experience per se, even one as reliable and persuasive as the daily experience of up-to-usness, is no guarantee of its ontological status. More formally we will therefore define free will as the 'the ability to intentionally choose between courses of action in ways that make us responsible for what we do', and it is this definition which will be discussed in Chapter 11 in the light of genetics, with all that this entails for moral responsibility, the criminal justice system and the structure of human society more widely. In the interim the book focuses on one main question: Are there particular genotypes, that is, sets of genetic variants, that correlate so tightly with certain displays of human behaviour that we are led to the conclusion that we really are 'constrained to follow one particular future'?

Genetic Determinism in Contemporary Discourse

With possible rare exceptions (Cashmore, 2010), there are today no 'hard genetic determinists' as previously defined within the academic biological research community, although there certainly have been in the past, and there are without doubt examples of 'hard genetic determinism' relating to medical pathologies, which will be discussed later. Generally, biologists today take great pains to highlight the role of both genes and the environment in reporting their work, and no one doubts that gene-environment interactions are critical for the development of all living organisms, not least the human. So are we then tilting at windmills by including the word 'determinism' in a book about contemporary genetics? I suspect not. There are two issues here, one relating to the biological research community and the other to the public communication and understanding of science. As far as the research community is concerned, despite the care taken by most biologists to place the publication of their genetic results within a rich discourse of gene-environment interaction, one cannot avoid the impression that for some geneticists, at least, it is the particular genetic variation which is providing the 'real story' as to what is going on in terms of the animal or human behaviour under investigation. It is both the power and the peril of the methodological reductionism – without which biological scientific enquiry would cease to function – that the success of genetics as an explanatory field can leave other valid and complementary levels of explanation unexamined or even unmentioned, thereby opening the way to an implicit, if not explicit, ontological reductionism, a philosophy parasitic on science, though not part of science. There can, therefore, be a creeping 'backdoor determinism' displayed in the language of some academic genetic discourse which gives primacy to the role of the genes almost as a matter of habit. 'Backdoor determinism' also crops up sometimes when geneticists are collaborating with economists, sociologists or criminologists. Academics not used to handling the complexities of quantitative genetics may be tempted to assign greater causative power to the genes than perhaps is warranted by the data.

A news feature in the scientific journal *Nature*, titled 'The Anatomy of Politics – from Genes to Hormone Levels, Biology May Help to Shape Political Behaviour' (Buchen, 2012),well illustrates this point. The author writes that '[a]n increasing number of studies suggest that biology can exert a significant influence on political beliefs and behaviours', reporting that 'genes could exert a pull on attitudes concerning topics such as abortion, immigration, the death penalty and pacifism'. In the article, John Hibbing, a political scientist at the University of Nebraska-Lincoln, is quoted as saying that 'it is difficult to change someone's mind about political issues because their reactions are rooted in their physiology'. We note the dualist language involved and its assumption of determinism. Genes and physiology are seen as something different from 'us' and 'our mind', and they seem to be controlling us, so we cannot even change our mind.

Political commentators and historians appear to find genetic explanations for cultural and political differences particularly alluring, perhaps because their grasp of the genetics does not match their expertise in other academic disciplines. In his book *A Farewell to Alms* (2007), the economic historian Gregory Clark argued that the English came to rule the world because the rich out-bred the poor, contributing more of their 'superior' genes to the conquering nation. In 2014, *A Troublesome Inheritance – Genes, Race and Human History*, by Nicholas Wade, stirred up a hornets' nest with its suggestion that genetic differences between 'the three major races' help explain economic differences between races and 'the rise of the West'.[1] *Plus ça change, plus c'est la même chose*, and Chapters 1 and 2 will track the long historical background that provides the cultural context for such fallacious claims.

But scientists are also sometimes guilty of hyperbole, particularly when it comes to the publicising of notable scientific advances. The publication of the full human genome DNA sequence in 2004 provided ample opportunity. Metaphors for the genome such as 'the Holy Grail', 'the Book of Life' and 'the Code of Codes' were all used. The 'blueprint' metaphor became very popular, replacing older, less deterministic terminology such as 'genetic lottery' (Condit et al., 2009). Walter Gilbert, who first used the phrase 'Holy Grail' to describe the genome at a conference at Los Alamos in 1986, and who was one of the foremost promoters of the Human Genome Project, described its potential with this graphic image: '[O]ne will be able to pull a CD out of one's pocket and say, "Here is a human being; it's me!" ... To recognize that we are determined, in a certain sense, by a finite collection of information that is knowable will change our view of ourselves. It is the closing of an intellectual frontier, with which we will have to come to terms' (Gilbert, 1992, p. 96). No equivocation there. In 2012 the first wave of thirty papers reporting the results of the Encyclopedia of DNA Elements (ENCODE) project were published. ENCODE 'aims to map all the functional sequences of the human genome. The main introductory paper in this series begins its Abstract by emphasising that the "human genome encodes the blueprint of life"' (Dunham et al., 2012). The genome in popular scientific literature is often referred to as 'an instruction manual', giving the impression that the human body is assembled from the manual much as you might put together a piece of furniture from the kit supplied.

Blatant narratives of genetic determinism are perhaps most clearly seen in the media reporting of the latest discoveries in genetics, and public assumptions about the deterministic roles of genes are proving remarkably resistant to change, even when those assumptions are no longer generally held within current academic biological discourse (Moore, 2008). The possible role of a variant gene in some variant human trait is often reported as the discovery of a gene 'for' this, that or the other – there are mean genes, gluttony genes, gangster genes, liberal genes which cause you to read *The Guardian* and even the whimsical suggestion of a 'geneticism gene' that predisposes some people to think that behaviour is caused by genes. Some sample media headlines illustrate the point: 'Binge-Drinking Gene Discovered';[2] 'Study Links Spread of Religion with "Believer Gene"';[3] 'Study Shows How to Tell If that Man in your Life Has Caring Genes';[4] 'Teen Survey Reveals Gene for Happiness';[5] 'The Science of Stress – Does Your Child Have the "Worrier" Gene?';[6] 'Happiness Gene Is in Britain's DNA';[7] 'Exam Success May Be Due to a Handful of Genes';[8] and so forth. An interview with the singer Sinead O'Connor was headlined with a quotation from the singer: 'I have no shame. I don't have an embarrassed gene'.[9] In 2006 an Australian Associated Press article began by stating that 'New Zealand

Maori carry a "warrior" gene which makes them more prone to violence, criminal acts and risky behaviour, a scientist has controversially claimed' (Kowal and Frederic, 2012). Even sober academic journals such as *Nature* can seemingly not resist the temptation to compress a complex genetic finding into such attention-grabbing headlines as 'Ruthlessness Gene Discovered'[10] or 'A Gene for Impulsivity',[11] even though the authors of the scientific papers whose work is being publicised studiously avoid such language. Discussing the tendency that many people drink alcohol at times of stress, *Newsweek* reassured readers that 'if this is you, don't blame yourself. Blame your DNA'.[12] Another widely read newspaper asks, 'Could it be that binge eaters really can't help themselves? A new study says that weak genes – not weak willpower – may be the reason some people compulsively overeat'.[13]

Despite sporadic protests made by scientists and science communicators concerning the continuing prevalence of such misleading language, the fight for readers and the general 'dumbing down' of media stories in a news-hungry world together suggests that such stories and headlines may be difficult to dislodge. The public language of genetic determinism may therefore partially be blamed on the present parlous state of the news media with all the accompanying pressures to hype up stories in misleading ways, sometimes dubbed 'genohype'. Some attempts have been made to assess whether the presence of 'genohype' in the media has itself been over-hyped. In one study, an investigation was carried out on 627 newspaper articles from the more serious broadsheets produced in four different countries reporting on 111 papers published in 24 scientific and medical journals during the period 1995–2001 (Bubela and Caulfield, 2004). The investigators assigned the newspaper articles to one of three categories: 'moderately to highly exaggerated claims', 'slightly exaggerated claims' or 'no exaggerated claims'. Only 11 per cent of the articles were categorised as having moderately to highly exaggerated claims, with 26 per cent assessed as being in the slightly exaggerated category. Also of interest in the present context was the finding that stories about behavioural genetics or neurogenetics were over-represented in newspaper articles, comprising 16 per cent of the articles compared to a roughly estimated 1 per cent of all academic press genetic articles at the time being on these topics. This might help explain why the reader gains the impression that much of genetics research is directed towards explaining human behaviour; such copy makes newsworthy stories for obvious reasons.

The Influence of the Media on Public Attitudes

Sociological attempts have been made to assess the impact of the media language of genetic determinism on public attitudes. The Condit research group and others have shown that the public's view on the causes of behaviour are often confused and contradictory, as they seek to incorporate a number of media narratives, life experiences and scientific findings into their worldview (Condit et al., 2009; Condit, 2010; Jayaratne et al., 2009). Much data suggest that the stories promulgated by the kind of 'elite media' stories cited previously do not act like 'magic bullets' to be instantly absorbed by the reader, but rather are resisted, critiqued or accepted depending on the reader's economic interests, health and social status and access to competing discourses. A recurring theme is that people display a 'two-track model' in which they can readily switch between more genetic deterministic explanations for disease or different behaviours and those which favour environmental factors or human choice (Condit et al., 2009). The two 'tracks' are often presented in contradictory ways: one moment respondents agree with statements that genes are completely determinative causes in conditions such as heart disease or diabetes, whereas the next they are willing to state that such conditions can be avoided by diet or exercise (Condit, 2010). When people are forced to think about specific cases of known links between genetics and disease, they test higher on subsequently administered genetic determinism questionnaires (Smerecnik et al., 2009). Generally such studies find little appreciation of gene-environment interactionism, the 'two-track' discourse tending to see the two kinds of explanation as mutually exclusive. There is also a tendency to see the determinative contribution of genetics as being greater for bodily traits, whereas environmental influences are seen as more dominant for matters relating to the mind (Condit and Shen, 2011). Figure 1 illustrates this tendency by combining results from three different surveys.

On the particular question of the impact of media stories that highlight genetic determinism, people seem to deploy elements of fatalism or determinism into their worldviews or life goals when they suit particular ends, either in ways that are thought to 'explain' why other groups are the way they are or in ways that lessen their sense of personal responsibility (Condit, 2011).

The proliferation of direct-to-consumer (DTC) genetic testing companies has also contributed to the idea that it is the genes that are pulling the strings of human destiny. On the whole, statements on genetics in relation to the environment are made on company websites in a reasonably

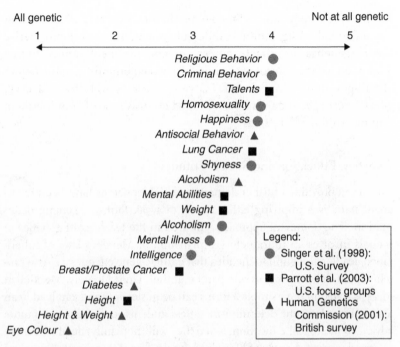

FIGURE 1. Results of three studies of public understandings of the role of genetics and non-genetic factors in the causation of human characteristics and behaviours transformed to a common standard to facilitate comparison. (From Condit, 2011 fig. 1, p. 625).

judicious way. But occasionally claims are made with distinctly deterministic overtones. As the Map My Gene website assures us: 'Genes have also been found to dictate the talents and abilities of people, which serve to explain why some people appear to be naturally-gifted in performing certain tasks while others apparently cannot get the same done despite persistent attempts'.[14] The idea of a genetically determined destiny is reinforced by sperm banks that suggest that prospective users should consider the donor's educational record, his athletic prowess, hobbies and favourite foods, as if these were somehow written into the genetic script provided by the sperm. Human eggs can likewise be purchased online with accompanying details about the donors.

Besides the company websites that provide information about DTC testing, traditional mass media are the other main source of information. 'Time and Newsweek magazines have devoted numerous covers to the subject, typically framing it as a miracle technology that will revolutionize the practice of medicine' (Rahm et al., 2012). News stories are presented using multiple different frames, such as a progressive frame (genetics as miracle cures), an empowerment frame (testing allows you to take control of your health) or a deterministic frame (genes are all-powerful): 'Studies of DTC genetic testing websites and DTC advertising of genetic tests have ... found excessive use of empowerment framing' (Rahm et al., 2012).

Genetics, Education and Social Attitudes

Education provides a further domain in which concerns have been raised about narratives implying either strong or weak forms of genetic determinism. The dominant approach still used in the teaching of genetics in schools involves an introductory explanation of Mendel's laws of inheritance, leading the pupil to the idea that a single gene encodes a single protein which in turn controls one particular trait. Unfortunately, Mendelian ideas as stated in their simplest forms can be misleading and can lead to an assumption of genetic determinism, unless students continue on with more advanced biological education, when they will (hopefully) acquire a more nuanced picture. A study of French textbooks 'found that direct, linear and causal genetic determinism is the interpretative model most often associated with genetic diseases', although more recent textbooks spoke more of polygenic disease and the role of the environment (Castera et al., 2008). A comparison of school pupils in France and Estonia in the sixteen to eighteen age range found that there was a greater level of genetically deterministic beliefs amongst the Estonian compared to the French students (Castera et al., 2013). For example, 32 per cent of the Estonians but only 10 per cent of the French 'agreed or rather agreed' with the statement that 'Ethnic groups are genetically different and that is why some are superior to others', whereas 40 per cent of Estonian students 'agreed or rather agreed' with the statement that '[i]t is for biological reasons that women more often than men take care of house-keeping' compared to 10 per cent for the French. In general the investigators found a correlation between genetically deterministic beliefs and traits such as sexism and racism, speculating that the lower correlations found in France may partly be explained by the fact that philosophy is a compulsory course in the last grade of French secondary school, a course that tackles the topic of determinism. An alternative genetic pedagogy has been suggested from within the UK educational system that might give a more nuanced perspective as to how genes work in the development of traits (Jamieson and Radick, 2013).

Similar correlations between beliefs in genetic determinism and intolerant attitudes have been reported amongst university students. Assessed according to a social dominance orientation (SDO) scale, which is said to measure 'the degree to which individuals desire and support group-based hierarchy and the domination of "inferior" groups by "superior" groups' (Sidanius and Pratto, 1999, p. 48), beliefs in genetic determinism ('geneticism') in a group of German university students was found to strongly correlate with a variety of ideologies including sexism, racism and high scores on the SDO scale (Keller, 2005). Participants who were psychologically primed to be more open to genetic explanations for human differences also displayed more prejudice and in-group bias. Similar results were reported based on studies of groups of students from Blaise Pascal University in France (Dambrun et al., 2009). The more the participants believed in genetic determinism, the higher their SDO scores and the greater was the correlation with prejudice towards Arabs and the poor, together with support for the death penalty. Because in the social sciences there is more of a focus on social and environmental factors compared to genetic factors, the investigators predicted that university exposure to a psychology course would lead to the perception that genetic variables play a less important role than environmental ones, and their results confirmed this hypothesis. While genetic determinism scores of first-year students were significantly higher in psychology than in biology, after three years of university exposure they were significantly lower in psychology than amongst those studying biology, for whom determinism scores stayed the same (Dambrun et al., 2009).

These and many other reports suggest that beliefs in the power of the genes to determine social identities and future destinies are not merely neutral, but correlate with a broad array of social and political attitudes. Of course, correlation should not be equated with causation, a recurring theme throughout this book, and one can argue that people with particular social and political beliefs may be attracted to genetic determinism precisely because it provides apparent justification for those beliefs. But results from the kind of longitudinal studies cited previously from the work of Dambrun et al. do support the idea that beliefs about genetic determinism play a causal role in changing social attitudes. On the other hand, people with racist attitudes can shift from genetic to cultural accounts for perceived differences among groups without decreasing their level of racism (Ramsey et al., 2001; Lynch et al., 2008), so if someone really wants to be a racist, then it seems that they will draw their justification from any convenient resources that may be available.

More powerful (in this context) than formal education is the process of 'cultural osmosis' whereby knowledge of, and attitudes towards, genetics are absorbed from many sources. In a critique of what he dubs 'genetic essentialism', one reviewer comments that '[l]earning about genetic attributions for various human conditions leads to a particular set of thoughts regarding those conditions: they are more likely to be perceived as (a) immutable and determined, (b) having a specific etiology, (c) homogeneous and discrete, and (d) natural, which can lead to the naturalistic fallacy' (Dunham et al., 2012). In a U.S. study based on focus groups, participants cited a wide range of media that impinged on their understanding of genetics (Bates, 2005). Participants drew, for example, on sci-fi texts. Others cited television. One participant reported that she had watched a TV news magazine show on genetics, saying, 'I think it was Dateline or something like that, and they were talking about genetic makeup, where they wanted the blue eyes, blonde hair.... They can make sure that that would happen. It's all taken care of. I mean you can determine what your child looks like.' Another participant explained that it was likely that parents would choose Nordic traits for their children and deselect other traits in a quest for 'perfection'. Certain films had also clearly influenced participants' opinions in the direction of genetic determinism. Several cited the film GATTACA, one participant remarking, 'I don't know if anyone saw the movie GATTACA, where, basically, if you aren't the best, then people start to manipulate their children's genes and almost order what they want, like a package deal.' The author of this study emphasises the nuanced way in which focus group participants weave genetic narratives out of a wealth of cultural referents which are processed to support their claims, not simply absorbed as 'bare facts', suggesting that linear assumptions of media influence on the public understanding of genetics may be overstated (Bates, 2005). Yet despite or perhaps because of such processing, many of the comments made did display a strong subtext of genetically deterministic thinking.

One indication of the iconic profile of DNA language in public cultural discourse is that the phrase 'it's in his/her DNA' has come into common usage in all kinds of contexts, some rather odd. As Brad Pitt told the *Daily Mail* in 2012 whilst discussing U.S. gun control: 'America is a country founded on guns. It's in our DNA' (Pitt, 2012). The Cloud Computing service provider Oxygen assures us that 'for Oxygen, security is in our DNA. The security of you and your company's data will always be our priority' (Mak, 2011). In 2012, the year of the London Olympics, the director of the Design Museum, Deyan Sudjic, suggested that 'London as a whole has been strengthened in its claims to be

Europe's only real world city. It's not the Olympics that have done that; it's London's 2000 years of urban DNA' (Sudjic, 2012). In commenting on its new TV drama series, the director-general of the BBC was quoted as saying that 'drama is something that is in the lifeblood of this country and in the DNA of the BBC too'.[15] The presumed implications of such language are clear: what is in the DNA must be immutable and unchangeable – somewhat missing the point that our DNA is undergoing a constant process of change and diversification.

Does all this outpouring of the language of DNA in popular culture act as a marker for deterministic thinking? It's hard to say. But at the least it should act as a reminder of the way in which the language of science can be absorbed into public discourse and be deployed in ways that lie well beyond science. Given the long history of the ideological abuse of genetics, a topic touched on in Chapter 2, one cannot necessarily assume that such misuse of language is merely benign. Cultural osmosis is a powerful process in shaping attitudes, be they expressed in the context of politics, social attitudes, economics, sport or religion. Although the main aim of this book is to investigate the role of genetic diversity in differential human behaviours, and whether, as a matter of fact, purported roles are validated by the available data, the considerable ideological investments often made in the outcomes of such assessments should act as a warning that, in this branch of science more than others, the scope for use and abuse remains particularly large. More examples illustrating this point will be given as different topics are addressed throughout the book, including intelligence testing, criminology, sexual orientation, religiosity and politics. The investigators who entitle their publication 'The Heritability of Foreign Policy Preferences' (Cranmer and Dawes, 2012) cannot seriously expect that their paper will be treated as 'pure science'.

Overall, therefore, 'genetic determinism', with all its various shades of meaning, remains a lively topic in public discourse, and the outcome of the discussion is not merely academic. In addition to the correlation of genetically deterministic beliefs with non-egalitarian attitudes, and the implications of discoveries about genetic variation for a wide range of social attitudes and practices, there is also abundant evidence that beliefs concerning the fixity of human identity, be it for perceived genetic or environmental reasons, have a remarkably negative impact on human flourishing. Of course the truth or falsity of beliefs does not hinge on their consequences, even though those may be negative. On the other hand, given the history of ideological abuse of genetics, it is as well to be very sure about scientific claims before making them available for public dissemination.

The Social Impact of Deterministic Beliefs

Psychologists have carried out many experiments in which subjects are influenced in subtle or not so subtle ways to disbelieve in free will, and the social consequences are then measured under controlled conditions. In one study, subjects started by reading either a text promoting the idea of determinism, claiming that scientists now realise that free will is an illusion, or a neutral text, before being tested for either passive or active cheating (Vohs and Schooler, 2008). The passive cheating opportunity arose from an experimental set-up in which the correct answer to a mathematical question appeared on a computer screen. Active cheating was measured by means of subjects paying themselves financial rewards on a difficult cognitive test when no one was looking. In both cases a higher level of cheating was measured amongst those with prior exposure to the 'free will is an illusion' text. In a separate study, three different experimental protocols revealed that subjects previously exposed to pro-determinism texts rather than pro-free-will texts were less likely to be pro-social and more likely to engage in antisocial behaviour (Baumeister et al., 2009; Stillman and Baumeister, 2010). In a further study, brain recordings were carried out on two groups of subjects who had previously been exposed to either pro-determinism or pro-free-will texts to measure the readiness potential which is associated with the decision-making process (Rigoni et al., 2011). The early phase of the readiness potential was found to be reduced in those subjects exposed to the 'free will is an illusion' text, consistent with the measured behavioural changes. Furthermore, undermining free will can degrade self-control, perhaps helping to explain the increase in antisocial behaviour (Rigoni et al., 2012).

It could be argued that controlled laboratory experiments do not accurately reflect the social consequences of beliefs about determinism and free will found in normal everyday life. However, studies investigating beliefs about free will have tended to confirm the conclusion that they correlate with generally positive social effects in daily life. For example, possessing a belief in free will, but not several other key social beliefs, predicted better career attitudes and actual job performance amongst workers as assessed by their supervisors (Stillman et al., 2011). Belief in free will also appears to influence people's willingness to think independently of group opinions – a stronger belief in free will was associated with a higher measure of non-conformity in one rather small online survey (Alquist et al., 2013). Using a similar online methodology, Vonasch and Baumeister (2013) found that belief in free will correlated with an increased belief in social mobility and greater sympathy for the disadvantaged. Investigations using a much larger online sample found that increased belief in free will moderately correlated with belief in a just world, intrinsic religiosity and more moral-based attitudes towards judging both self and others (Baumeister and Brewer, 2012; Carey and Paulhus, 2013). Parents who believe that parenting does not make much difference to their children's outcomes tend to have children with worse outcomes (Baumrind, 1993). Whereas such observations carry the usual caveat that correlation does not entail causation, a reasonable interpretation is that belief in free will supports convictions about non-conformity and personal moral responsibility, reflected in increased conscientiousness in the workplace and in parenting, together with a heightened sense of justice.

Experiments on individual or group labelling in ways that implant either aspirational or, conversely, fatalistic attitudes provide further data illustrating the power of suggestion to shape human behaviours. In a process known as 'stereotype threat', presenting an intellectual test as diagnostic of ability reduced the scores obtained from African American students as compared to white U.S. respondents, whereas the scores of African Americans were likewise lower than other African Americans who performed the same test after being told that it was non-diagnostic (Steele and Aronson, 1995). Similarly, the giving of more positive or negative labels to different groups of schoolchildren who had previously been shown to be equivalent in cognitive and academic testing revealed that they then performed according to the expectations implied in the label given to their group. 'Cognitive simulations in which individuals visualize themselves executing activities skillfully enhance subsequent performance' and the opposite is also the case, as every sports trainer knows (Bandura, 1989). Personal goal setting is influenced by self-appraisal of capabilities, and the higher that perception the firmer is the commitment to achieving the goal. When Decca records turned down a recording contract with the Beatles on the grounds that '[w]e don't like their sound. Groups of guitars are on the way out', it was fortunate that the Beatles had sufficient belief in their capabilities to persevere (Bandura, 1989).

'Genes, Determinism and God' – an Outline

The first two chapters of this book track dichotomous ideas about human nature all the way from the Babylonians, Aristotle and Plato, through to contemporary molecular biology and human behavioural genetics. Chapter 3 then provides a survey of the way in which it is recent discoveries in biology, more than anything, which subvert such dichotomous notions of personhood. Chapter 4 suggests an alternative framework, drawing on contemporary science, for understanding how human personhood develops through a complex multi-causal process. Chapter 5 illustrates the way in which similar developmental processes also operate in animals, even in those invisible to the naked eye. With this material providing the essential scientific background, Chapters 6–10 then explain the underlying methods and assumptions of quantitative human behavioural genetics, providing a sympathetic assessment along the way (Chapter 6) before delving into the molecular genetics of human behaviour in Chapter 7, and providing a critical survey of the applications of behavioural genetics in the study of intelligence, religion and politics (Chapter 8), sexual orientation (Chapter 9) and criminality (Chapter 10). Chapter 11 tackles some of the philosophical challenges of the topics surveyed, bringing biology into conversation with philosophy and arguing that both science and philosophy are consistent with the ontological reality of free will rather than providing support for any particular form of neuro- or genetic determinism. Finally, Chapter 12 brings genetics into conversation with theology, with the concept of humankind being made 'in the image of God' providing a bridge for the transmission of a positive and fertile interchange of ideas between science and religion.

This is a 'both-and' book. Those who prefer confrontational 'either-or' discourse should look elsewhere.

CHAPTER 1

Human Personhood Fragmented?

Nature-Nurture Discourse from Antiquity to Galton

> For good nurture and education implant good constitutions, and these good constitutions taking root in a good education improve more and more, and this improvement affects the breed in man as in other animals.
>
> Plato[1]

> Of all the vulgar modes of escaping from the consideration of the social and moral influences on the human mind, the most vulgar is that of attributing the diversities of conduct and character to inherent natural differences.
>
> John Stuart Mill, 1848 (2004)

> It is well ascertained that many persons are born with such natures that they are almost certain to become criminals. The instincts of most children are those of primaeval man; in many respects thoroughly savage, and such as would deliver an adult very quickly into the hands of the law. The natural criminal retains those same characteristics in adult life.
>
> Francis Galton (1890)

The idea that 'nature and nurture' or 'acquired and innate' provide an adequate framework for understanding human personhood was already being declared redundant by biologists within a few decades of the popularisation of such dichotomous language by Francis Galton in the 1870s. One such biologist was Leonard Carmichael whose academic career in psychology and biology included becoming president of Tufts University (1938–52). In a prescient paper published in 1925, Carmichael wrote:

In man, from the first environmental stimulation of the fertilized ovum until, it may be, well past three score and ten years, the human individual is not made up of two substances: one acquired, the other innate. The human organism and personality, rather, is a unity produced by both of these forces. The unique resulting total-ity cannot profitably be violated by a destroying analysis, and dichot-omized as part native and part acquired ... the question of how to separate the native from the acquired in the responses of man does not seem likely to be answered because the question is unintelligible. (Carmichael, 1925)

Unfortunately such insights were quickly lost in the subsequent battles to maintain the hegemony of either 'nature' or 'nurture', as if these were some reified forces struggling over the destiny of human personhood, as will be described further.

In the latter decades of the twentieth century, and to the present day, the death of the 'nature-nurture' debate was announced by biologists at regular intervals. Nevertheless, Galton's 'convenient jingle of words', as he dubbed his phrase 'nature and nurture', has proved remarkably impervi-ous to extinction in public discourse. Browse the library shelves or your favourite online bookseller and you will find a sizeable collection of recent books with titles such as *Nature or Nurture* (Dowling, 2004), *Nature and Nurture* (Keating, 2013), *Nature via Nurture* (Ridley, 2003), *Nurture the Nature* (Gurian, 2009), *The Nurture of Nature* (Wall, 2009), *The Nature of Nurture* (Wachs, 1992), *Nature versus Nurture* (Collins, 2002), *Nurture by Nature* (Tieger and Barbara, 1997), *Nurture through Nature* (Warden, 2007), *The Nature-Nurture Debate* (Brewer, 2006) and *The Nature and Nurture Debates – Bridging the Gap* (Goldhaber, 2012). Wikipedia reports the discussion in terms of 'Nature versus Nurture'.[2] One cannot avoid the impression that the possible ways of combining the two words might be close to saturation.

Historical searches for the first use of the words 'nature' and 'nurture' have become something of a cottage industry. Here we take the standard view that contemporary discussions about nature and nurture have their roots in the writings of Francis Galton in the late nineteenth century. But in any event, our main concern is not with that particular histori-cal question, but rather with the broader ideological concerns over the centuries that have rendered changing views concerning human identity either more less plausible. Does the historical debate help us understand the enduring appeal of describing humanity within dichotomous lin-guistic frameworks?

Astrology, Gods and God in the Ancient World

As far back as we go in human history we find a fascination with the factors that determine human fate. For the Babylonians their fates were in the stars together with the planets, the sun and the moon. As Babylonian texts from the seventh century BC point out: 'If the Northern Fish (Mercury) comes near the Great Dog (Venus), the king will be mighty and his enemies will be overwhelmed', or 'If Mars stands in the house of the Moon (and there is an eclipse), the king will die and the kingdom will become small' (Needham and Wang, 1956, p. 353); no equivocation here about the ultimate rulers of human destiny. And as Joseph Needham points out, the Babylonian texts have some remarkable parallels in ancient Chinese literature.

More than a thousand years later, Augustine of Hippo (354–430) critiqued the idea that the stars played a role in human destiny in one of the earliest texts that discusses the similarities between twins. Augustine wished to critique those who credited the success and longevity of the Roman Empire to the stars, instead ascribing this to God's universal providence. Those who wrote in the name of the famous physician Hippocrates had reported that the disease pattern in a pair of male twins was remarkably similar, an observation which the Stoic Posidonius then explained by the fact that they must have been conceived under the same star (at a time when the difference between identical and non-identical twins was unknown). Not so, said Augustine, in Book V of his *City of God*, for look at Jacob and Esau, the most famous twins of the Old Testament, how different were their lives, and look at how different are the lives of contemporary twin pairs (Augustine and Dods, 2009). By such arguments did Augustine subvert the claims of the astrologers, declaring that '[a]strology is a delusion' and explaining similarities between twins using the environmental argument that 'all that was necessary for the growth and development up till birth' of the twin fetuses was 'supplied from the body of the same mother' so that 'they might be born with like constitutions'.

Plato and Aristotle

Well before Augustine, Plato (427–347 BC) and then his pupil Aristotle (384–322 BC) were beginning to establish contrasting contours of the philosophy of mind in ways that continue to resonate down to the present day. In his dialogue *Phaedo*, Plato uses Socrates' 'theory of recollection' to strengthen his case for an immortal soul. This theory entails that the soul must obtain a priori knowledge before it becomes united with a new human body. And this, says Plato, can readily be demonstrated by interrogating people and finding that they have innate knowledge, which they can recall from memory (Plato and Bluck, 1955). The hard disk, as it were, comes pre-loaded with both software and content. By contrast Aristotle maintained that the soul, the organ for thinking, contains no innate thoughts and the mind only starts to exist when the soul starts thinking: 'What it thinks must be in it just as characters may be said to be on a writing tablet on which as yet nothing actually stands written: this is exactly what happens with mind' (Aristotle and Ross, 1961). The soul comes with the software, the potentiality, but no knowledge base. Thus was born the idea of the mind as a tabula rasa, a blank slate, a term found often in thirteenth-century European literature following the translation of Aristotle's *De Anima* into Latin from its Arabic version by Michael Scot in around the year 1220 (Aristotle, 1994).

A further contribution of Platonic thought to later debates on nature and nurture is his particular understanding of the words 'nature' or 'essence', virtually synonymous in his writings. For Plato the nature of something was a reflection of, or participation in, its Form or Idea, the real and transcendent essence of that kind of thing, existing in a celestial realm. In turn nature was linked to 'race' – the 'race of philosophers' are those who have 'philosophic natures'. The nature of a person should therefore define his class and occupation, for, as Plato says, 'I am myself reminded that we are not all alike; there are diversities of natures among us which are adapted to different occupations ... we must infer that all things are produced more plentifully and easily and of a better quality when one man does one thing which is natural to him and does it at the right time, and leaves other things' (Plato, 2012, p. 151). In his *Timaeus*, Plato places in the mouth of Socrates flattering words about his discussants 'that people of your class are the only ones remaining who are fitted by nature and education to take part at once both in politics and philosophy' (Plato and Zeyl, 2000). The phrase 'nature and education' here is sometimes translated as 'nature and nurture'. Today we speak of 'human nature' or a 'doggy nature' or the 'nature of mathematics' or the 'denaturation of proteins', perhaps unaware of the deep platonic roots of our language.

There were other aspects of Plato's thought that tended to be treated somewhat benignly in earlier centuries, but which took on a darker hue when viewed through the lens of twentieth-century eugenic history.[3] With royal blood in his veins, Plato was raised during the Peloponnesian War in which Athenian democracy was pitted against Sparta, the main city-state of the Peloponnese, still wedded to an ancient tribal aristocracy. The war brought famine, epidemics, the fall of Athens and its subsequent rule by two of Plato's uncles during a civil war in which both lost their lives. Small wonder that Plato's philosophy sought unchanging eternal realities to be reflected in an unchanging state free from the corruption of decay in which stability reigned and each class knew its rightful place. Indeed, Plato's vision of the perfect state equated to the original state from which all other states had descended, albeit in various degenerate forms. Plato's *Republic* is therefore a story not of progressive evolution towards a utopian future, but of restoration back towards an uncorrupted beginning (Plato, 2012). The original father of degenerate things in flux, such as states that have fallen into tyranny, is the Form or Idea which is perfect and does not perish. The physical world is composed of the shadows of the ideal Forms which represent the ultimate reality. The essence of 'sensible things' (what we can touch and feel) can only be discerned by reference to their original Forms.[4]

Plato's political challenge was therefore to envision a state that was so close to the original Form that it would not be subject to decay. Democracy was clearly out of the question because it was a recipe for change. Instead, Plato's city-state, so close to perfection that it would be unlikely to change, entailed a rigid class system in which there were just three classes: the guardians, vastly superior in race, education and values, and then the warriors and the working class. The guardians are philosopher-kings who would rule the state in wisdom. Class war is avoided by giving the ruling class such power that an uprising from the 'cattle', as Plato called the lower classes, would be impossible. The task of the philosopher-kings is to herd the cattle. Only the ruling class is permitted to carry arms and to receive an education. The role of workers is to supply the needs of the ruling class through trade and manual labour.

Plato's ideal city-state faced the challenge as to how to preserve the static nature of the class system and in particular the internal unity of the ruling class. This would happen by a strict programme of training, by the elimination of economic interests and by the destruction of family loyalties. The communal raising of children from birth would ensure that no member of the ruling class would be able to identify his children, nor his parents, and infanticide would be practised to help in preserving the purity of the class. As Plato comments in his summary of the *Republic* in *Timaeus*, 'each should look upon all as if belonging to one family' (Plato and Zeyl, 2000, 18 c/d). The nurture of young children of the ruling class was robust and they should be raised to experience the life of a warrior first-hand. Children of both sexes must be 'taken to see the battle on horseback; and ... if there were no danger they were to be brought close up and, like young hounds, have a taste of blood given them' (Plato, 2012, p. 266). No playing down of the role of nurture here. As Plato points out, with a spin that we would now call Lamarckian: 'For good nurture and education implant good constitutions, and these good constitutions taking root in a good education improve more and more, and this improvement affects the breed in man as in other animals' (Plato, 2012, p. 188).

As we preserve the purity of domestic animals by breeding, Plato reminds his readers, so we should not neglect the purity of our own race. To help people understand the origins of their class and to stay loyal to their city-state, Plato proposed to tell a big utilitarian lie, a Myth of Blood and Soil, something that he himself labelled as 'audacious fiction'. People were to be told that their upbringing and training was really a dream and that in reality they had been formed in the 'womb of the earth … where they themselves and their arms and appurtenances were manufactured; when they were completed, the earth, their mother, sent them up; and so, their country being their mother and also their nurse, they are bound to advise for her good, and to defend her against attacks, and her citizens they are to regard as children of the earth and their own brothers' (Plato, 2012, pp. 183–4). Plato's metaphor for pure class breeding was that 'God … has put gold into those who are capable of ruling, silver into the auxiliaries, and iron and copper into the peasants and the other producing classes' (Plato, 2012, p. 27).

Plato's audacious fable would hardly be worth mentioning but for the fact that it resonates so powerfully with the hereditarian and eugenic narratives that gained such traction from the late nineteenth century and onwards. Plato's own eugenic programme involved a further fabrication in which defective offspring arising from marriages arranged by the rulers would not be blamed on the organizers: 'We shall have to invent some ingenious kind of lots', writes Plato, 'which the less worthy may draw on each occasion of our bringing them together, and then they will accuse their own ill-luck and not the rulers' (Plato, 2012, p. 214). 'We do Plato the gravest of wrongs', says A. E. Taylor, 'if we forget that the *Republic* is no mere collection of theoretical discussions about government … but a serious project of practical reform put forward by an Athenian patriot, set on fire, like Shelley, with a "passion for reforming the world"' (Taylor, 1908, p. 122).

Aristotle's long shadow, like that of Plato, extended itself over the following millennia to provide much of the discourse and the very meanings of words that were later to frame the varied understandings of the human person, not least because of his huge influence in the early formation of what we now call the biological sciences. In a splendid piece of early Victorian hyperbole, the great anatomist Richard Owen pronounced in 1837 that 'Zoological Science sprang from his [Aristotle's] labours, we may almost say, like Minerva from the Head of Jove, in a state of noble and splendid maturity' (Owen and Sloan, 1992, p. 91).[5]

In Raphael's great painting *The School of Athens*, Plato points upwards to the unchanging Forms, not of this world, whilst Aristotle's hand is held flat, parallel with the surface of the earth in which he took such detailed interest. Though Aristotle's essentialism ran deep, it was both more nuanced and more practical than that of Plato, being 'spiritually' far less ambitious (Shields, 2012). The genuinely necessary features of reality can be uncovered. One category of properties of things is known as *idia* (or *Propria* in medieval Latin): 'non-essential properties which flow from the essence of a kind, such that they are necessary to that kind even without being essential' (Shields, 2012). So the *idia* of things can be described without any necessary reference to their ultimate nature or essence. This is what we find in Aristotle's *History of Animals* in which he organizes an immense array of observations about animal structure and behaviour without any recourse to terms such as 'nature', 'essence', 'form' or 'necessity' (Lennox, 2009, p. 425). It is a work of descriptive classification. But in Aristotle's explanatory work *On the Parts of Animals*, the approach is quite different. Here the language focuses on the functional necessities that flow from the nature of the organism's 'being what it is', full of essentialist and teleological language. 'Not all animals have a neck, but only those with the parts for the sake of which the neck is naturally present.... Now the larynx is present by nature for the sake of breathing ... for everything that Nature makes is means to an end' (Aristotle and Lennox, 2001). No one characteristic of an organism, says Aristotle, can in itself define the essence of an organism, but a list of many characteristics gets closer.

Bound up with such discussion is Aristotle's influential theory of causation, a topic which features multiple times in later chapters (especially Chapter 11). Aristotle maintained that a proper explanation for something entails four aspects:[6] material, formal, efficient and final, using the question, 'Why is there a statue?' to make his point. The statue being made of bronze is the material cause of the statue. The shape or form of the statue is its formal cause. Aristotle did not hold to Plato's eternal Forms, rather his formal cause refers to the being-in-itself (the 'constitution') of the statue. The form of statues as a class of things has certain characteristics and the sculptor made sure that his particular statue conformed to the class. Transferred to biological organisms, the idea of Aristotelian form then describes its being as a whole.[7] Living things are more than the sum of their parts.

The efficient cause of the statue is the sculptor, the one who makes the matter take the form that it has. Efficient causes are the movers and shakers, those people or things that make things happen. Out of Aristotle's four, they provide the main understanding of causation in contemporary scientific discourse. A does something to B which causes something else to happen. Aristotle, however, would have seen this type of explanation as impoverished without his notion of final cause. Why is the statue there? Because the sculpture wished to commemorate the wisdom of Socrates. This is the *telos* of things, their ultimate end or goal. Contemporary biology, unlike other branches of science, remains saturated with such teleological language. The beaver builds the dam to protect its home from predators. The male peacock displays its plumage to attract a mate. Natural selection has entailed that giraffe necks become longer for the purpose of feeding off high branches.

In an oft-quoted saying, 'Everyone by nature is a disciple either of Plato or of Aristotle'. Though the alternatives may not be quite that stark, the fact that this assertion is repeated so often is at least a reminder of the immense influence on the history of western thought of these two giants of Greek philosophy. Their contrasting views of the human person continue to resonate in biological understandings of personhood right down to the present day.

The Medieval and Early Modern Period

In the late eleventh century a Benedictine monk called Constantine (fl. 1065–85) made his way to the monastery of Monte Cassino in southern Italy. There he began to translate medical works from Arabic to Latin, the language of European academia – these included the works of Galen and the so-called Hippocratic School, so supplying the foundations of medical literature on which the West would build for many centuries.

These early translations whetted the European appetite for more. Beginning in the first half of the twelfth century, translation of ancient classical works into Latin – first from Arabic, later from the original Greek – became a major scholarly activity, with Spain as the main geographical focus. By the thirteenth century a flood of works by Aristotle were translated, and the European academic world reeled before this onslaught of new knowledge.

Ancient classical texts had if anything the effect of diversifying rather than unifying the varied understandings of words such as 'nature' and 'nurture' as they are encountered in the writers of the early modern period. Not until the nineteenth century did the question of innate human qualities, particularly mental abilities and their capacity for improvement, begin to be viewed as scientific rather than philosophical questions. Until that time it was mainly philosophers, educationalists and literary writers who pondered on the deep questions of human nature and human destiny.

The earliest text known so far which opposes the twin influences of nature and nurture on a single life comes from an unexpected source. In 1911 a thirteenth-century French novel was discovered in a locked box labelled 'Old Papers – No Value' in the house of a British Nobleman (Groff and McRae, 1998). The novel is entitled *Silence* and was written by an author, previously unknown, who designated himself as Heldriss of Cornwell (Heldriss of Cornwall, c. 1250–1300). In the story, as Groff recounts, King Evan of England has recently married the daughter of the king of Norway, thus ending a period of devastating warfare. Evan passes a law forbidding family inheritance to a woman. Shortly after passing this decree, the king is rescued from a passing dragon by a young knight, Cador of Cornwall. The king offers Cador a large estate plus the woman of his choice, who turns out to be his nurse Euphemie who has come to bind up his wounds. Their subsequent marital bliss is marred only by the fact that their first child is a daughter and so unable to benefit from their inheritance. To overcome this problem they christen their daughter Silentius, the masculine of 'Silence', and have her raised as a boy by two trusted servants. Being a medieval play life then gets very complicated although, dragons aside, the topics tackled feel remarkably modern, if not post-modern, focused as they are, on the question of gender identity. And it is on this topic in particular that the novel becomes relevant for our present context. Over the course of the novel there are two vigorous tussles between nature and nurture, personified, as was common in the literature of the time, as duelling combatants seeking hegemony over the life of the hapless Silentius. The first tussle occurs at puberty when nature seeks to persuade Silentius that she should return to the role appropriate for her gender:

> This is a fine state of affairs,
> You conducting yourself like a man,
> running about in the wind and scorching sun
> when I used a special mold for you,
> when I created you with my own hands,
> when I heaped all the beauty I had stored up
> upon you alone!
>
> (2502–9)

But Nurture will have none of it:

> Nature, leave my nursling alone,
> or I will put a curse on you!
> I have completely dis-natured her.
>
> (2593–5)

Nurture teams up with Reason, another personification, and together they come up with some convincing arguments for Silentius to stay as she is, not least the attractive reason that Silentius is actually quite ambitious and is looking forward to her eventual knighthood and inheritance. 'If I'm on top', says Silentius, 'why should I step down?' Fair point. After many further adventures (I am leaving out a lot), Silentius is sent on a quest to track down Merlin who has been living naked and feral in the woods. This leads to the second tussle between Nature and Nurture, but this time over Merlin's present state. Nurture uses the cunning theological argument that Adam and Eve had no parents, so lacked parental nurture, but were nevertheless capable of committing sin. Nature, equally cunning, turns the argument to his advantage, arguing that Adam and Eve were by nature without sin, and it was only the devious serpent that caused them to go against nature, and to fall. Irrespective of the respective persuasiveness of their positions, what is interesting about the novel is the way in which Nature and Nurture are viewed as distinct competing forces tussling for human hegemony. It is not really until we reach Galton six centuries later that nature and nurture are pitted against each other in quite the same way.[8]

The phrase 'nature and nurture' crops up frequently in the writings of Richard Mulcaster (1530–1611), particularly in his educational treatise *Elementarie* (1582; Teigen, 1984), although with a very different nuance from that found in *Silence*. Mulcaster was the headmaster of Merchant Taylors' and St. Paul's schools in London and a pioneer in educational theory and practice, emphasizing the education of girls, physical education and an appreciation of the English language which, he felt, had reached the zenith of its perfection in his own Elizabethan age. 'Nurture' for Mulcaster meant education. 'Nature' is what had been endowed to the children under his care by God. 'When I use the name of nature,' writes Mulcaster, 'I mean that power, which God has implanted in these his creatures both to continue their own kind, that it does not decay, and to answer that end, wherefore they were made' (Mulcaster, 1582, p. 32). Nature and nurture are complementary influences; nurture (i.e., education) has the power either to improve on or destroy innate abilities. In words with which every teacher can identify, Mulcaster asserted that, '[i]f nature in some children be not so pregnant, as these may take the full benefit of this whole train[ing] yet by applying it widely, there may be some good done, even in the heaviest wits and most un-apt bodies, though nothing so much as in the very quickest' (Mulcaster, 1582, pp. 31 and 41). As Mulcaster comments in his mellifluous prose: 'Nature makes the boy toward, nurture sees him forward' (Mulcaster, 1581, p. 35).

A few years after the publication of Mulcaster's *Elementarie*, Shakespeare started on his prodigious output of plays (with Parts 1, 2 and 3 of *Henry VI*, around 1590 onwards). There is no evidence that Shakespeare read Mulcaster, although it would not be surprising if he did, especially given that Mulcaster taught Shakespeare's contemporary Edmund Spenser (Teigen, 1984). Spenser in *The Faerie Queene* (1590) paints a picture of Aristotle's golden mean being banished by the extremes on either side (Johnson, 1810, p. 100):

> ... the face of golden Meane.
> Her sisters, two Extremities:
> strive her to banish cleane.

But there is no evidence that nature and nurture were generally seen at the time as two such 'extremities'. Shakespeare perhaps gets close in *The Tempest* in Prospero's curt dismissal of Caliban as

> A devil, a born devil, on whose nature
> Nurture can never stick[9],

but this is more a comment of despair about the character of an individual than about human nature in general. Through the medium of drama, Shakespeare provides us with an incredibly perceptive range of insights into human nature. In his *Cymbeline* (first performed 1611/12) is the King of Britain whose children (when aged two and three) are kidnapped by a nobleman called Belarius in retaliation for his unjust banishment. After observing their behaviour as they develop over the years, Belarius exclaims

> How hard it is to hide the sparks of nature!

Despite not knowing that they were sons of the King, Belarius surmises that

> The roofs of palaces, and nature prompts them
> In simple and low things to prince it much
> Beyond the trick of others.

Once a prince, always a prince. As Conley points out: 'The use of "nature" in the sense of innate qualities, especially those that are hereditary, is quite frequent in Shakespeare's plays, and it is impressive how strongly Shakespeare argues in favor of psychological hereditarianism' (Conley, 1984).

While Shakespeare was bringing to the attention of mass audiences meanings of 'nature' and 'nurture' that dramatically illustrate the immense range of complexity in human behaviours, the French philosopher René Descartes (1596–1650) was emphasising the platonic tradition of innate ideas. 'On first discovering them', wrote Descartes, echoing Plato, 'it seems that I am not so much learning something new as remembering what I knew before'.[10] Innate truths are accessed 'by the power of our own native intelligence, without any sensory experience'.[11]

But soon after Descartes, John Locke (1632–1704) powerfully re-expressed the tabula rasa tradition in his hugely influential *Essay Concerning Human Understanding* in which Locke starts his thesis by launching a frontal attack on the inheritance of innate ideas: 'If we will attentively consider new born children, we shall have *little* reason to think that they bring many ideas into the world with them', the child's mind being likened to an 'empty cabinet' or 'white paper', waiting to be written on by the language of experience. Knowledge is obtained by the direct perception of the senses. Locke's rejection of the inheritance of innate ideas coheres well with his espousal of the liberal society in which the virtuous citizen will freely exercise his reason to resist the tyranny of absolute monarchy. In Locke's view, no citizen is burdened by the inheritance of ideas that would subvert their use of reason and the acquisition of knowledge by experience.

The tabula rasa terminology has often been attributed to Locke, although, in fact, he himself never used the term in his *Essay* (as mentioned earlier, it was drawn from the Latin translation of Aristotle's *De Anima*). And in any case, whereas Locke certainly opposed the notion of innate ideas, he had no problem with innate predispositions, as becomes clear in *Some Thoughts Concerning Education* (1693) in which he exhorts educators of young children 'to make the best of what nature has given, to prevent the vices and faults to which such a constitution is most inclin'd.... Every one's natural genius should be carry'd as far as it could; but to attempt the putting another upon him, will be but labour in vain'. Here perhaps the clear ideas of the philosopher are tempered by the challenging realities of education. Locke's argument was not one of dichotomy and of disjunction as we find in the later nature-nurture debate, but rather about how education and an appropriate political system could best nurture nature. Human nature was malleable, open to social, economic and political change, even to revolution.

The musings of the philosophers, literary authors and political theorists were shaped by astute observations of character on the one hand and by clear ideological agendas on the other. Indeed, ideological agendas continue to influence the discussion to the present day. But it was in the transformation of views about heredity that science eventually began to gain traction and create an epistemic space in which the views of Galton would then provide the dominant framework for the earlier part of the twentieth century.

Changing Views on Heredity

Before the eighteenth century, living things did not reproduce, they were more often spoken of as being created. The generation of every plant and animal was a unique and isolated event, not some supernatural *creatio ex nihilo*, but rather an organized process whereby the parental organisms made their offspring. The process included sex, conception, pregnancy, birth and the period of weaning. Development and inheritance were seen as two aspects of the same process, the separation between the two not becoming firmly established in biology until the beginning of the twentieth century (Müller-Wille and Rheinberger, 2012, p. 16). Although John Ray (1627–1705) had introduced the species idea in the seventeenth century, most commentators did not envisage the characteristics of a species as being necessarily immutable. Instead the world was full of weird and wonderful transmutations (Daston and Park, 1998). Medieval texts reported odd unions between different species. John Locke was convinced that he had seen 'with his own eyes ... the Issue of a Cat and a Rat' (Locke, 1979, p. 451), though the most bizarre offspring to modern ears was the so-called Scythian Lamb, formed from a union between plant and animal and leading to pictures of lambs (with four legs) growing out of the ground on a kind of stalk.[12] Not until 1748 do we find the term 'reproduction' coming into the life sciences, introduced by Georges-Louis Leclerc, Comte de Buffon, the author of a thirty-six-volume natural history. The etymology is informative. Buffon borrowed the word 'reproduction' from contemporary theology where it meant 'resurrection'; therefore, applied to natural history, the word 'reproduction' in its initial usage carried the sense of a reconstitution from component parts (Müller-Wille and Rheinberger, 2012, p. 15).

The fusion of development and inheritance into a single process of propagation also allowed scope for offspring quite dissimilar to the parents. For example, if a woman was thinking of another man whilst having sex, her 'imaginative power' might influence her unborn embryo in such a way that the child might be quite unlike the father. This became relevant in paternity disputes in the courts where dissimilarity between father and child might not necessarily persuade the judge because it could be ascribed to the over-active imagination of the mother. Whilst the modern science of epigenetics has not yet taken the influence of the mother on her developing fetus quite that far – science that will be discussed further in following chapters – it is interesting to note that the idea of reproduction as an integrated process in which the mother's behaviour has measurable effects on the fetus is now back in fashion.

This impact of the environment, as we would now call it, played a central role in explaining the similarities between ancestors and their descendants well into the seventeenth century. Effects of climate, the economy, social factors and politics all came into the rich explanatory framework as to why an individual was or was not more or less similar to their parents and grandparents. If the weather changed at the wrong moment during the pregnancy, or the king was deposed and political tumult ensued, that could change everything. The current 'humours and dispositions' of the household were uppermost in people's minds. When people thought mechanistically at all, then it was often to alchemy or to machines that they looked for their metaphors (Müller-Wille and Rheinberger, 2012, p. 20). René Descartes was at least solidly biological in likening the act of procreation to the brewing of beer. In the formation of the embryo, wrote Descartes, the seed 'is fermented and well concocted by material heat, so that its parts are mixed together all the more thoroughly'. Male and female 'seminal materials' acted on each other 'as a kind of yeast', claimed Descartes, somewhat optimistically observing that the 'the scum formed on beer is able to serve as yeast for another brew'.

As with the modern idea of reproduction, it is often not realized how recent is the idea that 'inheritance' can refer to a substance that is passed down from parents to children, a theory that did not gain traction until the mid-nineteenth century, even though the mechanism of inheritance still remained quite unknown at that time. The closely linked noun 'heredity' as used with reference to the propagation of living things only started appearing around 1800 (Müller-Wille and Rheinberger, 2012, p. 41). The word *hérédité* in French and 'heredity' in English both derive from the Latin *hereditas* meaning 'inheritance or succession'. The French psychiatrist Prosper Lucas popularized *hérédité* with its biomedical meaning in his *Traite philosophique et physiologique de l' hérédité naturelle* (1847–50), and the word 'heredity' came into the English language in its modern biological sense in chapter 8 of Herbert Spencer's *Principles of Biology* (1864, vol. 1) and then Francis Galton's *Hereditary Genius* (1869). The transition of the idea of heredity from European property legislation into inheritance in biology was a lengthy process. The early laws of primogeniture ensured that property was passed down to the firstborn. Later inheritance laws broadened the inheritance to other family members in varying proportions. The important point for its subsequent usage in biology was that inheritance was carried out in an orderly and law-like manner in which property was distributed to the next generation in certain proportions. Property law created an epistemic space in which biological ideas of heredity could begin to flourish.

The word for heredity in German was *Vererbung*, which began to be used in Immanuel Kant's (1724–1804) anthropological writings during the 1770s and 1780s, although *Vererbung* still retained its legal connotations, in contrast to *hérédité* and 'heredity', which lost their legal meanings completely as the words rapidly began to be used in new biological contexts. Kant provides an interesting example of the way in which thinkers of the time sought to address not just the permanence of characteristics within family histories, but also the variation, the unexpected appearance of 'throw-backs' to previous generations, together with diseases and 'monstrosities'. Kant discussed what he called 'races' by which he meant human varieties that 'bred true', being impressed by the fact that Portuguese living in Africa continued to produce white children, whereas Africans who came to live in Europe continued to have black children, even though in both cases the climates were very different. Clearly the environment (our term, not Kant's) could not trump all other considerations. But to explain these observations Kant did not invoke physical particles that were mechanistically conveyed from one generation to the next, but rather the power of natural law and the particular circumstances of family history. In Kant's view, the *Anlagen* or dispositions that were destined to characterize the different races were present in the ancestral roots of humanity. Once these traits had been expressed in response to varying environments, they then became fixed and were transmitted down the generations. The environment exerts its influence on its inhabitants, defining the properties of their lineage, which then breeds true.

As far as the individual is concerned, Kant was on the side of freedom as spontaneous originality, not bound by external causes, but leading to its own causal chain entailing genuine moral responsibility (what we would now refer to as the libertarian notion of free will) (Kant et al., 1998, Book II, chap. 2, sec. 2). Kantian freedom was derived from the moral law – it is this which entails genuine moral responsibility (Kant et al., 1998, Book III, chap. 2, sec. 9).

Writing against the background of social ferment in which individuals were rising up to change their own destinies, Kant's influential contemporaries, whether writing on economics or politics, likewise tended to place their focus on individual freedom and people's ability to rise above the constraints that their upbringing might have bestowed on them. As Adam Smith commented in *The Wealth of Nations* (1776): 'The difference between the most dissimilar characters, between a philosopher and a common street porter, for example, seems to arise not so much from nature, as from habit, custom, and education' (Smith and Skinner, 1982, p. 120). A similar narrative extends through to John Stuart Mill (1806–73) in the nineteenth century, Lord Rector of St. Andrews University, whose Scottish philosopher father James Mill trained his son to follow in the footsteps of his friend Jeremy Bentham by starting him on Greek at the age of three and Latin, Euclid and Algebra by the age of eight. Certainly Mill himself experienced no lack of nurture, later writing in his *Principles of Political Economy* (1824) that '[o]f all the vulgar modes of escaping from the consideration of the social and moral influences on the human mind, the most vulgar is that of attributing the diversities of conduct and character to inherent natural differences'. Mill's whole programme of political reform depended on the malleability of social and economic structures, claiming that '[t]he power of education is almost boundless: there is not one natural inclination which is not strong enough to coerce, and if needful, to destroy by disuse' (Mill, 1969).

A major factor that began to change such convictions was the widespread medical observations of human disease, including mental disease, that increasingly highlighted the role of heredity in its biological sense during the course of the nineteenth century. Until then, Galenic scholastic medicine still held sway. Bloodletting reached its peak in the eighteenth century but was already in sharp decline by the nineteenth century (Warner, 1997). As the sun finally set on Galenic medicine, so the idea of hereditarian disease gained greater traction, and the determining power of 'nature' became more common in the discourse of medical practitioners. The hereditarian ideas of Charles Darwin, son of a doctor, and of his first cousin Francis Galton did not spring into being de novo but were nurtured within a cultural background strongly influenced by medical opinion that was becoming increasingly deterministic in tone. Back in the eighteenth century, *Chambers' Cyclopedia* (1738) had referred to heredity as the inheritance of 'good or evil qualities', and Edinburgh Professor of

Medicine John Gregory had written of the transmission of 'madness, folly and the most unworthy dispositions' (1765), maintaining, in the words of a later commentator, that criminals 'have hidden in their organism the germs of their fatal disposition of which they are the victims', a notion supported by phrenology (Waller, 2001, p. 462). Victorian medical textbooks and journals likewise exhorted people to breed well and avoid consanguinity not just for the health of their offspring but to prevent national degeneration. Eugenic ideas were already in the air long before Francis Galton invented the word 'eugenics' ('born well'). In his book *Practical Moral and Political Economy* (1828), the Radical Thomas Edmonds maintained that '[t]he minds of a people may be improved by selecting for propagation those people who excel in the more useful qualities of mind, as justice, judgment, imagination, benevolence, &c.; and not permitting ideots or madmen, or people approaching to such, to propagate' (Edmonds, 1828, p. 269). Victorian physicians had data to back up their claims about the heredity of disease. In 1828 George Burrows concluded that as many as 85 per cent of his patients had diseases tainted by heredity, and a decade later Sir Henry Ellis, the superintendent of Wakefield asylum, calculated that approximately 15 per cent of the insane in his charge had inherited their malady (Waller, 2001, p. 460). A critic of the theory of hereditary disease, the London surgeon Benjamin Phillips, declared in 1846 that 'the advocates of the opinion that no diseases are hereditary, though able men, are few in number' (Phillips, 1846, p. 101). The notion that acquired characteristics could be inherited held sway throughout this era, so all physicians assumed that diseases acquired by parents could be inherited at the point of conception by their offspring, along with their constitutional endowments and accumulated experiences. It was the health and wealth of the nation that was at stake, especially the wealth of the growing professional middle class. The Scottish essayist William Greg in 1868 condemned the 'pernicious social system' that allowed inferior people to reproduce, writing that '[a] republic is conceivable in which paupers should be forbidden to propagate'. As John Waller reports, 'by the mid-Victorian years the heritability of "mental aberrations" was as fully taken for granted as the inheritance of bodily diseases and infirmities' (Waller, 2001, p. 460). Heredity came to imply fixity and lack of hope for medical intervention.

Lamarck, Darwin and Galton

A trajectory running from Lamarck via Darwin to Galton sets the parameters for the nineteenth-century debates about nature and nurture that were destined to dominate the following century. Jean-Baptiste Lamarck (1744–1829) believed in the fixity of species until 1797 but then became an evolutionist, as he stated in his introductory lecture for his new post as Professor of Lower Animals at the newly founded Natural History Museum in Paris in 1800. Lamarck soon set out his new theory of evolution in works such as his *Philosophie zoologique* (1809). Strongly progressionist in tone, Lamarck envisaged the continuous spontaneous generation of new species which then move up the escalator of life, with all steps occupied at all moments. This is the primary process and it is then the differing circumstances on each step that lead to different adaptations and consequent variations. The acquired characteristics obtained by adaptation were passed on to subsequent generations, greatly speeding progress up the escalator. The inheritance of acquired characteristics is the theory for which Lamarck is now best remembered, but, in fact, for Lamarck it was a secondary mechanism, the main driver being some ill-defined 'intrinsic power' that pushed species ever onwards and upwards (Burkhardt, 2015). The Lamarckian scheme was deployed to highlight the importance of nurture for more than a century.

Darwin's theory of evolution published in 1859 in *On the Origin of Species* was strikingly different from that of Lamarck – as Darwin was fond of pointing out – and yet retained far more of Lamarck than is often recognized (Burkhardt, 2015). In place of a rather vague 'intrinsic power', there was now an actual biological mechanism in the form of natural selection that explained how organisms best suited to their environments had a greater chance of passing on their characteristics to subsequent generations, a process that eventually came to be known as 'reproductive success'. Nevertheless Darwin repeatedly cited the inheritance of acquired characteristics as a secondary evolutionary mechanism, so his theory remained infused with Lamarckian ideas throughout his life, and indeed for several decades afterwards.

Be that as it may, there is no doubt that Darwin's proffered mechanism of inheritance lent itself to Lamarckian interpretations, proposing that

> I venture to advance the hypothesis of Pangenesis, which implies that every separate part of the whole organisation reproduces itself. So that ovules, spermatozoa, and pollen-grains, – the fertilised egg or seed, as well as buds, – include and consist of a multitude of germs thrown off from each separate part or unit. (Darwin, 1868)

Darwin gave the name 'gemmules' to these hypothetical physical units that were gathered up from all parts of an organism and 'packaged' in some way in the eggs and sperm, and from there passed on to the offspring.

Although Darwin described the mechanism for the inheritance of acquired characteristics as 'most perplexing', he supposed that the tissues that were affected by environmental changes 'could throw off gemmules endowed with all the qualities which they have acquired', ensuring that the newly endowed benefits of adapting to an environment could be transmitted to the offspring. The same mechanism could also potentially explain a blending theory of heredity, to which Darwin stayed loyal in his published work, being quite unaware of the ground-breaking experiments of the Augustinian Moravian monk Gregor Mendel (1822–84) who had published in German in 1866 his experiments on the inheritance of differing traits in peas in the *Proceedings of the Natural History Society of Brünn*. But 'blending inheritance' in Darwin's terminology tended to refer to visible characteristics, the consequences of inheritance rather than its mechanism. When writing of heredity, Darwin often expressed the idea as the blending of hereditary 'blood' in fractions, an idea drawn from colonial inheritance laws relating to the half-caste offspring of black and white parents (Olby, 2013).

Despite the fact that Darwin remained as puzzled about the mechanisms of inheritance as did his contemporaries, his espousal of blending inheritance and of the possibility of the inheritance of acquired characteristics should not be portrayed as a weak view of heredity, as Darwin returned again and again to examples of domestic breeding. Referring to the pedigrees of Shorthorned cattle and the 'more recent' Hereford breed, Darwin asked rhetorically, '[I]s it an illusion that these recently improved animals safely transmit their excellent qualities even when crossed with other breeds?' He then responded with his own knock-down argument: 'Hard cash paid down, over and over again, is an excellent test of inherited superiority' (Darwin, 1868, 2:3). Darwin also understood the inheritance of the instinctive behaviours of the 'lower animals' as beyond doubt, launching a robust critique of John Stuart Mill's position, as he clearly stated the point in *The Descent of Man* (1871a, p. 98, fn 5): 'It is with hesitation', writes Darwin, who often hesitated, 'that I venture to differ from so profound a thinker, but it can hardly be disputed that the social feelings are instinctive or innate in the lower animals; and why should they not be so in man?' By the second edition of *The Descent of Man* in 1874, p. 67, Darwin has built up a greater head of steam, all hesitation now gone, adding that '[t]he ignoring of all transmitted mental qualities will, as it seems to me, be hereafter judged as a most serious blemish in the works of Mr. Mill'. Strong words, indeed, from the gentlemanly Darwin.

Unlike Darwin, who never wrote a book on heredity, his cousin Francis Galton (1822–1911) wrote several and indeed did much to make the word more common in its biological sense in the English language, though it was through Darwin that he first became interested in the subject, writing that it was 'the publication in 1859 of *Origin of Species*', which 'made a marked epoch in my own mental development, as it did in that of human thought generally' (Galton, 1908). Looking back in 1908 over his early work on heredity, a few years before he died, Galton reflected that '[i]t seems hardly credible now that even the word heredity was then considered fanciful and unusual. I was chaffed by a cultured friend for adopting it from the French' (Galton, 1908, p. 288).

Born in 1822, the same year as Gregor Mendel, Francis Galton was the youngest of seven children, learning to read at the age of two. After an abortive attempt at studying medicine at King's College London, Galton went instead to Cambridge where he read mathematics, suffering a nervous breakdown in his third year brought on by overwork and failure to achieve a first, resulting in a pass degree (Kevles, 1985, p. 9). After Galton had unwillingly returned to medicine for a while, his father died in 1844, leaving him a large estate, which was sufficient to provide an independent income for the rest of his life. The young Francis dedicated himself to travelling, visiting Egypt, the Sudan, Jerusalem and Beirut and learning Arabic along the way. Galton also managed to pick up venereal disease during his travels, which may partly explain why he himself later failed to produce any children of his own. His later extended explorations in South Africa, coupled with his successful efforts to pacify local warring tribes, earned him a gold medal from the Royal Geographical Society and election to the Royal Society. Marriage to the daughter of the headmaster of Harrow, then Dean of Peterborough, together with membership of the Athenaeum Club and a move to a posh house near Hyde Park in London, ensured that Galton would be counted amongst Britain's 'eminent men'.

This personal background is relevant to what follows. Galton was self-driven, nervously fragile, independent of thought, elitist, somewhat eccentric in his ways and an inventor. His many inventions included underwater reading glasses, a sand-glass speedometer for cyclists and a self-tipping top hat. More useful, Galton introduced fingerprinting for identifying criminals. He counted everything. At the Epsom races on Derby Day, Galton surveyed the rows of white faces in the stand opposite with his opera glasses, noting what a 'capital idea it afforded of the average tint of the complexion of the British upper classes', comparing this to the average tint after the race had passed by, which was 'uniformly suffused with a strong pink' (Kevles, 1985, p. 7). Galton's passion for observing and measuring things was important because it was in this way that he began to apply the nascent approach of statistics to questions of human biology. As Galton once commented, 'Some people hate the very name of statistics, but I find them full of beauty and interest' (Kevles, 1985, p. 17). At the same time Galton shared with Darwin a social concern, common to most Victorian gentlemen of the time, that the British stock of 'eminent men' was being diluted by their marriage to those of less ability, leading to 'racial degeneration'. The difference in their approaches was that Galton was considerably more robust than his cousin in suggesting solutions. Darwin was cautious and tolerant, worried about state interference in the private affairs of individuals. Galton was outspoken and assertive in suggesting that the state could and should become involved in arranging the 'best' types of marriage that would improve the stock of the nation. In reading Galton's publications with the benefit of hindsight, it is difficult to avoid the conclusion that he 'discovered' what he knew must already be the case.

Galton's initial foray into heredity was made in a two-part article published in *Macmillan's Magazine* in 1865, subsequently published as his first book, *Hereditary Genius* (1869), in which he demonstrated, to his satisfaction at least, that high achievement ran in families. Based on a statistical analysis of 500 families with several 'eminent' members, the methodology was flawed, but nonetheless represented the first attempt to introduce scientific rigour to theories of human behaviour. The social influence of families might be important for the army or legal profession, thought Galton, but this was not the case for men distinguished in science, literature or the law, whose inborn abilities would outweigh any hindrances to success that they might encounter. 'I have no patience', wrote Galton, 'with the hypothesis occasionally expressed, and often implied, especially in tales written to teach children to be good, that babies are born pretty much alike, and that the sole agencies in creating differences between boy and boy, and man and man, are steady application and moral effort.' Unlike Darwin, Galton did not criticize Mill by name, but he very likely had Mill's educational philosophy in mind as he wrote. The scientific reviews of *Hereditary Genius* were generally enthusiastic, those in literary and religious journals much less so, making the point that Galton had overstated the case for heredity (Paul, 1995, p. 31). But Darwin was in the enthusiastic camp, writing to Galton after his wife Emma had read aloud to him *Hereditary Genius*, 'You have made a convert of an opponent in one sense, for I have always maintained that, excepting fools, men did not differ much in intellect, only in zeal & hard work; & I still think there is an eminently important difference.'[13] And by the time he came to write *The Descent of Man* (1871a), we find Darwin writing that '[w]e now know, through the admirable labours of Mr Galton, that genius which implies a wonderfully complex combination of high faculties, tends to be inherited; and, on the other hand, it is too certain that insanity and deteriorated mental powers likewise run in families'.

It was one particularly potent critique that led Galton eventually to start using the phrase 'nature and nurture'. This came from the Swiss botanist Alphonse de Candolle (1806–93) who felt that Galton had been far too dismissive of the differing circumstances of his subjects (Fancher, 1984; Ridley, 2003). In his *Histoire des Sciences et des Savants depuis Deux Siècles* (1873), de Candolle presented a survey of 300 eminent European scientists and argued that external stimuli including climate, population size, language, religion and wealth had influenced their rise to eminence in the sciences. He specifically took aim at Galton's assertion that an eminent man would rise regardless of circumstance (Fancher, 1983). Galton's response to de Candolle was to conduct his own survey of English scientists based on data from questionnaires sent to 180 distinguished members of the Royal Society, men remarkably like himself, of whom results from 100 were accepted into his statistics. The results were published in *English Men of Science: Their Nature and Nurture* (1874), launching Galton's first use of the phrase 'nature-nurture'. As Galton pointed out 'the phrase "nature and nurture" is a convenient jingle of words', useful because 'it separates under two distinct heads the innumerable elements of which personality is composed', yet not forgetting to add that 'no carefulness of nurture can overcome the evil tendencies of an intrinsically bad physique, weak brain, or brutal disposition' (Galton, 1874, pp. 12–13). Lest the point be missed, Galton goes on to recount the story about a 'child who had developed into manhood, along a predestined course laid out in his nature' (Galton, 1874, p. 15). Galton then published an article entitled *The History of Twins as a Criterion of the Relative Powers of Nature and Nurture* in *Popular Science Monthly* (Vol. 8, 1876), with a particular emphasis on mental heredity, in which Galton concluded that 'Nature is far stronger than Nurture within the limited range that I have been careful to assign the latter'.

Despite impressions to the contrary, neither de Candolle nor Galton belonged solely to the nature or nurture camps. Galton's survey of English scientists prompted by de Candolle helped convince him that external stimuli, such as wealth, had played a part in the expression of genius. He also pointed out that he and de Candolle agreed that genius was to some extent hereditary, particularly as shown by the scientific aptitudes of different races (as was thought at the time, Galton referring to Negroes as a 'subrace'). Although Galton and de Candolle can be said to be the 'founding fathers' of the nature and nurture camps, respectively, their opinions often overlapped and they had a long and friendly correspondence, suggesting that the nature versus nurture dichotomy was not as polarised in the late Victorian period as it would afterwards become (Fancher, 1983). Having said that, there is little doubt that on average Galton tended to weight his conclusions heavily biased towards the power of heredity. His own theory for the mechanism of inheritance supported such conclusions. In his essay 'Hereditary Talent and Character' Galton suggested that we are 'no more than passive transmitters of a nature we have received, and which we have no power to modify' (Galton, 1865).

Over the period 1869–71 Galton tested Darwin's theory of gemmules by transfusing blood samples from one rabbit to another, finding that this did not result in the transfer of traits, so at least the gemmules could not be in the blood. Darwin responded in *Nature*, slightly grumpily, that he had never said that the gemmules were transported in the blood anyway (Darwin, 1871b), Galton responding obsequiously a week later to claim that it was all based on an unfortunate semantic misunderstanding (Galton, 1871), and there the matter rested. Meanwhile, Galton busily pursued ideas of heredity significantly different from those of his cousin, summarized in his 1876 paper 'A Theory of Heredity' (Galton, 1876). Rejecting Darwin's theory of pangenesis, Galton argued that all living things were derived from organic units that in turn originated from common germinal material, which he called the 'stirp', from the Latin *stirps*, descendant or stock, a line descending from a common ancestor. In Galton's definition, the stirp was the 'sum-total of the germs, gemmules, or whatever they may be called, which are to be found … in the newly fertilised ovum' (Galton, 1876, p. 330). Far from being a novel creation of the bodies of the parents, the stirp in Galton's view maintained its ancestral characteristics, although the sum of the germs making up the stirp of a fertilised egg is much larger than the total number of germs that actually develop in an individual, those that remain forming a 'residue' that is later passed on to the progeny. In Galton's hands the units of inheritance hardened and became invariant, fixed entities passed down the generations, independent of environment. No longer was there scope for the inheritance of acquired characteristics, a possibility left open by Darwin's theory of pangenesis.

Galton is best known not for his theory of heredity but for his invention of the word 'eugenics' and his subsequent promotion of eugenic ideas. The history of eugenics has been well recounted by others (for example Kevles, 1985; Paul, 1998; Müller-Wille and Rheinberger, 2012), and it is not the intention in what follows to reiterate this literature. Nevertheless, eugenics became so entangled with ideas of nature and nurture and with the science of genetics in the late nineteenth and well into the twentieth century that it is impossible to ignore the topic altogether. Galton introduced the idea in his *Inquiries into Human Faculty and its Development* (1883) where he suggested that his theory of heredity could be applied to improving the human race using a 'science of improving stock'. Earlier Galton had considered the word *viriculture* (the 'cultivation of men') as an alternative. The aim was clear: to prevent racial degeneration and improve the human stock, mimicking the work of cattle and horse breeders. Galton's idea was not so much to weed out the unfit as to promote the marriage of the high-born. Celibacy and philanthropy would lead to the degeneration of populations, whereas a meritocracy with careful selection (an unlikely combination with the benefit of hindsight) would improve the race. Eugenics began to attract utopian aspirations in Galton's thinking. As he wrote in his autobiography: 'I take Eugenics very seriously, feeling that its principles ought to become one of the dominant motives in a civilized nation, much as if they were one of its religious tenets' (Galton, 1908). Galton's goal, he said, was 'to replace Natural Selection by other processes that are more merciful and not less effective. This is precisely the aim of Eugenics' (Galton, 1908, p. 323). At the time the views were seen as generally benign, albeit distasteful by some; only now in retrospect, with the history of the twentieth century in view, do the words 'other processes' take on a more sinister tone.

CHAPTER 2

Reifying the Fragments?

Nature-Nurture Discourse from Galton to the Twenty-First Century

> *Mistaken regard for what are believed to be divine laws and a sentimental belief in the sanctity of human life tend to prevent both the elimination of defective infants and the sterilization of such adults as are themselves of no value to the community. The laws of nature require the obliteration of the unfit, and human life is valuable only when it is of use to the community or race.*
>
> Madison Grant, President of the New York Zoological Society (Grant and Osborn, 1916, pp. 44–5).

> *Children inherit their minds and dispositions in the same way and to the same degree as they inherit their bodies.*
>
> *The Fruit of the Family Tree* (Wiggam, 1925)

> *We are forced to conclude that human nature is almost unbelievably malleable, responding accurately and contrastingly to contrasting cultural conditions.*
>
> Margaret Mead, 1935 (Mead, 1963, p. 280).

> *We used to think that our fate was in our stars. Now we know, in large part, that our fate is in our genes.*
>
> James Watson, 1989[1]

Galton's jingly phrase – 'nature and nurture' – established the parameters for the discussion throughout the century that followed, with great swings of academic opinion regarding the relative roles of one or the other remaining a characteristic feature. And although science played a much larger part in the debate than in previous centuries, it is nonetheless striking that ideological, political and social factors continued to inform and shape the debate in ways that go well beyond science right throughout the twentieth century and indeed to the present day.

Genetics, Instincts and the Power of Nature

In 1893 August Weismann (1834–1914) showed that only germ cells, not the somatic cells which comprise the rest of the body, can be agents of heredity and noted that the two types of cell replicated in different ways (Churchill, 2015). Somatic cells came from germ cells, but not vice versa. Weismann's finding contradicted the theory of pangenesis and represented a significant blow to Lamarckian ideas of the inheritance of acquired characteristics. Ironically, however, Weismann's own theory of 'germinal selection' suggested that some changes in the germ line could be brought about by beneficial changes imposed by the environment that could then be inherited. Lamarckian ideas turned out to be very persistent.

Mendel's particulate theory of inheritance eventually came to dominate the science of genetics, but the route there was long and winding. It is not the case that Mendel's work was lost to view during the thirty years after its publication in 1866, but his paper remained unknown to Darwin and Galton (Kampourakis, 2015). At the turn of the century Mendel's seminal work was rediscovered and extended by three fellow plant breeders: Hugo de Vries (1848–1935) in Amsterdam, son of a Mennonite deacon who later became prime minister of the Netherlands; Carl Correns (1864–1933) in Tübingen, who was encouraged to study botany by a correspondent of Mendel; and Erik von Tschermak (1871–1962) in Ghent, whose grandfather had taught Mendel during his time in Vienna. All three had been using different plant breeding systems to investigate inheritance, and each confirmed the famous three-to-one ratio between dominant and recessive traits in his own system, as originally published by Mendel. With varying degrees of alacrity, they recognised that Mendel's work had foreshadowed their own, and together they helped launch Mendel to the central place that he still enjoys in the history of genetics. William Bateson, first professor of genetics at Cambridge University, widely publicised these breakthroughs in his monumental work *Mendel's Principles of Heredity – a Defense* (1902), in which Bateson spelled out the importance of Gregor Mendel's rediscovered ideas (Bateson, 1909). In the very same year, 1902, Sir Archibald Edward Garrod made the first detailed biochemical description of several inborn errors of metabolism, using Mendel's results to explain the pattern of heredity (Garrod, 1902). It was Bateson who first coined the word 'genetics' (from the Greek *gennō, γεννάω;* 'to give birth').[2] All these results provided striking confirmation of the particulate theory of inheritance. In 1909 the Danish botanist Wilhelm L. Johannsen (1857–1927) introduced the term 'gene' to replace older terms like factor, trait and character: the word was deliberately chosen to contrast with 'pangene,' the earlier term associated with the now discredited ideas of pangenesis. Johannsen also introduced the terms 'genotype' and 'phenotype' to refer to the genetic information contained within germ cells and the visible characteristics of an organism, respectively.

The first few decades of the twentieth century were marked by a range of insights into the newly discovered explanatory powers of genetics. A team of researchers at Columbia University led by Thomas Hunt Morgan, the first native-born American biologist to receive the Nobel Prize, pioneered genetic studies using the fruit fly *Drosophila*. The studies met with great success, showing that genes were strung out on chromosomes, as Morgan put it, 'like beads on a string'. Mutant genes for eye colour followed a recessive pattern of inheritance, so Morgan was able to show that Mendelian inheritance applied as much to animals like *Drosophila* as it did to plants. In 1915 Morgan co-authored with three other collaborators the famous book *The Mechanism of Mendelian Inheritance*, which completed a revolution in scientific thought by placing genes at the centre of biologists' ideas about heredity.

Given these breakthroughs, it might have been thought that Mendelian genetics would immediately come to take centre-stage as *the* mechanism to explain both the evolutionary theory of natural selection as well as the heredity of human traits. As far as the former is concerned, this did not occur until natural selection was fused with population genetics during the 1920s and 1930s. With regard to the latter, early twentieth-century studies of heredity tended to investigate the transmission and development of traits within a given environment as the organism matured (Cooke, 1998). Neo-Lamarckian ideas still remained popular until around 1920, though they were reformulated by arguing, for example, that the 'germinal material' could in some way be influenced by environmental changes.

Overall, during the first few decades of the twentieth century, the new genetics became associated with a steady hardening of hereditarian views and increasingly became entangled with political and social issues via the path provided by eugenics. It was widely accepted that most human mental traits and human behaviours had a hereditary component (Gillette, 2011). 'The number of characteristics which eugenicists believed could be transmitted genetically was particularly all-embracing. They included not only such defects as insanity, mental deficiency and epilepsy, but also unemployment, alcoholism, pauperism and criminality. Therefore, by the eugenic definition, almost the entire urban poor could be classified as "degenerate"' (Woodhouse, 1982). Two infamous family studies mark the transition from a Lamarckian view holding out the possibility of change, to the hard hereditarianism which typified the early twentieth century, the first being published by Richard Dugdale in 1874, and the second by the psychologist Henry Goddard in 1912.

In 'The Jukes': A Study in Crime, Pauperism, Disease, and Heredity, Dugdale described his discovery of a criminal family, called the Jukes, whose criminal careers could be tracked back through several generations (Paul, 1995, pp. 43–4). Dugdale reported that '[t]hey had lived in the same locality [in upstate New York] for generations, and were so despised by the reputable community that their family name had come to be used generically as a term of reproach' (Dugdale, 1877, p. 8). But Dugdale was optimistic, maintaining on Lamarckian grounds that the 'environment tends to produce habits which may become hereditary', arguing therefore that 'public health and infant education ... are the two legs upon which the general morality of the future must travel' (Dugdale, 1877, pp. 66 and 119). One answer was prison reform, so that the criminal's naturally enterprising nature could be channelled in better directions. But in 1915 Arthur Estabrook (1885–1944), working out of the Eugenics Record Office in New York, revisited Dugdale's data on the Jukes family, adding some further results of his own, now arguing that such families should be prevented from reproducing given that their genetic susceptibility to criminality proffered no hope for change based on environmental factors (Estabrook, 1916). Such conclusions aligned closely with those of Henry Goddard, director of research at the Training School for Backward and Feeble-Minded Children in Vineland, New Jersey. Goddard had been impressed with the work of a French psychologist called Alfred Binet who had developed a test for mentally deficient French children in 1905, and Goddard soon developed his own version (later to develop into the IQ test). This came in useful in his study of The Kallikak Family: A Study in the Heredity of Feeble-Mindedness (1912), a book that eventually went through twelve editions (the last in 1939) and was immensely influential. The Kallikaks were a real family, but their fictitious name was made up by Goddard by joining the Greek words for 'beautiful' (kalos) and 'bad' (kakos). As Goddard recounts, the family were descended from a Martin Kallikak who had made the mistake of having a child with a mentally defective young woman that he had met in a tavern during the time of the Civil War. The offspring of that union had, according to Goddard, 480 descendants of which, for 143, 'we have conclusive proof, were or are feeble-minded, while only forty-six have been found normal. The rest are unknown or doubtful' (Goddard, 1927, p. 13). The conclusions were clear, suggested Goddard, since 'in the light of present-day knowledge of the sciences of criminology and biology, there is every reason to conclude that criminals are made and not born. The best material out of which to make criminals ... is feeble-mindedness'. Families like the Jukes and the Kallikaks, claimed Goddard, are the 'true parasites' of society and, as they cannot be killed off like lower animals, should therefore be prevented from breeding (Paul, 1995, p. 50).

Goddard's new test for mental feeble-mindedness, together with Stanford psychologist Lewis Terman's further revised version of the Binet test in 1916, provided useful tools for studying such families. It was found that between a third and a half of all delinquents, prostitutes and criminals were feeble-minded, or 'morons' as Goddard defined them, referring to those who tested with a mental age between eight and twelve, those below this range being referred to as 'imbeciles', with 'idiots' yet further down Goddard's range (Goddard, 1910). Studies by Goddard and others created a rising sense of public concern about national degeneracy. It was thought that the large number of children being born to working-class families would soon overwhelm the 'fitter' children of the middle classes by sheer numbers.

Along with family studies and the testing of ever larger cohorts of the population, biological theories of 'instincts' in both animal and human behaviours contributed to the sense that nature was more powerful in its effects than nurture. Following the publication of William James' *Principles of Psychology* in 1890, the doctrine of instinctivism became dominant in psychology, replacing John Stuart Mill's associationism as the primary theory of mind. Instinctivism argued that most human behaviours are instinctive or innate, with little or no role for the environment in how a behaviour presents. Through the 1900s and 1910s, this viewpoint was shared by most animal behaviourists, who proposed an increasingly extensive number of behaviours as instinctive – well over a thousand by 1920. One of the most influential promoters of the importance of instincts was the psychologist William McDougall who received his early education in psychology at Cambridge, but then moved to the United States where his textbooks of psychology, such as *Introduction to Social Psychology* (1909), became very influential. Whilst a professor of psychology at Harvard, McDougall promoted the idea 'of instincts in the human species as innate tendencies to pursue by purposive actions certain biological ends'. These instincts included the gregarious instinct, dominance, companionship and 'purposive striving' (Degler, 1991, pp. 34–5).

It was within a cultural climate in which hereditarian ideas were beginning to harden and the power of instinctual behaviours was widely touted that eugenic ideas began to flourish and, eventually, to affect state legislation. The Society for Racial Hygiene was founded in Germany in 1905, The British Eugenics Education Society in 1907, and the American Eugenics Society in 1923. The first organisational effort in America to apply eugenics had been much earlier. In 1906 the American Breeders' Association set up a committee to 'investigate and report on heredity in the human race', and the committee was expected to make clear 'the value of superior blood and the menace to society of inferior blood' (Kimmelman, 1983). The eugenic societies never had huge memberships but were widely supported by a largely middle-class clientele of professionals. They were fed by worries about national decline and, especially in America, by concerns about mass immigration diluting the 'national stock'. The population of the United States was exploding. In 1870 its population was hardly more than that of Germany, but by 1890 more people lived in America than all the European nation-states put together, including Russia (Müller-Wille and Rheinberger, 2012, p. 98). The apparently high birth rate of the feeble-minded was also of concern to many. The eugenic movement drew support from socialists and from conservatives, from the religious and from the anti-religious, from politicians and intellectual elites, and from academic biologists as well as those in the humanities. Many prominent eugenicists, especially in Britain, were Marxists and socialist radicals, including J. B. S. Haldane, George Bernard Shaw and Hermann Muller. Eugenic goals could be ideologically packaged in such a way as to make them acceptable to almost any social or political cause. But not quite: Catholics often opposed eugenics because of eugenic arguments against the reproduction of the 'unfit' and eugenic opposition to celibacy. No eugenic legislation was ever passed in any state or country where Catholics wielded significant political influence (Larson, 2010). Opposition also came from the Labour Party in Britain, and from some critical individual libertarians, such as Britain's G. K. Chesterton. Most social reformers, however, saw eugenics as a key way of improving the health, intelligence and well-being of the nation. 'More children from the fit, less from the unfit, that is the chief issue of birth control' declared the feminist campaigner Margaret Sanger in 1919 (Paul, 1995, p. 20). Membership of the eugenic societies was rather fashionable at the time. The first International Congress of Eugenics, held in London in 1912, had Winston Churchill as one of its English vice-presidents, and amongst its American vice-presidents was Charles Eliot, president of Harvard University. Had he still been alive to see the day, Galton would have been pleased.

Eugenic ideas drew upon the latest science for their support. 'In the 1910s and 1920s, eugenics was simply considered as applied human genetics' (Paul, 1995, p. 4). Every member of the first editorial board of the American journal *Genetics*, founded in 1916, supported the eugenics movement (Ludmerer, 1972). Eugenics was taught at most major American universities, or became a standard part of genetics courses, and was endorsed in more than 90 per cent of U.S. high school textbooks (Selden, 1989). At the First International Eugenics Congress held in London in 1912, Professor Punnett stated that 'feeble-mindedness' was a case of 'single Mendelian inheritance' and that '[t]here was every reason to expect that a policy of strict segregation would rapidly bring about the rapid elimination of this character'. Samuel J. Holmes, professor of zoology at the University of California at Berkeley, informed his readers in his book *Studies in Evolution and Eugenics* (1923), that anyone familiar with genetics could in a few generations 'breed a race of idiots, a race of dwarfs, a race of giants, an albino race, an insane race, a race of moral imbeciles ... a race of pre-eminent mental ability, or a race of unusual artistic talent'. There was no excuse to allow 'degenerate human beings' to reproduce (Paul, 1995, p. 1). As Holmes was keen to emphasise in his much later book *Life and Morals* (1948, p. 41), Darwinian evolution 'does not *logically* compel me to adopt any one standard of conduct rather than another'.

The emerging field of statistics was also deployed in the pursuit of eugenic ideals, most strikingly by Galton's successor, the biometrician Karl Pearson (Kevles, 1985, pp. 20–40). 'Biometry' was a new discipline that Pearson himself had created, together with his close collaborator Walter Weldon, being the statistical study of evolution and of heredity. Like Galton, Pearson read mathematics at Cambridge. Kevles paints him as 'the rationalist and lonely romantic' who 'tended to love people more easily in the abstract group than in the particular flesh and blood'. After a complex political pilgrimage, Pearson came to 'equate morality with the advancement of social revolution, the outcome of the Darwinian struggle with the ascendancy of the fittest nation, and the achievement of fitness with a nationalist socialism' (Kevles, 1985, p. 23). Pearson became professor of mathematics at University College London, where he invented a new theory of correlation and used it to study 4,000 pairs of school-children siblings, concluding that the correlation coefficients for physical characteristics as well as for intelligence all equalled about 0.5. As Pearson announced in his 1903 Huxley Lecture to the Anthropological Institute, 'We are forced, I think literally forced, to the general conclusion that the physical and psychical characters in man are inherited within broad lines in the same manner, and with the same intensity.... We inherit our parents' tempers, our parents' conscientiousness, shyness and ability, even as we inherit their stature, forearm and span'. Concerned that Britain was in a state of national deterioration, Pearson saw no point in expanding schools, declaring in the same lecture that 'No training or education can *create* [intelligence], you must breed it'. Pearson spent much time with Galton – ever the dutiful son to the proud father-figure – telling Galton privately that charities for mentally defective children were a 'national curse and not a blessing'. The answer was to restore natural selection and restrict 'reproductive selection', which favoured the most fertile over the most fit. Galton's endowment of a new Galton Eugenics Professorship at University College, that bastion of progressive education, following his death in 1911, found its obvious first appointee in Pearson, who established the Galton Laboratory. This carried out extensive statistical surveys on the relationships between physique, intelligence and disease, focusing its attention on studies in national deterioration. In studying ancestral inheritance statistically, Pearson was interested in the variation in a population, the forerunner of methods still used by behavioural geneticists today (Chapter 6), whereas the geneticists like Bateson were interested in the phenotypic characteristics of individuals in relation to their particular family histories. The distinction was and is a crucial one.

In light of the overwhelming public support for the eugenics movement, amplified by well-orchestrated campaigns, the legislation of sterilisation for the unfit seemed to be a very rational next move. Coercive moves to ensure such an outcome were first made in the United States. In 1896 Connecticut passed a law stating that 'no man and woman either of whom is epileptic, imbecile, or feeble-minded' could 'inter-marry, or live together as husband and wife, when the woman is under forty-five years of age'.[3] In 1907 Indiana passed the first law for the compulsory sterilisation of 'confirmed criminals, idiots, rapists and imbeciles' whose condition was pronounced incurable by a committee of three medical doctors (Larson, 2010). By 1914, over half of the states had imposed new restrictions on the marriage of persons afflicted with mental defects. The seminal Supreme Court case, Buck v Bell, made sterilisation of the feeble-minded on eugenic grounds by U.S. states constitutional in 1927, and by 1930 more than thirty U.S. states had compulsory sterilisation statutes, although the degree to which these were enforced differed greatly depending on the state. The laws were generally viewed as both reasonable and reformist. By the 1930s around 12,000 eugenic sterilisations had been performed in the United States of which 7,500 had been carried out in the most progressive state of all, California. By the 1960s this number had reached 63,000 (Larson, 2010). In Europe, Denmark passed a law allowing the sterilisation of defectives in 1929, followed by the other Scandinavian countries and then finally Germany, in 1933, even before Hitler came to power (Degler, 1991, p. 46). Sterilisation was seen as progressive and an obvious responsibility for a state-organised society with a social conscience. A Swedish doctor writing in 1934 stated that '[t]he idea of reducing the number of carriers of bad genes is entirely reasonable. It will naturally be considered within the preventative health measures in socialist community life' (Burleigh, 2000, p. 366). Sterilisation laws in Sweden stayed in place until the 1970s. Based on a solid biological basis in the power of nature over nurture, eugenics represented the rational response of progressive science-based state control in the light of the social problems contributed by the unfit and the feeble-minded. The Nobel prize-winning Marxist geneticist Hermann Muller (1890–1967) declared to the Third International Congress of Eugenics in 1932 '[t]hat imbeciles should be sterilized is of course unquestionable'. In 1935 Muller envisaged that, through selective breeding, within a century most people could have 'the innate qualities of such men as Lenin, Newton, Leonardo, Pasteur, Beethoven, Omar Khayyam, Pushkin, Sun Yat Sen, Marx, or even to possess their varied faculties combined' (Muller, 1936, p. 113).

Nurture Fights Back

Soon after the First World War, the discussion between nature and nurture began to split along disciplinary lines. Psychology and the newly emerging field of social anthropology both began, in increasingly strident tones, to highlight the power of nurture to shape human destiny, leading eventually to a sharp division with the hereditarians and the eugenicists (Cooke, 1998; Logan and Johnston, 2007). The two poles of nature and nurture became defined and entrenched (Paul, 1998; Gillette, 2011).

In psychology the pendulum had already begun to swing with Edward Thorndike (1874–1949) at Columbia University in the early 1900s as psychologists began to argue for the primacy of learning. Thorndike, working with rats, was the first to strictly divide behaviours into 'innate' and 'learned' and to demonstrate that a learned behaviour had no heritable basis. Nature for the behavioural psychologists was now redefined as instinct and nurture as learning. Thorndike himself was a moderate eugenicist, who concluded that there were instinctive behaviours in both humans and animals, although he chose to study learned behaviours (Gillette, 2011). Knight Dunlap, a psychologist working at Johns Hopkins University, published a paper in 1919 entitled 'Are There Any Instincts?' thereby opening the floodgates for a string of papers attacking the concept of instinct in both humans and animals; during the 1920s the concept became increasingly unacceptable to psychologists, and behavioural psychology became influential (Boakes, 1984). Behaviourists did not generally completely reject instinct as an explanation for behaviour, but considered it far less important than learning, especially in humans (Logan and Johnston, 2007). Thorndike's ideological successors, such as J. B. Watson and B. F. Skinner, took his distinction between innate and learned behaviours much further than had Thorndike himself. Watson was a vigorous proponent of the primacy of learned behaviour, arguing in his seminal works *Behavior as a Psychologist Sees It* (1913) and *Behaviorism* (1924) that there were only a very few human instincts. The deep divide between nurture and nature reached its zenith in the 1930s, when Skinner proposed his theory of 'radical behaviorism', concluding that humans had only one instinctive behaviour, the ability to learn, and that all other behaviours could be considered reactions learned in response to environmental stimuli and the perception of punishment and reward for actions. As Skinner famously declared: 'the nature of human nature is that humans have no nature' (Gillette, 2011). Genetic determinism was replaced by environmental determinism, with Watson claiming that 'I should like to go one step further now and say, "give me a dozen healthy infants, well-formed, and my own specified world to bring them up in and I'll guarantee to take anyone at random and train him to become any type of specialist I might select – doctor, lawyer, artist, merchant-chief and, yes, even beggar-man and thief, regardless of his talents, penchants, tendencies, abilities, vocation, and race of his ancestors."' Watson was realist enough to add: 'I am going beyond my facts and I admit it, but so have the advocates of the contrary and they have been doing it for many thousand of years' (Watson, 1925).

In parallel with behavioural psychology, cultural anthropology was also moving in an increasingly environmental direction over the same period. Instituted as a discipline by the German-American Franz Boas (1858–1942) at the turn of the century, it underwent a rapid rise, and Boas and his students took over many university departments and academic journals, expanding their reach and influence (Boas and Stocking, 1974). The value of cultural diversity and a strong antipathy to the use of genetics in identifying racial types were traits nurtured by Boas's own life experiences, having been raised by liberal Jews in Germany in the latter decades of the nineteenth century in an era marked by growing nationalism and anti-Semitism. In *The Mind of Primitive Man* (1911) Boas downplayed

development of the organism, Hogben pointed out, that was the key to understanding biological variation.

In parallel with the impact of the neo-Darwinian synthesis and of discussions about developmental biology, many of the other pillars on which eugenic ideas had previously flourished, such as strongly hereditarian views on IQ, were now steadily being dismantled. For example, the notable eugenicist Charles Davenport's 1929 study, *Race Crossing in Jamaica*, which had been commissioned to reveal the differences in IQ between the black, white and mixed-race populations on the island, in fact showed that there was very little difference in IQ scores between groups, with substantial overlap between them (Gillette, 2011). Studies carried out by the Iowa Child Welfare Research Station in the 1930s, where feeble-minded children from an orphanage were placed in 'normal' homes, showed that feeble-mindedness was not so absolutely determined as had been assumed just a decade earlier when it was upheld in the courts as grounds for compulsory sterilisation (Degler, 1991). Children whose pedigree should have made them feeble-minded were found to score within the normal range for IQ. These results helped cause a major backlash against sterilisation in the mid-1930s, in both Britain and America, where Parliamentary and Congressional committees 'flatly declared that there was no established case for compulsory sterilization, eugenic or otherwise. Both [committees] observed that sterilization might be warranted for the few disorders that were demonstrably genetic in origin. Both insisted that any such sterilization should be entirely voluntary' (Kevles, 1985).

Eugenics and the Third Reich

The backlash against eugenics came sadly too late to have any effect in slowing down its most tragic applications in the hands of the architects of German National Socialism. Eugenic policies, albeit relatively benign compared with what was to come under the Nazis, became well established in Germany during the Weimar Republic (1919–33) (Weindling, 2010). The importance of controlling reproduction was advocated by the 'social hygienist' Alfred Grotjahn, with the aim of weeding out the *Minderwertigen* – 'the inferior ones' – and by contrast providing incentives to promote the reproduction of the *Kinderreichen* – the 'child rich'. Biology lay at the core of the Nazi racial culture. The weak were a burden holding back the renewal of primitive Aryan racial vigour. Already, by 1929, Hitler had demanded euthanasia for the mentally ill. Drawing upon Californian legislation, compulsory sterilisation was introduced in 1933 in the Law for the Prevention of Hereditarily Diseased Progeny. The law specified eight (supposedly) hereditary illnesses that would justify sterilisation, including congenital feeble-mindedness, manic-depressive illness and epilepsy. Chronic alcoholics were also sterilised, as were homosexuals. Applications for sterilisation were made to the newly established Hereditary Health Courts, their judges being informed that they 'must always bear in mind Hitler's words that "the right to personal freedom always gives way to the duty of preserving the race"' (Burleigh, 2000, pp. 357–8). One of the Law's architects was psychiatric geneticist Ernst Rüdin, director of one of the first eugenics research institutes. At least 375,000 individuals were sterilised by the German authorities under this law, and there were an estimated 5,000 deaths from complications (Weindling, 2010). Of these around 60 per cent were defined as feeble-minded, many coming from psychiatric institutions. The basis for being feeble-minded was often established by IQ tests, which tested knowledge on such luminaries as Christopher Columbus, Luther and Bismarck. 'We must see to it that these inferior people do not procreate,' proclaimed the well-known German biologist Erwin Baur. 'No one approves of the new sterilization laws more than I do, but I must repeat over and over that they *constitute only a beginning*' (Larson, 2010). Those applying the sterilisation laws were helped in their task by newly established university chairs of racial hygiene (Proctor, 1988). Public opinion was swayed by a barrage of Nazi state propaganda which 'encouraged and incited people to doubt venerable religious precepts, or to entertain thoughts which under normal circumstances they might have remained blissfully ignorant of' (Burleigh, 2000, p. 361). Pre-war compulsory eugenic sterilisation then merged with the broader aims of the Holocaust entailing the extermination of Jews, gypsies, the mentally ill, homosexuals, criminals and anti-socials. Many detailed scholarly accounts of the horrors that followed have been published elsewhere (Burleigh, 2000, Weindling, 2010).

Dominant Nurture Makes Way for Resurgent Nature

In 1949 the American psychologist Nicholas Pastore published a book entitled *The Nature-Nurture Controversy* in which he surveyed twenty-four English and American scientists 'prominent in the nature-nurture controversy in the period 1900–1940' with the aim of relating their views on this topic to their attitudes toward social, political and economic questions. Goodwin Watson of Columbia University declared in the foreword that '[t]he instinct theories which played a great role in psychology at the beginning of the twentieth century have been very largely abandoned as a result of increased knowledge of cultural anthropology' (page vii). Nurture reigned supreme in 1949. The book was a pioneering work of its kind, with Pastore categorising twelve of his samples as environmental-ists, of whom eleven were classified as either liberals or radicals, with a single conservative (J. B. Watson), whereas the other twelve were catego-rised as hereditarians, with all being classified as conservative, except for one lone liberal. The author concluded, not without good reason, that the controversy was as much sociological in nature as it was scientific. But what is perhaps most striking about the book with the benefit of hindsight is its failure in the author's introductory and summary sections to even mention the holocaust or to make any effort to link the central theme of the book with the eugenic horrors that had taken place such a short time previously.

It did indeed take some years before the full scale of the Holocaust became widely known following the end of the Second World War in 1945. As this happened, genetics and eugenics became tarred with the same brush and linked to the evils of genocide – Auschwitz and Buchenwald. This helped promote the strong move towards environmentalism that, outside of Germany, had already been in full swing in the pre-war years. Genetic determinism became replaced by an almost equally strong envi-ronmental determinism, promoted by Boas and his students. In his *Social Darwinism in American Thought* (1945), American historian Richard Hofstadter commented that biological ideas 'are utterly useless in attempt-ing to understand society' and that 'the life of man in society, while it is incidentally a biological fact, has characteristics which are not reducible to biology and must be explained in the distinctive terms of a cultural analysis'. UNESCO statements (1951 and 1952) declared that race was a sociocultural construct and disavowed any connection between biological characteristics, including genes, and racial categories (Muller-Wille et al., 2008). Skinner was a well-known figure, writing extensively for the popular press (Rutherford, 2004). The anthropologist Margaret Mead was widely read: 'In the period from the onset of World War II through perhaps the end of the Vietnam War and the height of the Civil Rights movement, a liberal political imagery, based in considerable part on the anthropological view of the world, dominated public thinking in America. It was central in debate and formulation of public policy. Mead became an American icon' (Greenfield, 2001). This popularity helped promote environmentalism in the public consciousness, to the exclusion of any other explanation. At the same time, there was no viable explanatory framework for how genes might influence behaviour; the 'one gene, one trait' model had been shown to be too simplistic, but complex multigenic traits were still poorly understood, leaving environmentalism to dominate the field.

Environmental determinism at the time seemed more in harmony with social and political attitudes, making it more attractive as an explanatory framework than biological determinism. In the aftermath of the war, people were keen to embrace an explanation that promised the possibility of change, rather than one that limited opportunity by arguing for a biological destiny (Keller, 1992). Furthermore, new social movements, such as civil rights, women's liberation and the anti-war movement, emphasised individuality, autonomy and equality, qualities that were seemingly antithetical to an unchangeable genetically determined future. '[N]ot surprisingly, feminists usually oppose ... biological determinism, for so often it seems to fix the status quo. To be determined by biology is to surrender to limitations, to deny the possibility of change' (Birke, 1999). Reproductive rights and the right of bodily autonomy became important issues, and the idea of state control of reproduction became repugnant. Liberalism in social attitudes prevented a return, at the time, to the early twentieth century attitudes about heredity.

In the immediate aftermath of the War, key figures moved swiftly to depoliticise human genetics. What remained of eugenic organisations in America, such as the Eugenics Record Office, were quietly closed down by the president of the American Eugenics Society, Frederick Osborn (Gillette, 2011), and funding by groups such as the Rockefeller Foundation was discontinued (Biehn, 2009). 'Most of the research work on human behavior accomplished by scientists identified as eugenicists was forgotten, regardless of whether the work was eugenic in nature or not' (Gillette, 2011). The 1949 UNESCO Statement on the 'Nature of Race and Race Differences by Physical Anthropologists and Geneticists' officially asserted the condemnatory position of newly neutral geneticists on eugenics and any tight linkage between race and genetics.

Genetic determinism was now seen as politically suspect, whereas environmental determinism seemed not to have any political ramifications. '[T]he UNESCO statement marked the explicit declaration of environmentalism as the politically and intellectually sanctioned approach in opposition to biological determinism' (Weingart et al., 1997). It also entailed dropping the focus on human behaviour, superficially at least, leaving it to the social sciences and drawing firm boundaries for the limits of genetics: '[T]he force of genetics was confined to purely physiological attributes, while behaviour came increasingly to be seen as belonging to the domain of culture' (Keller, 1992). In 1945 the eugenics critic Lionel

Penrose became Galton Professor and head of the Galton Laboratory of National Eugenics at University College London, renaming his chair in 1951 the Galton Professorship of Human Genetics and changing the title of the laboratory's journal in 1953 from *Annals of Eugenics* to *Annals of Human Genetics*.

But whilst this distancing of genetics from eugenics was in full swing, a quiet revolution was brewing, one that in the end was destined to swing the pendulum back in a way that rendered biological explanations once again more acceptable and which continues to dominate discussions concerning human nature to the present day. In 1944, DNA, rather than protein as was earlier thought, was identified as the genetic material by Oswald Avery working at the Rockefeller Institute in New York. In 1953 Watson and Crick published their famous one-page paper in *Nature* describing the double-helical structure of DNA (Watson and Crick, 1953). Soon the genetic code was broken, and by the 1960s molecular biology began to dominate a broad swathe of biological research. At the beginning the emphasis was on the molecular genetic analysis and understanding of bacteria, viruses and animals. The language of molecular biology became saturated with powerful metaphors drawn from linguistics and cybernetics: 'transcription', 'translation', 'encode', 'replication', 'open reading frame' and more. It did not take long before geneticists turned their attention once more to questions of human behaviour. In 1960 John L. Fuller and William R. Thompson published their textbook *Behavior Genetics*, the first of its kind, and in 1970 the journal *Behavior Genetics* was established together with the founding of the Behavior Genetics Association with Russian-American geneticist Theodosius Dobzhansky as its first president (Griffiths and Tabery, 2008).

In parallel with the emergence of molecular biology, the new science of ethology was established by Niko Tinbergen and Konrad Lorenz, amongst others, who defined it as 'the biological study of behaviour' (Burkhardt, 1999). With its roots in the pre-war years, post-war ethology became more of a professional discipline and set out to study behaviours of animals only in natural or semi-natural environments, in contrast to the earlier behaviourist focus of research on animal behaviour in captivity (Mitman, 1992). The concept of instinct was central to early ethology. The beginnings of ethology were not far distant in time from the heyday of genetic determinism, and Tinbergen was significantly influenced by William MacDougall's work of twenty years earlier (Gillette, 2011). However, the notion of instinct now became more sophisticated, being rooted in neural mechanisms, even though those mechanisms still required elucidation. Lorenz insisted on a strict division between behaviours that are instinctive and those that are learned, arguing that behaviours could never cross from one to the other, a stance repeated in Tinbergen's 1951 book *The Study of Instinct* and expounded by some British ethologists (Griffiths, 2004). This position was attacked by Daniel Lehrman in his 1953 paper 'A Critique of Konrad Lorenz's Theory of Instinctive Behavior'. Lehrman maintained that making sharp distinction between instinctive and acquired behaviour was untenable in the light of what was already known about behavioural development (Griffiths, 2004). The critique had some effect and British ethologists in particular subsequently toned down the attempt to distinguish between instinctive and learned behaviours, with Tinbergen's 1963 book *On the Aims and Methods of Ethology* now describing the term 'innate' as 'heuristically harmful' (Griffiths, 2004). But ethology as an independent discipline had a relatively brief life. Population genetics and evolutionary theory took over in the field of animal behaviour, and after 1975 the discipline was swallowed up by the emerging field of sociobiology.

The heyday of ethology coincided with the decline in popularity of the behaviourist psychology of Watson and Skinner, a decline hastened by the inability of researchers to find support for some of the field's most sacred tenets. For example, John Garcia found that rats have an instinctual avoidance of foods that led to nausea, his work initially being rejected for publication because it seemed so at odds with the then current view that behaviour avoidance was the result of conditioning (Gillette, 2011). In 1961, Keller and Breland reported that their attempts to train thousands of animals using operant conditioning had primarily ended in failure and in reversion to 'instinctual' behaviours (Degler, 1991). Behaviourism as a model for humans also came under increasing attack from the late 1950s onwards. A seminal moment occurred in 1959, when Noam Chomsky published a harsh review of Skinner's *Verbal Behavior* (1957), convincing many psychologists that behaviourism was insufficient to explain linguistics and that Skinner was making claims that went well beyond the data (Degler, 1991). Behaviourism was also challenged by 'cognitivism', which emphasised the role of the mind and its relation to the brain in behaviour. The brain came to be seen as an information processor, based on computing analogies, with a focus on the biological functioning of different areas of the brain, a very different model from that provided by the behaviourists (Greenwood, 1999). However, cognitivism never really replaced behaviourism, but rather developed alongside it, both disciplines influencing the other to adopt less radical positions (Greenwood, 1999; Watrin and Darwich, 2012). Although cognitivists in the 1960s did not directly assert that the functioning of the brain, and consequently, the behaviour it caused, might be genetically determined, they allowed for the possibility that this might be so. The determinist possibilities of cognitivism would not be picked up and made explicit, however, for another three decades, when evolutionary psychology brought the cognitivist framework within the scope of evolutionary genetics.

In 1969 Arthur Jensen published an infamous article in the *Harvard Educational Review* entitled 'How Much Can We Boost IQ and Scholastic Achievement?', making proposals on educational policy based on a deterministic view of IQ levels as measured by twin studies (Jensen, 1969). This truly put the genetic cat amongst the environmentalist pigeons and Jensen's claims were met with a barrage of hostile critiques. Prominent amongst these were the writings of Richard Lewontin (Lewontin, 1975), who had carried out his graduate studies under the tutelage of Theodosius Dobzhansky at Columbia University in the early 1950s. Lewontin recruited like-minded scientists to his left-wing organisation *Science for the People* and in the process the science became thoroughly mixed up with politics. Left-leaning British scientists such as Steven Rose likewise joined the chorus of protest against ultra-reductionist and deterministic views of human biology, generating a valuable body of literature on this topic in the process (Rose et al., 1984; Rose, 1997).

The publication in 1975 of E. O. Wilson's *The New Synthesis* represents a pivotal moment in the history of nature and nurture (Segerstråle, 2000). The inordinate amount of press attention gained by the controversy that surrounded its publication, together with the controversial views of educational determinists such as Arthur Jensen, marked the beginning of the modern media emphasis on the nature-nurture controversy and the re-emergence of genetic determinism as an academically viable concept after long years as a fringe theory indelibly associated with dubious politics. Despite the title, much of the book was not actually new, but primarily aimed to draw together all the work on animal social behaviour that had been separately taking place in a number of disciplines. Wilson's genius was to see that these various strands could be united within an evolutionary framework, and to give the field a name (Griffiths et al., 2008; Schifellite, 2011). Sociobiology drew heavily on ethology, basically replacing the field (discussed previously), and on the theoretical work of Hamilton, Trivers and Price in evolutionary game theory, inclusive fitness and altruism, as well as using population genetics and Wilson's specialism on eusociality. The discussion of human sociobiology took up less than five per cent of Wilson's book, but it was this small percentage that aroused the most controversy. The aim for sociobiology, especially human sociobiology, was hegemonic in that it claimed to provide a better way of understanding human societies than the social sciences currently provided. Within this framework, those behaviours, cultures, social relations and formations, which had been the sole purview of the social sciences, were given genetic and evolutionary explanations (Schifellite, 2011).

The early sociobiologists certainly sounded like genetic determinists. As Wilson wrote: 'The individual organism is only the vehicle (of genes), part of an elaborate device to preserve and spread them with the least possible biochemical perturbation.... The organism is only DNA's way of making more DNA' (Wilson, 1975a, p. 3), and 'The genes hold culture on a leash. The leash is very long, but inevitably values will be constrained in accordance with their effects on the human gene pool' (Wilson, 1975a, p. 167). As Wilson assured his readers, in a passage which he may later have regretted: 'In hunter-gatherer societies, men hunt and women stay at home. This strong bias persists in most agricultural and industrial societies and, on that ground alone, appears to have a genetic origin.... My own guess is that the genetic bias is intense enough to cause a substantial division of labour' (Wilson, 1975b). Such deterministic language had not been heard since the strict hereditarians held sway in the early decades of the century, although in later works such as his *On Human Nature* (1978), Wilson sought to redress the balance by highlighting human freedom. In general, the following wave of sociobiological literature was somewhat more judicious in its claims, not least in response to feminists concerned about the deterministic stereotypes of female roles that were rife in such literature, by neo-Marxists concerned that the forces of economic change were being drowned in a wave of genetic determinism and by anthropologists who thought sociobiologists were being naive about the power of culture. The new wave was expressed most typically by C. J. Lumsden and E. O. Wilson in their book *Genes, Mind, and Culture* (1981), which allowed for a much larger role for cultural inheritance than found in the earlier sociobiological writings. If culture was normally held on a rather tight genetic leash in the first wave of sociobiology, in the later version the leash was envisaged as considerably more flexible. Richard Dawkins' *The Selfish Gene* (1976) sounded like genetic determinism given the powerful metaphor utilised in the title, but the author was also careful to point out in the book's concluding pages that memes ('culturgens' in the language of Wilson) as cultural units of replication had now taken over from genes in shaping culture, writing that 'we are built as gene machines and cultured as meme machines, but we have the power to turn against our creators. We, alone on earth, can rebel against the tyranny of the selfish replicators' (Dawkins, 1976, p. 215). What is striking in the language of both Wilson and Dawkins is the dualist framework in which reified genes, or even reified memes, are perceived as being in a tug-of-war with the human will.

A decade after Wilson's *Sociobiology: The New Synthesis*, the field of sociobiology was absorbed into evolutionary psychology, which now claimed to explain not only human behaviours but also the human mind. In the writings of Tooby and Cosmides, the 'adapted mind' was perceived as a modular structure, each module having been selected for during evolution to cope with a particular challenge in humanity's evolutionary past (Tooby and Cosmides, 1989, 1990). Ipso facto, claimed the evolutionary psychologists, there must be a set of genes involved in the development of such modules that was selected for in order to generate precisely such a 'module'. It was against this background of a renewed interest in gene-driven 'mind modules' that Herrnstein and Murray published *The Bell Curve* with its thesis that intelligence was primarily genetic, fixed and immutable (Herrnstein and Murray, 1994). With time the claims for the genetically driven explanatory power for such 'units' of human behaviour became ever more ambitious. *The Dark Side of Man: Tracing the Origins of Male Violence* (Ghiglieri, 1999) maintained that human males are by nature rapists, murderers, warriors and perpetrators of genocide, whereas in *A Natural History of Rape: Biological Bases of Sexual Coercion* (Thornhill and Palmer, 2000), the authors argued that rape is an adaptation to increase the reproductive success of men who would otherwise have little sexual access to women. The revival of biological explanations for human nature has often been associated with attacks on social anthropologists and sociologists as if they remained purveyors of a 'blank slate' view of the human mind. For example, in his book *The Blank Slate: The Modern Denial of Human Nature* (reviewed by Patrick Bateson under the title 'The Corpse of a Wearisome Debate'), Steven Pinker writes that '[m]y goal in this book is not to argue that genes are everything and culture is nothing – no one believes that – but to explore why the extreme position (that culture is everything) is so often seen as moderate, and the moderate position is seen as extreme' (Pinker, 2002, p. ix).

The increasing medicalisation of genetics, powered by the new and powerful techniques of molecular biology, also led to narratives in which the genes appeared to be in control. The teaching of Mendelian genetics in schools lent itself readily to the idea of 'one gene – one trait', with many students never being exposed to the fact that the vast majority of diseases with genetic involvement are polygenic in character. The identification of some 5,000 genetic diseases following a Mendelian pattern of inheritance appeared to support such a straightforward understanding, again the point somewhat being lost that, as a group, such diseases represent only a few per cent of the total disease load, with polygenic disease representing a much higher proportion.

The Human Genome Project also served to give genetics a central role in the public consciousness as a controller of human destiny. Launched in 1990, this $3 billion project to sequence the human genome reached its culmination in 2004 with the publishing of the full sequence (Consortium, 2004). By 2016 the cost of sequencing a human genome had fallen dramatically to around $1,000. The huge cost of the very first sequence was justified on medical grounds. As already mentioned in the Introduction, metaphors for the genome such as 'the Holy Grail', 'the Book of Life' and 'the Code of Codes' were all used. The 'blueprint' metaphor became very popular, replacing older, less deterministic terminology such as 'genetic lottery' (Condit, 1999). As some of the hype associated with the earlier phase of the Human Genome Project gathered steam in the 1990s, so it was met with a veritable barrage of books expressing concerns that biology was reverting to the bad old days of the stringent hereditarianism of the early twentieth century. Books in this genre included *Exploding the Gene Myth* by Ruth Hubbard and Elijah Wald (1993), *The DNA Mystique* by Dorothy Nelkin and Susan Lindee (1995), *The Rise of Neurogenetic Determinism* by Stephen and Hilary Rose (1995) and *Genetic Maps and Human Imaginations* by Barbara Katz Rothman (1998), all arguing that the 1980s had seen a major upswing in deterministic messages in popular media associated with the new genetics. In turn, this wave of literature was met with the response that the public did not as a matter of fact consistently interpret the new genetics according to the deterministic media stereotypes (Condit et al., 1998; Hedgecoe, 1998), a response already briefly reviewed in the Introduction.

The 1990s also saw a burgeoning of results emanating from new behavioural genetic research programmes which had been initiated in the 1980s with the inauguration of several very large-scale twin studies/databases, including the Australian Twin Registry in 1980, the Swedish Adoption/Twin Study of Aging in 1984, the Virginia 30,000 in 1986 and the Minnesota Twin and Family Study and Minnesota Study of Twins Reared Apart in 1989. These studies all tended to highlight gene-environment interactions as critical in the development of human behavioural traits. When a group of professional UK geneticists was interviewed in the late 1990s, the researchers reported that

> our respondents resorted to the use of what we have called the popular parable of the nature-nurture pendulum. According to this story, early geneticists, who were also eugenicists, discounted nurture in favor of nature. Following the Nazi atrocities of World War II and public distaste for eugenics, the pendulum is said to have swung in favor of nurture, and social scientists had a larger role in the study of behavior and disease. Now, after the supposed failure of this approach, the pendulum is at equilibrium, where new geneticists are melding together an understanding of nature and nurture. (Kerr et al., 1998)

Overall there seems to be a considerable amount of truth in that narrative. We started Chapter 1 by posing the question: Does the historical debate help us understand the enduring appeal of describing humanity within dichotomous linguistic frameworks? Considering the brief overview in these two chapters of the discourse between nature and nurture, covering a span of more than 2,000 years, what is most striking is the extreme swings of opinion that have occurred on the matter. As the discussion became more informed by science from the nineteenth century onwards, the swings became, if anything, more dramatic, not less. The scientific discussion has been marked by a dialectical approach in which strongly contrasting positions have been pitted against each other, the strength of opinion nowhere near being justified by the data available at the time. Nevertheless, by the late twentieth century the consensus for most people was settling down into some form of nature-nurture/gene-environment interactionism. But in that very dichotomous language lies a problem, as it tends to reify these interacting components as if there are two separate forces competing in dualistic fashion for their influence over the identity of human personhood. Fortunately the science of the past few decades has now begun to paint a very different picture of the human person that thoroughly subverts all forms of dichotomous description, and it is the aim of the rest of this book to paint that picture as accurately as the most recent scientific literature will allow.

CHAPTER 3

The Impact of the New Genetics?

How Contemporary Biology is Changing the Landscape of Ideas

> *Concepts that have proven useful in ordering things easily achieve*
> *such authority over us that we forget their earthly origins and accept*
> *them as unalterable givens. Thus they come to be stamped as 'neces-*
> *sities of thought', 'a priori givens', etc. The path of scientific progress*
> *is often made impassible for a long time by such errors.*
>
> Albert Einstein (1916)

> *Since it has become evident that genes interact with their environ-*
> *ment at all levels, including the molecular, there is virtually no inter-*
> *esting aspect of development that is strictly 'genetic', at least in the*
> *sense that it is exclusively a product of information contained within*
> *the genes.*
>
> J. L. Elman (1996, p. 21)

> *The gene does not lead. It follows.*
>
> M. J. West-Eberhard (2003)

There are two powerful trends in contemporary biological research programmes. The first is arch-reductionism. Over the course of the twentieth century when physicists were giving up on the Newtonian universe
in favour of less mechanistic and less deterministic understandings of the
properties of matter, biochemists and molecular biologists were going in
the opposite direction, dissecting living matter at the molecular level with
techniques of ever increasing sophistication. In my own field of molecular immunology, in which the immune system is broken down into its
molecular components, models of the signalling pathways that regulate the
immune system often look more like Lego kit constructions than anything
resembling a living organism. The same is often the case with diagrams
illustrating the information flow leading from the genome to the many

intermediary molecules that then regulate the cell's growth, metabolism and replication.

But the second trend, becoming increasingly powerful as a research strategy, is that of the systems approach in which the living cell or whole living organism is investigated as a complex system in which the myriad components are completely integrated. The massive increase in the use of genetically modified animals, especially mice, in biological research, has opened up new ways of investigating the role of molecules in the system as a whole. In parallel, it is increasingly possible to model the workings of at least some biological systems in silico using powerful computing techniques. At first glance, it might seem that the systems biology approach must exist in some kind of tension with the strictly reductionist approach, but in practice this is not the case, or certainly need not be the case, because it is precisely the better understanding of the components of the system which then leads to better approaches to considering the system as a whole.

In this chapter we will see how contemporary biology is helping view living organisms as integrated complex systems. In particular we will introduce eleven different biological concepts and findings, some of them very recent, which are relevant to the way in which genetic information is processed and utilised within living systems. In the following chapter we will then introduce a matrix for understanding the development of a human being, from the fertilised egg all the way to the grave, to see how these new biological insights shed light on the way in which human personhood emerges. Those already familiar with DNA and the molecular understanding of the gene should have no problem with what follows. Others who wish more background information may find it in introductory textbooks of genetics or, more briefly, in *The Language of Genetics*(Alexander, 2011).

Eleven Key Aspects of Contemporary Biology

It is easy to receive the impression from biology textbooks that the genome is like a recipe and that the flow of information is one-way – the nuclear DNA as the centre of power, ruling over the functions of the cell and, thereby, the organism as a whole. Nothing could be further from the truth. When I was reading biochemistry in the Oxford of the mid-60s, the so-called central dogma, introduced by Francis Crick in 1958, reigned supreme. This is the idea that the flow of information is from genes to proteins, but not the other way round. Crick was not sufficiently sensitised to the nuances of the word 'dogma', but quickly became so, later confessing that its use caused 'almost more trouble than it was worth'. The phrase has, in any event, long since disappeared from the molecular biological literature as the sheer complexity of the system has become more apparent. Much of the discovery of this complexity was itself triggered by the advent of genomics, with its focus on DNA sequencing, which in turn led to a much greater appreciation of the role of the non-protein-coding sections of the genome.

Not one but many dogmas have been scattered to the winds by recent biological discoveries. We now know that the size and complexity of the genome has no obvious relationship to the size and complexity of the organism. We also know that a mere 21,000 human genes can encode for more than a million different proteins by mechanisms described later in this chapter (Mueller et al., 2007). The genome of each cell in the body is very likely to be slightly different and identical twins are not really identical. Genetic information from a single gene is not necessarily converted into the same protein information. Environmentally induced information can even be inherited across several generations. These and other fascinating findings from contemporary biology will be described further. The aim in presenting the following list is not to discuss each point in any detail, but rather to see how the genome is embedded as but one component in a complex system in which each component contributes a critical functionality to the system as a whole, a system which is highly sensitive to environmental inputs.

Development

The theme of development is one of the oldest topics in biology, going back to Aristotle's wonderful embryological observations of chick development described in his *Historia Animalium* in the fourth century BC. But occasionally the theme of biological development has been eclipsed for a while, or at least come in conflict with other powerful theories. For example, in the first half of the eighteenth century there were vigorous debates about reproduction between the so-called preformationists and epigeneticists (Roe, 2010). 'Preformationism' referred to the belief that new organisms came from 'germs' which represented preformed organised matter that had originally been brought into being by divine creation, and the idea gained popularity as a reaction against Cartesian mechanical views of reproduction. By contrast the epigeneticists, who supported what we would now call developmental biology, saw the generation of each new organism during reproduction as entailing the formation of new order out of disorganised matter. As embryology became better understood during the nineteenth and early twentieth centuries, epigenesis became the view that 'the characters of the adult do not exist already in the newly fertilized germ, but on the contrary arise gradually through a series of causal interactions between the comparatively simple elements of which the egg is initially composed' (Waddington, 1952, p. 156). In the early part of the twentieth century this view was clearly being overshadowed by the rise of genetic determinism, leading Leonard Carmichael to protest, in words already quoted, that because of the developmental process 'the question of how to separate the native from the acquired in the responses of man does not seem likely to be answered because the question is unintelligible' (Carmichael, 1925). By 1952, with the post-War suspicion of genetic determinism at its height, we find the biologist Conrad Waddington commenting that '[t]here can be no doubt nowadays that this epigenetic point of view is correct', although the word 'epigenetic' has once again changed its meaning since Waddington's time, as will be discussed further. In any event, a mere decade later molecular biology became the dominant voice in the biological sciences, and once again the central importance of development in defining the characteristics of adult organisms became somewhat lost in the language of 'genetic programmes' and 'central dogma', with a revival of the preformationist idea that the adult organism was somehow a direct phenotypic read-out of the genomic instruction manual. I suspect that the biologist Gilbert Gottlieb (1929–2006) may have sometimes felt that he was ploughing a lonely furrow as, from the 1970s onwards, he began to insist that a developmental systems approach was the only way to understand living organisms in which genetic, neural, behavioural and environmental contributions to growth and development were all integrated into one complex system in ways that defied the (then) central dogma of genetics (Gottlieb, 1992). Indeed, despite his huge influence on the field, Gottlieb is quoted as saying, shortly before his death, that 'getting across the developmental point of view has been the largest failure of my career' (Keller, 2010a, pp. xi–xii), though in retrospect, given Gottlieb's major influence on the field, few now would accept the validity of that comment.

Today such ideas are commonplace – not least due to the impact of such pioneers in the field – but the landscape has in any case changed for, I suggest, four main reasons. The first is the way in which evolutionary developmental biology, evo-devo, has come to occupy such a prominent position in evolutionary thinking, spilling over to impact more broadly on the role of genes in developmental processes (Muller, 2007). The second, ironically enough, is that the methods of molecular biology have now been applied to investigate the molecular mechanisms of developmental biology with startling success, in particular demonstrating the various ways in which environmental inputs integrate with the genomic system to bring about developmental change. The third reason is that an increasing understanding of genetic regulation by proteins and by epigenetic modification (as will be discussed later) has highlighted the idea of development as a complex system in which DNA plays an important role, but by no means the only role. And the fourth reason is the recognition that differences in the environment during early human development make measurable differences to the health and well-being of the adult individual decades later.

The development of biological organisms is a continuum from fertilisation to death. Around one million of our own cells die every second, fortunately to be replaced (Gregory, 2009). Living organisms are always in the process of becoming: they are never static. Development is the key to the subversion of all forms of dichotomous language and is a theme that will crop up repeatedly through this book. In the following chapter we consider human development in greater detail. For the moment we will continue with our list of the eleven ways in which the genome is embedded as but one component in a complex system.

The Variable Products of the Genes

In the earlier understanding of the genome as a recipe, the products of the genes were 'read off' as if each gene had one fixed and determinate product. We now know that this is not the case: most genes can generate several different proteins and their amino acid sequence is not invariably specified by the sequence of base-pairs in the DNA. Our understanding of the genome has now changed radically from a 'read-only' concept, in which the genome was perceived as a static repository of information, to a 'read-write' model in which the genome is under constant dynamic regulation and modification (Shapiro, 2013). Five of the main mechanisms involved will serve to illustrate this point.

Alternative Splicing

A protein-coding gene encodes both introns (for *intragenic* sequences) and exons (for *expressed* sequences). The exons refer to the DNA sequences within the gene that end up encoding the protein, whereas the introns are 'spliced out' and their sequence does not encode any of the protein

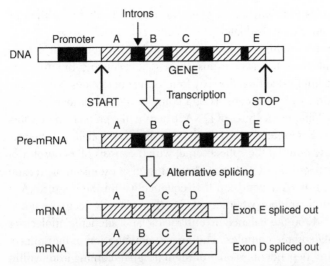

FIGURE 2 Alternative splicing of one gene to generate two different mRNA products. The hatched regions represent the exons.

sequence. So first the whole gene is transcribed into pre-mRNA and then the introns are spliced out to form the mature mRNA as illustrated.

Figure 2 shows five exons in the gene, labelled A to E. Four black introns appear between the exons. The exons contain the coded nucleotide sequence that will eventually be translated into the amino acid sequence in the protein. The pre-mRNA is processed whilst still in the nucleus to splice out the introns, and the exon sequences are then joined up together to form the mature mRNA, which then transits to the cytoplasm (the part of the cell outside the nucleus) to be used as a template for translation into proteins. Once the exons are joined up, they form a correct and continuous sequence of triplet codons. We can say that the protein-encoding sequence of the mature mRNA message can be read as a complete sentence in one breath, without any commas or semicolons.

The beauty of this system is that it allows for different combinations of exons to generate different mRNAs, and therefore different proteins (known as 'isoforms') with different functions. In the example shown, exons A to C are invariant – they are always used – whereas exons D and E are variable. Two mRNAs can be formed that contain A to C plus either D or E, and these mRNAs are then translated into two different proteins. In reality, data from the ENCODE project show that many genes generate far more than two alternative proteins: around 95 per cent of the 21,000 human protein-coding genes undergo alternative splicing, generating about 100,000 different proteins from our 21,000 genes (a number further amplified by other mechanisms described later) (Bernstein et al., 2012).

The genome contains a code within a code in the sense that certain features identify the introns and exons where splicing occurs. But this code is a lot more complicated than the sixty-four triplet nucleotides that encode the amino acids. Instead it consists of a complex algorithm that includes more than 200 different features of DNA structure that predict where splicing will occur (Barash et al., 2010; Ledford, 2010). The splicing of the pre-mRNA is carried out by the spliceosome, which consists of a complex of small RNA molecules and proteins. The inclusion or exclusion of an exon in the mature mRNA depends on the sequence within the pre-mRNA at the intron-exon boundary; whether splicing then actually occurs or not at that site depends on the balance of 'enhancing' and 'silencing' molecules within the spliceosome complex. The precise composition of the spliceosome is, in turn, dependent on environmental signals coming from within the cell and from the wider environment of the individual. The important point here is that the information regarding the structure and function of the protein that is produced is only partially dependent on the DNA sequence of the gene. It is the environmental context within the cell which imposes splice site specificity via the spliceosome.

RNA Editing

Whereas alternative splicing works by re-assembling and swapping around segments of primary DNA sequence, RNA editing is in a sense more radical in that it changes the primary sequence of the mRNA during or after its transcription, and this can happen also in many other types of RNA. This happens by the insertion, deletion or substitution of one of the four nucleotides in RNA at selected sites. The RNA so produced is called a 'cryptogene', so-called because the 'image' in the DNA of the resulting products is now no longer recognisable in the DNA.

RNA editing, though less well-known than alternative splicing, is by no means a rare esoteric mechanism, but rather reflects the normal functioning of both plant and animal cells. More than 100 million editing sites have been reported in human RNA (Bazak et al., 2014), although only a relatively few of these cause a change in the amino acid sequence following translation (Ulbricht and Emeson, 2014). The majority lie within ion channels and transporter proteins that convey essential chemicals across the membranes that enclose cells and are of particular significance in the functioning of nervous tissue.

One example of great relevance to the animal brain, including the human, will serve to illustrate (Holmgren and Rosenthal, 2015). Glutamate is a key brain neurotransmitter that communicates fast excitatory signals across the synaptic junctions that functionally connect one neuron (nerve cell) to another. Like other transmitters, glutamate is secreted by one neuron into the small gap at the synaptic junction that separates it from the neuron to which it communicates. Crossing this small gap (the 'synaptic cleft') the glutamate then binds to a specific receptor that opens a channel in the membrane through which generally sodium, but sometimes calcium, then flows. This then modulates the functions of the neuron that receives the message. In one particular glutamate receptor (there are several types), at the 586th codon in the gene that encodes the receptor, RNA editing of the mRNA results in the amino acid glutamine being replaced with an arginine in virtually all the receptors of that type. This tiny change is sufficient to change completely the properties of the receptor. The receptor now no longer transports calcium, and its efficiency of conductance is reduced by tenfold, whereas other glutamate receptors retain the amino acid glutamine at the 586th codon and are fully functional. What is the point of rendering the RNA-edited glutamate receptor non-functional in this way? The answer seems to be that the arginine-containing 'blocked' receptor is dominant in its effects, competing out the other functional glutamate receptors. So the edited receptor seems to play a critical role in 'quietening down' the excitatory functions of glutamate. How vital this is for normal brain function has been demonstrated in many different ways. For example, if a version of the receptor is made non-editable in transgenic mice, this leads to epileptic seizures. In fact, expressing different levels of the non-editable version in mouse brain led to a range of behaviours, from lethargy to hyperactivity, depending on the expression level of the mutated gene. In some humans suffering from amyotrophic lateral sclerosis (a disease made famous by one of its high-profile sufferers, Prof. Stephen Hawking), a similar mutation occurs in spinal neurons and may contribute to the selective neuronal death observed in these patients (Kawahara et al., 2004), raising some interesting therapeutic possibilities (Yamashita and Kwak, 2013).

Translational Recoding

Translational recoding represents a further mechanism whereby mRNA can be 'reinterpreted' by cellular machinery (Dinman, 2012). For example, the translational system which uses the mRNA to synthesise proteins starts at a different nucleotide from the normal one to start the translation process. This 'frameshift' would normally lead to nonsense because the codons are read as triplets and a shift of one or two nucleotides will put the sequence out of frame so that it no longer makes sense. But in rare cases, such a shift creates a whole new frame, generating a meaningful sequence of codons, and a completely different protein is made in place of the usual one.

RNA Modifications

Cellular RNA molecules can be chemically modified more than a hundred different ways, often in response to environmental cues and in a reversible manner (Wang and He, 2014), thereby introducing multiple extra levels of regulation.[1] For example, pseudouridine is the most common modified nucleotide across all RNA species, in which an enzymatically regulated chemical modification is made to uridine, one of the four nucleotides that make up the structure of RNA. The presence of pseudouridine in mRNA leads to several changes in RNA structure that are thought to impact on functionality, including translational efficiency and localisation in the cell. Intriguingly, the presence of pseudouridine can also change the way in which the translational machinery 'reads' the codon that contains the pseudouridine, so that the mRNA is 'interpreted' in such a way that there is a different amino acid in the resulting protein (Carlile et al., 2014). Such changes can occur in response to environmental inputs, results which 'suggest a mechanism for the rapid and regulated rewiring of the genetic code through inducible mRNA modifications' (Carlile et al., 2014).

Post-translational Protein Modifications

Once a protein has been synthesised, there are at least a dozen different ways in which chemicals can be attached to one or more amino acids within the protein, thereby markedly changing its properties. These are dynamic changes which can be reversed in seconds in most cases, thereby providing many finely tuned levels of control and further ways in which the environment integrates with cell physiology.

Phosphorylation of proteins, entailing the attachment of a phosphate group to one or more amino acids within a protein, is one of the most important modifications. More than half of all human proteins are regulated by phosphorylation, often leading to striking changes in the functions of the protein (Wilhelm et al., 2014). A further critical modification involves glycosylation whereby carbohydrate structures are added to the protein, often dramatically changing its properties. For example, the insulin given to patients suffering from Type 1 diabetes needs to be correctly glycosylated to exert its full pharmaceutical effect. Glycosylated proteins play a critical role in the development of the correct synaptic architecture of the brain. The attachment of fatty acids to specific amino acids represents another important post-translational protein modification as this renders the proteins more hydrophobic ('water-hating') and can relocate them into the lipid-rich plasma membrane (Charollais and Van Der Goot, 2009). Loss of lipid modifications in humans due to mutations can lead to severe disorders, such as disruption of normal sex development (Callier et al., 2014), and such modifications are vital for the normal functioning of brain cells (Kang et al., 2008).

It has already been mentioned that about 100,000 different proteins can be generated from our 21,000 genes by means of alternative splicing. Now add in RNA editing, translational recoding and pseudouridylation, together with the many ways in which post-translational modifications can change the functions of the resulting proteins and it becomes clear that such combinatorial mechanisms can easily generate more than a million human proteins with varying functions. An apparently rather modest number of protein-coding genes within a genome can therefore give a rather misleading impression as to its eventual scope. It should also be noted that, whereas the molecular machinery that results in post-translational protein modifications is itself encoded by the genome, the dynamic regulation of protein functions also depends on multiple physical parameters that are not encoded by the genome, such as their precise conformation, how hydrophobic or hydrophilic they are and the more acidic or more alkaline nature of their immediate environment, all of which in turn impacts upon protein functionality.

Transcriptional Regulation

The proteins that regulate gene expression – the switching on and off of genes – are known as 'transcription factors'. Their name depends on the fact that they regulate which genes are transcribed into mRNA. Transcription factors bind to specific DNA sequences located near to protein-coding genes known as 'promoters', which guide the transcriptional molecular machinery to start transcribing the gene at the 'transcriptional start site'. Other DNA sequences known as 'enhancers' and 'silencers' are also regulated by the binding of transcription factors, some located quite far from the gene that they help regulate. Many enhancers may be involved in regulating a single gene (Spitz and Furlong, 2008). Some promoters and other regulatory sequences require the binding of a whole cluster of different transcription factors before the gene is in the fully 'on' position. Transcription factors can also exert either a positive or a negative effect on the regulatory sequences. In some cases the binding factors increase in number until a certain critical threshold is reached at which point the gene flips to the 'on' position (Spitz and Furlong, 2008). So the actions of transcription factors need to be understood in an integrated, combinatorial manner.

The ENCODE project aims to delineate all the information-containing sequences encoded in the human genome. The initial wave of results from this project has identified around 400,000 regions with enhancer-like features in the human genome, together with about 70,000 promoter-like regions. In all, 636,336 binding regions representing 8.1 per cent of the genome are enriched for sequences recognised by DNA-binding proteins, indicating the huge extent of DNA regulation by proteins (Bernstein et al., 2012). Around 560 different transcription factors have been identified in human cells at the protein level, although the eventual total number is likely to be higher than this (Wilhelm et al., 2014). At any one time there are some 250–300,000 transcription factor molecules in every eukaryotic cell – that is, cells (like our own) that contain a nucleus (Simicevic et al., 2013). Every aspect of the living organism's existence is regulated by the gene regulatory networks within which transcription factors play a vital part, be it development, brain function, reproduction, digestion, muscle action or the functioning of the immune system which protects our bodies against foreign invaders such as bacteria and viruses.

There is a constant traffic of information coming from the environment and impacting the expression of either a few highly selective genes or hundreds of genes depending on the signals involved. When we choose our body state, whether it be by lack of sleep, too much exercise or too little, over-indulgence in food or drink, starvation, risk of infection by unclean habits or the choice to live angry lives, we are ipso facto influencing the status of our genomic system. For example, when participants in a study were exposed to just one week of insufficient sleep (average 5.7 hours per night) and compared to a control group who slept an average 8.5 hours sleep per night, no less than 711 genes were found to be expressed at either higher or lower levels in the sleep-deprived compared to the control group (Moller-Levet et al., 2013). The genes affected included many involved in circadian rhythms, metabolism, stress and immune responses. Of course not all the regulation of gene expression in such a study can be ascribed to transcription factors alone – non-coding RNA, micro-RNAs and other RNA regulators are all very likely involved in the complex web comprising the regulatory network that switches genes on and off.

A second example will also help illustrate the way in which the environment 'talks' to the genome by means of transcription factors. Let us imagine that you choose to take a high-flying job in the city which is highly paid but intensely stressful. As a result your adrenal cortex, located just above the kidneys, will very likely start to secrete more of a steroid hormone called cortisol (one of a class of chemicals known as 'glucocorticoids'). A region of the brain known as the hypothalamus controls the level of cortisol secretion, increasing it in response to stress. The good news is that those in leadership positions have been reported to have lower levels of cortisol (Sherman et al., 2012), so perhaps the best strategy is to get to the top as quickly as possible. Cortisol has numerous effects on the body, including raising blood sugar, modulating metabolism and suppressing the immune system, useful from an evolutionary perspective in terms of preparing the body for stressful situations in which a good supply of energy is important, and less useful perhaps in preparing the body for a day in the office, especially considering that cortisol is implicated in increased blood pressure. Cortisol enters its target cells to bind to glucocorticoid receptors found in the cytoplasm – receptors can be located within the cell as well as on the cell surface. The glucocorticoid receptor is itself a transcription factor. Upon cortisol binding the protein unfolds to reveal nuclear localisation sequences which then target the receptor to the nucleus. Once there it binds to enhancer and promoter regions where it forms a complex with a collection of other specific proteins that together ensure efficient binding and initiation of gene transcription (Ratman et al., 2013). Depending on the exact composition of this complex, the various forms of the glucocorticoid receptor can exert both positive and negative effects on gene expression, helping explain the very broad range of outcomes in different cells that occur following cortisol binding to its receptor.

So taking that stressful job in the city can have a major impact on the regulation of your genome, in turn leading to changes taking place in billions of your body's cells. Such examples could be multiplied 100-fold, illustrating the many different ways in which environmental information is conveyed to the genome, with combinatorial control by a myriad transcription factors and their associated co-factors regulating the fine-tuning of gene expression. The idea of linear signalling pathways inside cells has been replaced by a scenario in which vast interacting communication networks operate to regulate the genome, and it is the system as a whole which functions like an orchestra to produce the 'right' music.

Epigenetics

The term 'epigenesis' and its earlier eighteenth-century interactions with the 'preformationists' have already been discussed. In more recent decades the term 'epigenetic' was used to describe the overall events that occur during the development of an organism, often referring to 'epigenetic rules' with its implicit assumption of ignorance concerning the underlying molecular events. All that has now changed. Although the term is still in use with this more historical meaning, a far more common contemporary usage is in reference to the molecular changes that lead to the regulation of gene expression. Epigenetic regulation therefore represents a quite distinct level of gene regulation from that involving transcription factors, of equal or even greater significance as a way in which the environment communicates with the genome (Crews et al., 2014).

The term 'epigenetics' derives from combining the Greek *epi*, meaning 'over' or 'above', with 'genetics'. Today epigenetics refers to all those inheritable changes that occur in chromosomes that bring about changes in the phenotype without any changes occurring in the DNA nucleotide sequence. 'Inheritable' in this context may simply refer to the transmission of information from parent cells to daughter cells during replication, or it can refer to inheritance across generations. If the DNA sequence is the musical score on the page, then epigenetics is all those Italian words (*fortissimo* and the like) that give instructions as to how the notes should be played.

There are three known mechanisms that mediate epigenetic changes – DNA methylation, chromatin modification and microRNA – and the resulting chemical changes in the DNA are known as 'epigenetic marks'. Methylation involves the transfer of a methyl group to a cytosine nucleotide in the DNA. When cytosines in a gene promoter are methylated, this often causes the gene, in vertebrates at least, to be 'silenced' or 'switched off' (Schubeler, 2015). If a cytosine (C) nucleotide has a guanosine (G) nucleotide next to it (known as a CpG pair), it is more likely to be recognised as a site for methylation by the transfer machinery. Gene promoters often contain lengthy clusters of CpG repeats known as 'CpG islands': in the human genome, approximately half of gene-promoter regions contain stretches of sequence that are CpG-rich. One way in which methylation down-regulates gene expression is by preventing the binding of transcription factors to promoters. Conversely, removal of the methyl groups, known as demethylation, can be used to switch a gene on again. In mammals, 70–80 per cent of CpG units are methylated at any one time, whereas only 2–5 per cent of all cytosines are methylated.

To understand the second epigenetic mechanism, involving chromatin modification, it is first necessary to appreciate that the DNA is tightly inter-twined with proteins inside the cell nucleus, most of which are histones. Whether the chromatin is in a very tightly bound 'closed state' or more loosely bound 'open state' determines whether transcription factors have access to genes to switch them on and off. In eukaryotes, 147 base-pairs of DNA are wrapped around eight molecules of histone that come in pairs. This little unit is known as a nucleosome. Chromatin consists of lots of nucleosomes packaged together. Each histone has a tail that sticks out from the nucleosome. That is really significant because it enables enzymes to make chemical modifications to the tail that in turn change the structure of the nucleosome. Acetylation involves the enzyme-catalysed transfer of an acetyl group to an amino acid called lysine in the histone tail. Histone acetylation causes the nucleosome to loosen its tight grip on the DNA, so facilitating increased gene transcription by enabling transcription factors to bind to their promoters. Enzyme-catalysed removal of the acetyl groups, known as deacetylation, causes the reverse: silencing of the genes. As with DNA, histone tails can also be methylated. In general, methylation has the opposite effect to acetylation: histone methylation leads to gene silencing and demethylation leads to more gene transcription. The sum total of the specific combinations of histone modifications at a promoter is referred to as the 'histone code'.

The third epigenetic modification besides CpG methylation and histone modifications that is receiving increasing attention involves micro-RNAs, and their role seems to be that of 'directors' as they steer the methylation and acetylation molecular machinery to the appropriate genes. Further integration is mediated between these mechanisms by the fact that cytosine methylation attracts 'repressor complexes' that contain the enzyme that deacetylates histones, so increasing the 'closed chromatin' state in which accessibility of transcription factors to activating regulators is restricted. In general, cytosine methylation contributes to long-lasting changes in gene expression that can last for the life of the organism – like permanently crossing out lines of text in a book that you read on your laptop – whereas regulation by histone acetylation and methylation can be more dynamic, responding to short-term changes in the environment, as if you highlighted sentences in the book but then removed the highlighting the following day.

Epigenetic marks are important for regulating genes. It is of great significance that these epigenetic marks are preserved when DNA is replicated during cell division. When the DNA replicates during cell division, each strand serves as a template for the synthesis of a new daughter strand. The template strand contains the methylation pattern, and certain enzymes specialise in recognizing methylated cytosines in the template. As soon as the new CpG unit is made in the daughter strand, the enzymes simply place a methyl group onto the newly made CpG in exactly the right spot, so that the new strand is not only the exact replica of the template at the nucleotide level but also contains methylated cytosines in the right places as well. But the fidelity of this replication mechanism is not nearly as accurate as for replication of the DNA double-helix itself. The error rate is about one in a thousand for the transfer of epigenetic marks through a cell division, whereas for the nucleotide sequence, the error rate is more like one in a million (Petronis, 2010). A more complex mechanism ensures that acetylation marks, likewise, are copied to the next generation of cells.

That epigenetic marks can be passed from a parent cell to its daughters when it divides is important for maintaining differences among cells in your body. When cells begin to differentiate into specialised cells during early embryonic development to become brain cells, skin cells, liver cells and so forth, this commitment is a one-way street – fortunately for us. This means that when liver or skin cells divide, as they do throughout life, they will still retain their specialised functions. At a molecular level this can only happen because every cell of the body (except red blood cells) contains a complete genome, but different cohorts of genes become switched off, the particular cohort determining the properties of the cell as it specialises, and they remain permanently in the 'off' position for the rest of life. Passing epigenetic marks on in the DNA of each generation of specialised cells helps preserve the characteristics of their particular lineage.

Given the role of epigenetic 'marks' to maintain the stability of cell lineage commitments, even through many cycles of replication, it might not at first seem obvious that epigenetic regulation of gene expression is also a dynamic environmentally influenced process, but this is indeed the case. First, the development of the fetus involves waves of epigenetic modifications that can be modulated by environmental inputs in ways that impact on the health and well-being of the individual for the rest of their life. Some examples will be provided in the next chapter when discussing early human development. Second, dynamic epigenetic regulation of gene expression is important in the daily life of the organism in response to environmental inputs. Third, environmentally induced epigenetic modifications in an individual organism can be transmitted transgenerationally, in some cases for many generations in the case of certain animals and plants (Jablonka and Raz, 2009).

One animal model that well illustrates the way in which post-natal environmental differences lead to permanent behavioural changes in the adult via epigenetic regulation is provided by the maternal care of rat pups (Meaney, 2010). Rat pups like lots of licking and grooming by their mothers during the first week of life. More surprising is the fact that the effects of that one week of post-natal experience has an impact on the rat that remains for the rest of its life. Some mothers are naturally good at such care for all their pups, others less so. When the rats are later exposed in adulthood to mildly stressful experiences, the ones who received the most licking and grooming during the first week of life remain the most calm, having a lower startle response and less cortisol production (increased cortisol is a sign of stress, in the rat as in the human). Conversely, the ones who received poor maternal care overreact to such stressful stimuli. By swapping litters of pups between the 'caring' and 'uncaring' mothers on post-natal day one, it has been shown that such behavioural traits in adulthood are predicted by the licking and grooming received during week one, not by the genetic inheritance received from the parents. And by week two, the maternal care effects disappear – it is grooming and licking in week one that counts, not during week two. The caring or non-caring traits persist on into the next generation, because the females demonstrate either caring or non-caring responses to their own new litters in turn.

So how does this work at the molecular level? Epigenetics plays a key role in connecting the maternally received care or indifference of post-natal week one with the rat's responses to stress in later life. Licking and grooming stimulates the rise of the neurotransmitter serotonin in the hypothalamus, the area responsible for converting short-term into long-term memories. Serotonin in turn causes increased expression of the glucocorticoid receptor right into adulthood (a receptor which we have already encountered earlier whilst discussing the role of transcription factors). Epigenetics comes into the story when serotonin triggers a chain of events inside the cells of the hypothalamus (the details need not detain us) that cause a marked reduction in the methylation of the promoter for the glucocorticoid receptor gene, so increasing receptor levels. The reduction in methylation is highly specific to that region of the DNA and is only induced by high grooming and licking in post-natal week one, not during week two. But in the pups exposed to less grooming and licking, the methylation level remains relatively high. The fact that the environmental difference during the first post-natal week can induce such long-lasting changes in a small region of DNA that in turn impacts so greatly on differential behaviour in adulthood is indeed remarkable. Yet, at the same time it needs to be emphasised that multiple pathways are likely to be involved in such responses. The expression of hundreds of genes is different between high-care and low-care rat offspring, with measurable epigenetic changes at multiple promoters and enhancers (McGowan et al., 2011).

Although hundreds of such plant and animal studies could be cited to illustrate the environmental impact on the epigenetic status of the adult organism and their offspring, such studies are more difficult to reproduce in humans due to the inaccessibility of relevant tissues. In practice, changes generally have to be investigated in saliva samples or blood cells – a complex mixture of different cell types that may not reflect changes happening elsewhere in the body – although sometimes harmless biopsy of other tissues, such as muscle and adipose (fat), are feasible. Some intriguing results have been reported. For example, smoking in a male cohort was associated with lowered methylation of a regulatory region of a gene that encodes a receptor that mediates the detoxification of the components of tobacco smoke (Philibert et al., 2013). Furthermore, this same receptor gene was also under-methylated in the offspring of mothers who had smoked during pregnancy (Joubert et al., 2012). Vigorous exercise leads to reduced methylation in the promoters of a wide range of genes in the muscles with their concomitant increased expression (Barres et al., 2012). Body mass index, a measure of an individual's weight, correlates with differences in the methylation status of a particular gene in adipose tissue, although in such studies it is difficult to discern cause and effect (Dick et al., 2014). An 'epigenetic clock', based on the varying methylation status of 353 CpG sites in tissues from people aged one day to ninety-nine years, matches remarkably well with the actual chronological age of the tissue donor (Horvath, 2013).

Human psychological states also appear to correlate with marked differences in epigenetic status and gene expression. Extended periods of stress, threat or uncertainty lead to epigenetic changes which result in increased expression of genes involved in inflammation and a decreased expression of genes involved in antiviral responses. Furthermore, people who have high levels of eudemonic well-being, the kind that stems from having a deep sense of purpose and meaning in life, also have a favourable profile of gene expression in their immune cells, with low levels of inflammatory gene expression and high expression of antiviral and antibody genes. Conversely, those with high levels of hedonic well-being, arising from continual self-gratification, show just the opposite (Fredrickson et al., 2013).[2]

The transgenerational inheritance of epigenetic modifications has received much attention (Lim and Brunet, 2013; Heard and Martienssen, 2014; Szyf, 2015). In this context, it is important to distinguish parental effects that are intergenerational (for example, malnutrition of the fetus in utero leading to epigenetically mediated effects in the adult offspring) and genuinely transgenerational effects that are found in the offspring of those adults who have not been exposed to the original trigger or environmental change. Long-lasting transgenerational effects are best known in plants. An epigenetic change, which may be triggered by an environmental change, is known as an epiallele. In plants, such epialleles can be transmitted down the generations for centuries. For example, the famous naturalist Carl Linnaeus first described a variant of a wild flower called the toadflax back in the eighteenth century. In this variant the symmetry of the flower is changed. More recently it has been shown that the silencing of a specific gene by methylation explains the observed toadflax variation (Cubas et al., 1999). Could such long-lasting changes occur in animals also? In small animals, maybe. For example, in the nematode worm C. elegans, which will be receiving more attention in Chapter 5, exposure to certain smells in early development affects adult responses to those same smells, a behavioural change that can be transmitted over more than forty generations (Remy, 2010). But in mammals it is very unlikely that such long-term environmentally induced epialleles will ever be detected over more than a few generations for the simple reason that the great majority (though not all) of the epigenetic modifications gained during life are wiped from the genome during transmission of parental DNA to the next generation. Having said that, in rodents there is a growing number of examples in which epigenetically mediated behavioural or other changes are transmitted in the gametes (sex cells) over at least two generations (Radford et al., 2014). In one experiment, genetic manipulation of the sperm in a male mouse resulted in loss of histone methylation associated with multiple genes leading to impaired development and survivability in the offspring (Siklenka et al., 2015). Even though the genetic manipulation that changed the methylation status was not present in the offspring, nor in subsequent generations, the impaired development was still observed in two further generations, demonstrating transgenerational transmission.

In humans, intergenerational epigenetic changes have been studied in the offspring of parents who were undergoing environmental stress during the time of fertilisation and/or pregnancy. A famous example involves the children born of parents who suffered malnutrition in utero due to the Dutch Hunger Winter of 1944/45, children who were found to be fatter during their early years of life and suffered a greater measure of ill-health in later life compared to controls (Painter et al., 2008). The offspring of pre-natally undernourished fathers, but not mothers, were heavier and more obese than offspring of fathers and mothers who had not been undernourished prenatally (Veenendaal et al., 2013). Individuals who were prenatally exposed to famine had, six decades later, less DNA methylation of a specific gene, relevant to metabolism, compared with their unexposed, same-sex siblings (Heijmans et al., 2008), but it is not (as yet) known whether that methylation difference is transmitted to the next generation.

Claims for transgenerational epiallele inheritance in humans are difficult to demonstrate unequivocally. Nevertheless, there is a growing accumulation of observations in the literature that point in this direction, albeit based on plausible explanations rather than direct evidence. For example, extensive studies have been carried out on the Överkalix cohort, a group of three-generation families from northern Sweden in which the grandparents experienced alternating periods of feast and famine (Pembrey et al., 2006). It was found that grandsons lived shorter lives if their paternal grandfathers experienced abundant harvests in their youth (Pembrey et al., 2006; Bygren et al., 2014). In fact, they died on average much earlier than grandsons whose paternal grandfathers had been raised in times of famine. Longevity trends were also seen in grand-daughters whose paternal grandmothers had been exposed to poor nutrition versus years of plenty. A lower incidence of cardiovascular disease seems to have been a significant factor in those who lived longer and a higher incidence of diabetes in those who lived shorter lives – the incidence of diabetes was fourfold higher in those whose grandfathers had experienced abundant eating (Kaati et al., 2002). The results could not be explained by 'longevity genes' in one set of grandfathers compared to another because, irrespective of their diet, the average age of the grand-fathers when they died was remarkably similar. Somehow the 'memory' of an environmental influence is being transmitted from grandparents to grandchildren, although whether this is 'epigenetic memory' remains to be seen. There are many caveats in such studies (Lim and Brunet, 2013), and there may be other mechanisms for transgenerational inheritance besides epigenetic, perhaps arising from regulatory RNA molecules that are found in the cytoplasm of both sperm and eggs and which are therefore passed on to potentially influence the development of the offspring (Daxinger and Whitelaw, 2014).

In the present context, whether genuine transgenerational epigenetic inheritance occurs in humans or not – and the jury is still out – makes no difference to the overall message. What is quite clear is that parental behaviour impacts on the epigenetic status of human offspring in ways that have effects for the rest of their own lives. Epigenetics provides a key mechanism whereby the properties of the living organism integrate with changing environments, during development and for the rest of life (Charney, 2012).

Transposons

No less than 48 per cent of the human genome is composed of 'transposable elements', or 'jumping genes', also called 'transposons' (Feschotte and Pritham, 2007). These represent 'copy-and-paste' segments of DNA and the great majority arise by a molecular mechanism whereby a segment of DNA is copied into RNA and then randomly inserted into some other location in the genome by converting the RNA 'message' back into DNA again to make a second copy. Clearly if this happened too often, then our genomes would soon fill up with so-called parasitic DNA, and so it's not surprising that cells have mechanisms to protect their DNA from too much copying. One kind of protection involves methylation of the transposon gene promoters. Although most of the random insertions make no difference to genome function, occasionally the insertion event occurs within a gene or in one of its regulatory regions in which case the consequences can be severe, such as turning a neighbouring gene 'on' when it should really be in the 'off' position. Indeed, there are more than seventy reported cases of newly inserted transposons causing disease in humans (Cordaux and Batzer, 2009). Transposons are added to the human gene pool at a frequency of one new insert every ten births (Cordaux et al., 2006). Some new insertions have been generated so recently that they are found in only a single individual and are designated 'private de novo' insertions. Around 600 million of these private germ-line insertions (insertions that occur in cells that make eggs or sperm) have been generated in human genomes throughout the world, generating a large amount of 'extra' genetic diversity, some of it of functional significance (Mills et al., 2007; Cordaux and Batzer, 2009).

The particular reason for discussing transposons in the present context is the recent, surprising finding that transposon insertion events occur not only in the germ-line cell DNA that is then passed on to the next generation but also in somatic cells in response to environmental inputs. This is particularly the case within the brain, with the hippocampus being especially marked in this respect (Baillie et al., 2011), whereas new insertional events occur much less frequently in the cerebral cortex (Evrony et al., 2012). Much of the brain insertional activity occurs during development when neurons are actively dividing, but some also continues to occur in later life. The implication is that the genomes of such brain regions are actually a mosaic in which each cell lineage may have its own genome sequence, although this does not necessarily mean different functional properties, though it might. Functional impact is suggested by the observation that transposon insertions in the adult brain more frequently involve protein-coding genes in comparison with those that occur during early development (Baillie et al., 2011). Brain neurogenesis in the adult brain, referring to the ongoing production of neurons, is especially marked in the hippocampus, increasing in response to physical exercise, environmental enrichment, stress and ageing (Charney, 2012). New neurons are integrated by their synaptic connections into the ever changing architecture of the hippocampus. It is intriguing that many of the environmental inputs that stimulate neurogenesis also increase the rate of transposon insertion into the genomes of hippocampal neurons. The implication is that even the brains of genetically identical (monozygotic) twins are different – the transposon insertion events in the neurons of their hippocampus will be different for each twin (Singer et al., 2010). Differences in lifestyle between the twins – for example, one exercises more than the other – might well produce environmentally induced differences in their hippocampal genomic sequences. As the writers of a review describing transposon insertions in brain events comment: 'Many such insertions in many different cells would be expected to yield subtle or not so subtle differences in cognitive abilities, personality traits and susceptibility to neurological problems' (Gage and Muotri, 2013).

It is unfortunate, but true, that very often it is neurological and psychiatric disorders that allow investigators to uncover the involvement of a genetic abnormality. The effects of transposon insertion in the brain are no different. For example, Rett syndrome is a severe developmental disorder that affects girls in particular. A gene called *MeCP2* is often mutated in cases of Rett syndrome and other mental disorders. MeCP2 protein is involved in epigenetic silencing of genes: it binds methylated DNA and in turn can recruit proteins that deacetylate histone proteins. But mutations in the *MeCP2* gene can have another effect: in the brains of both mice and people the mutant protein increases the number of transposon insertions, events very likely to be involved in the aetiology of the disease. An increased number of transposon insertions has also been described in the frontal cortex regions of the brains from schizophrenics (Gage and Muotri, 2013), and one particular gene which encodes a transcription factor implicated in schizophrenia has been described with a transposon insertion (Baillie et al., 2011). As Gage and Muotri (2013) comment, 'The new discoveries will make it ever harder to disentangle the relative effects of nature and nurture on our psyches'.

The Microbiome

The microbiome refers to the number of species and genetic diversity of the bacteria that occupy our various tissues. There are around 4×10^{13} bacteria in our bodies, called the 'microbiota', which is roughly equivalent to the number of our own cells (Sender et al., 2016) and taken together contain around 100 microbial genes for each human gene. One gram of human gut tissue contains 60 billion bacteria, around tenfold higher in number than the human population of this planet. The distal region of our gastrointestinal system contains around 4,000 different microbial species with 800,000 different genes (Relman, 2012). Our microbiomes are different in each of the eighteen sample sites on our bodies that have been examined in detail so far and everyone has a different microbiome, which also varies with time and place in the same individual (Schloissnig et al., 2013). They even undergo diurnal rhythms, changing in relatively predictable ways over a twenty-four-hour period, rhythms that are disturbed by jet lag (Thaiss et al., 2015). The microbiomes of identical twins appear to be somewhat more similar than are those of non-identical twins, suggesting that the genetic profile of an individual impacts on their gut bacterial population (Goodrich et al., 2014). The microbial populations of our tissues help our digestion, supply several of our dietary needs and have profound effects on human health, including immunity, inflammatory disease and obesity, in turn impacting on our decisions in life and on our general well-being (Velasquez-Manoff, 2015b).

The development of fast genomic sequencing has opened up new avenues of research to investigate the functional roles of our microbiomes. As with any new research field that tends to capture the public imagination, the field has sometimes been marred, in its public reporting at least, by exaggerated reports of its functional significance (Hanage, 2014). The leap to infer causation from correlation without further supporting data is a tempting one. Having made that caveat, there are some interesting experiments suggesting a causative link between the varying composition of the microbiome and human body mass index (Wallis, 2014). Body mass index in itself impacts on human identity and behaviour.

But some findings in both animals and humans point to an even closer link between the gut, its microbiome and mental well-being and function. The gut is served by the enteric nervous system, sometimes referred to as the 'second brain' because of its size and complexity, containing 200–600 million neurons, equivalent to the number of neurons in the spinal cord. The enteric nervous system represents the main two-way pathways of communication between gut and brain. It is unsurprising that its communication pathways are so complex when one considers that the intestinal surface area is about 100 times greater than the surface of the skin, so the gut represents the body's main interface with the 'semi-outside' world. At such interfaces the body's defence system is critical. In fact, the gut-associated immune system contains more than two-thirds of the body's immune cells and a complex web of hormone-secreting cells, involving more than twenty different hormones (Mayer, 2011). A huge amount of information is constantly being relayed from the gut to the brain, resulting in changes in memory formation, emotional arousal and mood (Berntson et al., 2003).

In one study carried out on mice, evidence was presented to suggest that germ-free animals in which the gut microbiome had been artificially removed display reduced anxiety but an exaggerated stress response which correlates with the elevation of certain key neurotransmitters in the hippocampus (Clarke et al., 2012). Reduced anxiety in the germ-free animals was normalised following restoration of the intestinal microbiota. Gut microbes can produce hormones and neurotransmitters that are identical to those produced by humans, and they can directly stimulate neurons of the enteric nervous system to send signals to the brain via the vagus nerve (Forsythe et al., 2014). The influence of varying microbiomes on anxiety and depression remains an ongoing area of active research (Schmidt, 2015), and intriguing correlations between the composition of the microbiome and autism have also been described (Kang et al., 2013).

Given the complexity of the microbiome itself and its many faceted interactions with the body's nervous, immune and hormonal systems, it is clearly going to take much more work to elucidate the various ways in which gut microbes can influence human emotional and cognitive states (Reardon, 2014a). Nevertheless, even the preliminary data are sufficient to give a new twist to the phrase, 'I have a gut feeling that something is the case'. The microbiome once again illustrates the way in which all of the components of biological organisms are tightly integrated, often in unexpected ways. We cannot therefore divorce our microbiomes and their diversity from our own human identity, as we are inextricably bound together in one huge and, on the whole, mutually beneficial cellular community.

The Virome

The world of viruses has not escaped the recent trend to add 'ome' or 'omic' to almost every conceivable biological entity. The 'virome' refers to the sum collection of viruses in a particular organism or location and contains the largest and fastest mutating set of genetic components on earth. The number of viruses on earth is estimated at a staggering 10^{31} – that is a 1 followed by 31 noughts (Virgin, 2014) – and it has been estimated that only about 1 per cent of this huge storehouse of genetic diversity has yet been described (Mokili et al., 2012). Cosmologists are not the only scientists who handle very large numbers. Formally the virome is a subset of the microbiome, but requires some separate comment here. Our own virome refers to the number of strains and genetic diversity of the viruses in our bodies which display even more diversity between human individuals than the microbiome. We contain roughly ten times more viruses in our bodies than bacteria (human faeces alone contain about 10^{8}–10^{9} viruses per gram), so around ten times more viruses than the number of our own cells (Mokili et al., 2012). Many of these viruses, called phages, live inside the bacteria, so bacterial composition determines what viruses will be present also, and viruses are constantly mutating, so increasing the overall genetic diversity of the viral cohort (Minot et al., 2013). Longitudinal studies on identical twins and their mothers have revealed that whereas their viromes are very different, within one individual their signature virome remains rather stable in composition over time (Reyes et al., 2010). Taking the microbiome and virome data together, it is clear that the genetic diversity expressed in these communities is far higher than the diversity of our own genomes.

Fortunately for us, only a tiny percentage of our viromes are pathogenic, either potentially or in practice. Indeed, there is growing evidence that the virome is involved in 'fine-tuning' the basal responses of our immune systems in ways that are positive for our health and well-being, as is also certainly the case with our bacterial inhabitants (Virgin, 2014). Furthermore, studies in mice have shown that infection with the herpes virus leads to increased expression of around 500 different genes in different organs, including the brain (Canny et al., 2013). This has potential human relevance because we often carry chronic herpes virus infections for life, which can cause shingles amongst other nasty conditions. Whereas the study of pathogenic viruses has a very long history, investigation of the more positive influences of our giant viral load remains in its infancy. The virome is mentioned here because it represents such a huge storehouse of extra genetic diversity cohabiting within our own bodies and, therefore, it should not be ignored as we consider the many layers of biological regulation that together contribute to individual human identity.

Integrating the Genome

We have now considered eleven different factors or mechanisms that need to be kept in mind when considering how the genome integrates with other cellular mechanisms to generate biological identity. Not all of these mechanisms are of equal weight in their contributions to the daily lives of biological organisms, and the list ranges from the very well established to the speculative, but the point in presenting the list in this way is simply to highlight the poverty of metaphors presenting the genome as an 'instruction manual' or, even worse, a 'recipe'. It is nothing of the kind, but rather an important resource that contributes along with many other components to generate a highly complex information system which underlies the development and daily dynamic existence of biological organisms. Our account therefore lies well within the tradition of developmental biology nurtured and expanded in particular by the late Gilbert Gottlieb (1998) and by Kenneth Schaffner (1998).

The following chapter will focus on the ways in which the molecular mechanisms are integrated with environmental inputs during human development.

CHAPTER 4

Reshaping the Matrix?

Integrating the Human in Contemporary Biology

> *When a blind beetle crawls over the surface of a curved branch, it doesn't notice that the track it has covered is indeed curved. I was lucky enough to notice what the beetle didn't notice.*
>
> Albert Einstein (Smith, 2014)

> *The whole is more than the sum of its parts.*
>
> Aristotle, *Metaphysics*

> *It is our choices, Harry, that show what we truly are, far more than our abilities.*
>
> Dumbledore in J. K. Rowling, *Harry Potter and the Chamber of Secrets*, 1998

Human Development

Given that large volumes are written on this single topic, it should be emphasised that no attempt is made here to be comprehensive. As an aide memoire during this description we will utilise a new acronym, DICI (pronounced 'dicey'), which stands for Developmental Integrated Complementary Interactionism.[1] Each word in the acronym provides a key to understanding the roots of biological identity: the acronym serves as a collection of four pegs on which to hang ideas. The DICI theme will crop up repeatedly in the chapters that follow.

Fetal Development

We start with a newly fertilised human egg – a zygote. What is inherited from the parents is not naked DNA, which by itself can do nothing, but a complex system of DNA, RNA, proteins and nutrients that together operate to regulate cell growth and division. The human egg just prior to fertilisation contains at least 3,000 different proteins, 7,500 different mRNA molecules and many thousands more small non-coding RNA molecules involved in regulating gene expression. By itself DNA would be as useless as a piece of software without any computer to run it on.

From the very beginning a range of environments is in operation: the immediate molecular microenvironment of the DNA within the cell and then the macroenvironment which includes the fallopian tube down which the zygote soon starts travelling, as well as the mother and the mother's own broader environment. The DNA and these two types of environment are completely integrated from the moment of fertilisation. Genes, like people, function according to the company they keep. And in the early zygote it's not the DNA which causes development to begin, but rather proteins inside the zygote, inherited from the mother's egg, which regulate which genes in the DNA are switched on and off. It takes a couple of days before the new DNA inherited from both parents begins to generate the proteins that ensure its own continuing regulation. Hundreds of genes and proteins are involved in this transition from the life of the egg to the life of the zygote, known as the 'maternal-to-zygotic transition', resulting in rapid cell replication now organised by the new genomic system (Xue et al., 2013; Lee et al., 2014). Proteins are the players in the DNA orchestra that cause the genes to play an integrated symphony of life. Causal networks operate in all directions.

Around six to twelve days after fertilisation, the growing embryo, now known as a 'blastocyst,' implants in the wall of the uterus. The blastocyst is a hollow ball of cells, with a special group of cells, known as the 'inner cell mass,' at one end. These cells form most of the tissues that will generate the baby; the other cells in the blastocyst are largely devoted to producing the special support tissues that allow placental mammals like humans to attach to the mother's uterus. It is the size of the uterus not the size of the parents that regulates the size of the eventual baby, as shown by studies on surrogate mothers. Once implanted, the cells derived from the inner cell mass flatten to form the embryonic disk, around 1 mm in diameter, with its outer surface layer of cells known as the 'ectoderm'. Over the next few days the ectoderm begins to receive chemical signals from nearby cells in the embryo that cause it to form the 'neural plate'. As the embryo continues to grow, the edges of the neural plate curl around to form a tube which ultimately becomes the nervous system with the brain at one end and the spinal cord at the other. At this stage it contains around 125,000 cells, but these are not neurons, mature brain cells, but rather immature precursor cells which have the potential to develop into either neurons or glial cells (a crucial cellular supporting cast within the central nervous system).

At this stage it is worth pausing to jump ahead to consider the structure of the fully developed adult brain, as we are then in a better position to appreciate the dramatic developmental transitions that lead from a mere 125,000 precursor cells to the mature product. The adult brain contains around 100 billion neurons (10^{11}) and one trillion glial cells (10^{12}). If we take the current world population as 7.5 billion, then this means that you have enough neurons and glial cells to give each living human around 13 and 130 of them, respectively. Neurons are the brain cells that carry out the signalling functions of the brain; glial cells provide support and renewal for the neurons, but also engage in some forms of signalling as well. A neuron is activated by a wave of ionic exchange across its cell membrane ('depolarisation') which travels like a mini-tsunami along the axon of the cell to the synapse, there to release chemical messengers known as 'neurotransmitters' into the synaptic cleft. These then bind to receptors on the surface of the membrane of the next neuron along the chain, thereby triggering either negative or positive signals to this neuron. Each neuron receives, on average, 5,000 synaptic connections, though the range is very broad, and some neurons connect up to other neurons by as many as 200,000 synapses. Since there are around 100 billion neurons in the adult brain, this implies that the brain contains a staggering 500 trillion synapses (5×10^{14}), making the human brain the most complex entity in the known universe.

The transition from a 1 mm diameter neural plate to a brain with 500 trillion connections provides plenty of ways to illustrate DICI. By around one month after fertilisation, the nascent embryonic nervous system is growing at breathless speed: about 250,000 new cells are being generated every minute during the first half of the pregnancy. This rapid cell division is regulated by the products of a range of at least nine different genes, as illustrated by the consequences of mutations in one of them, the ASPM gene[2], leading to microcephaly, a severe syndrome in which the brain develops to only around 30 per cent of its normal size with associated mental retardation (Bond et al., 2002).

How do neurons in the developing fetal brain know how to connect up with the millions of other neurons being brought into existence each day of a pregnancy? The human genome encodes a paltry 21,000 proteins, although, as already noted, the final number of functionally distinct proteins may well be more than one million based on mechanisms already described. Even so, there is clearly insufficient information in the genome per se to specify the billions of specific synaptic connections that characterise the architecture of the mature brain. Many details have yet to be worked out, but the broad outlines as to how this happens are now clear. In the early months of fetal development, genetic information, in coordination with input from the fetal microenvironments, is dominant in terms of establishing the broad structures and areas of the brain. But increasingly during the final three months of pregnancy, at a time when the fetus can hear and feel, and then even more so in the immediate post-natal years, it is the wider macroenvironment that plays a critical role in constructing the developing brain, particularly in specifying its synaptic architecture. The early to mid months of fetal brain development lay down intrinsic 'learning structures' that are then primed to respond to environmental inputs at certain later sensitive periods, these in turn then leading to the next stage in the development of specific areas. Over the eighteen- to twenty-three–week gestational period, 76 per cent of human genes are expressed in the fetal brain and 44 per cent are dynamically regulated (not just permanently in the 'on' position) (Johnson et al., 2009). Genetic information and environmental information are integrated at every stage.

Genetic contributions are now much easier to investigate than before the genomic era, particularly by the use of Genome Wide Association Studies (GWAS), an approach that will be explained more fully in Chapter 7. For the moment all one needs to know is that it's a way of identifying small sequences of DNA, which may or may not contain known genes, that correlate with the variation in different traits as assessed in large populations. Of particular interest in this context is a study that entailed imaging the brains of 30,717 individuals to assess the variation in volume in seven specific regions (Hibar et al., 2015). A range of gene variants were identified that, taken together, explained 7–15 per cent of the variance noted in brain substructure volumes in this adult population. This underlines the point that it is genetic and environmental information taken together that shape the anatomy of the developing brain.

In the early fetus, specialised centres for precursor brain cell production are found near the ventricles, fluid-filled spaces in the centre of the brain. From there the cells move to their designated areas. Specific transcription factors switch on an array of gene products that include morphogens (chemicals with concentration gradients that regulate development) and signalling proteins that help guide the neuronal precursors to their designated locations (O'Leary et al., 2007). Negative signalling proteins that prevent cells moving to the wrong location play an important part in this process. Experiments involving the genetic manipulation of mice show that changing the expression of even a single transcription factor can profoundly perturb the normal construction of basic brain areas (Cholfin and Rubenstein, 2007). To construct those parts of the brain that have layered structures, such as the cortex and cerebellum, neurons crawl along scaffolds formed by a special class of glial cells that extend from the ventricles to the brain surface (Linden, 2007). The six-cell layered neocortex forms orderly radial columns that then form connections with the columns on either side. The neurons that form the base of the columns are generated first, and then newcomers are added layer upon layer to build up each column (Fox et al., 2010). Even at these early stages, when the broad outlines of brain areas are being established, there is considerable variation in their relative size, which is then reflected in the adult brain (O'Leary et al., 2007). So great are the individual differences in brain structure and function by adulthood that a research team was able to identify individuals out of a group with more than 90 per cent accuracy simply by inspection of a map of brain regions that are active during mental activity, giving a new meaning to the term 'finger-print' (Finn et al., 2015).

The emerging fetal nervous system is highly sensitive to environmental inputs and insults, particularly during the early months of pregnancy, the consequences of the insult being strikingly stage-specific. In general, disruption of neural development in the first trimester is more deleterious than in the third trimester (Schuurmans and Kurrasch, 2013). Fetuses in their tenth to twentieth week of gestation suffered severe brain development problems following the dropping of the atom bomb on Hiroshima, whereas such problems were less prevalent in those a little younger or older (Klingberg, 2013). Aberrant neuronal migration can result in cerebral palsy, mental retardation and epilepsy in adulthood (Linden, 2007). Ingestion of certain drugs, excessive alcohol or smoking by the pregnant mother, or severe inflammatory responses, can all impact negatively on the emerging fetal neuronal system in ways that affect later behavioural development (Fox et al., 2010; Alati et al., 2013). As is true for radiation, there are windows of sensitivity during which exposure to such agents, called 'teratogens,' is particularly damaging to the fetus.

The possible effects of 'maternal distress', a general term including a wide range of psychological stressors such as anxiety and depression, on the developing fetal nervous system has also received extensive attention (Rice et al., 2007), with well over 5,000 publications addressing this issue over the period 2008–13 (Schuurmans and Kurrasch, 2013). The effects of acute environmental traumas such as disasters and wars have likewise been intensively investigated. Some studies have assessed infant performance on the Bayley Scales of Infant Development (a widely used measure of cognitive development) in the first two years of life. Overall, these show poorer scores on the Bayley scale for prenatally stressed children (Van den Bergh et al., 2005; Talge et al., 2007). Such effects have been observed for both chronic maternal anxiety stressors and acute stressors such as the 1998 Quebec Ice Storm (Weinstock, 2008). The Avon Longitudinal Study of Parents and Children (ALSPAC) continues to be of particular relevance in such studies (Pearson, 2012). In 1990 researchers started collecting specimens together with social and medical data from 14,500 pregnant women in the Avon area of Western Britain, tracking the subsequent health and sociological char-acteristics of their offspring, a study which remains ongoing. One of the many findings from the ALSPAC study was a higher incidence of antisocial behaviour and higher anxiety at ages four and seven for prenatally stressed children after controlling for obstetric risks, psychosocial disadvantage and post-natal anxiety and depression (O'Connor et al., 2003). The fact that the effect persisted between the ages of four and seven suggests that the influence remains unaltered by changing life stresses on the child, as this period includes the beginning of formal schooling. In a different longi-tudinal study, this time based on a U.S. cohort, greater anxiety was noted in prenatally stressed eight- and nine-year-olds, again after controlling for other confounding factors (Van den Bergh and Marcoen, 2004). By 2010 an extensive review was reporting that

> [i]n the last 10 years several independent prospective studies have exam-ined the longer lasting effects of prenatal stress, anxiety or depression on the behavioural, emotional and cognitive outcomes during childhood. Even though these studies used a wide range of different methods, both for measuring prenatal stress or anxiety, and for assessing the child, *they all support a link between prenatal mood and changes in outcome that sug-gest changes in neurodevelopment*'[my italics]. (Glover et al., 2010)

Overall, acute stress associated with trauma (war, bereavement, natu-ral disaster) appears to be particularly disadvantageous early in pregnancy, while stress associated with maternal anxiety and depression has a greater effect later in pregnancy (Rice et al., 2007).

An obvious explanation for such findings could be a genetic profile in which mothers genetically predisposed to anxiety or depression give birth to offspring with similar characteristics. It is therefore significant that increased anxiety and antisocial behaviour at ages four to ten has also been noted in prenatally stressed children born via IVF treatment (Rice et al., 2010). This study examined children carried by their genetic mother and compared them with children born via egg and embryo donation, who share a uterine environment but no genetic link with the mother. There was no difference in the association between maternal stress and child outcome depending on whether the child was genetically related to the mother, suggesting that it is the environmental stress that plays a key role in this instance.

In all such human studies, there is always the proviso that correlation does not equate with causation, and even when other confounding factors have been excluded by statistical analysis of the data, it is difficult to exclude the possibility that other causal factors, known or unknown, may be involved. For example, malnutrition is known to affect fetal brain structure and function in a manner analogous to prenatal stress, and maternal malnutrition during pregnancy is often associated with anxiety and depression, so in the same maternal stress studies the stress may be acting as a proxy for malnutrition (Monk et al., 2013). It is also worth emphasising that effect sizes for prenatal stress as measured in the subsequent offspring tend to be small, albeit significant, averaging between 3 per cent and 20 per cent of the variance in phenotype; although this is fairly low, an effect size of 15 per cent is of clinical significance (Van den Bergh et al., 2005; Talge et al., 2007).

Overall, the impact of environmental factors on the developing fetal neuronal system provides many good illustrations of the DICI principle. Signals from the macro- and microenvironments interact and integrate with genomics and cell biology to influence the developing system in ways that have measurable cognitive and behavioural effects decades postnatally. In the process, many different complementary narratives can be told. There is the narrative of maternal choice: whether to smoke and drink excess alcohol during pregnancy, or not. There is a narrative in which a couple splits during the pregnancy, leading in turn to maternal distress and a greater probability of negative behavioural and psychiatric outcomes in later life. There is a narrative of war or natural disaster in which maternal stress is again mediated to the fetus through no fault of the mother. Further complex complementary stories could be told, beyond our present scope, of the hormonal changes in the mother in response to stress and of the ways in which these impact on the synaptic organisation of the emerging fetal brain via the placenta (Buss et al., 2012; Painter et al., 2012). And another complementary narrative would certainly include the array of epigenetic changes, switching genes on and off in response to the latest maternal messages, often in ways that lead to long-term epigenetic differences in the offspring. The important point is that none of these narratives, and indeed the many others that could be told, are in any kind of rivalry with each other, even though the narratives belong to very different disciplines, often using methodologically radically different approaches. At every level of discourse, be it sociological, anatomical, hormonal, genetic or epigenetic, there is complete integration of signals and influences, so that differences in fetal development between individuals become like the thousand different nuanced ways in which a Mozart symphony can be performed depending on how the notes on the page are interpreted using different instruments and conductors.

With new automated technologies for measuring epigenetic 'marks' on the genome or its regulating proteins, the epigenetic changes that correlate with the changing fortunes of the developing fetus are now gradually being mapped. Three 'health warnings' are necessary in interpreting such data. First, due to the ease of automation, the measured changes tend to be dominated by results from the methylome rather than by changes in histone acetylation: gaining the 'total epigenetic picture' in large-scale studies is currently challenging. Second, for obvious ethical reasons it is rare to be able to measure epigenetic changes in the most relevant tissue of the newborn, namely, neuronal tissue. So in practice, epigenetic markers tend to be assessed in the cord blood that is otherwise discarded with the placenta at birth, or in white blood cells or hair follicles. Such measurements can at least act as 'proxy markers' for the real McKoy. Third, when epigenetic differences are reported, it should not be assumed that all are functionally significant. However, many undoubtedly will be, particularly when studies focus on the methylation of genes that are known or thought to be involved in influencing the trait in question. On the other hand, comparison of total methylomes in a particular tissue between different individuals will undoubtedly contain much background 'stochastic noise'. The major challenge is then to select out the epigenetic differences that are likely to be of greatest functional significance in terms of regulating gene expression.

Each human sperm and egg appears to be epigenetically unique, displaying far more epigenetic variation than genetic variation (Petronis, 2010). It is therefore not surprising that epigenetic differences continue to amplify as the pregnancy continues, influenced by differing environmental inputs. A mother's diet during pregnancy can be sufficient to change the ways in which her baby's genes are expressed. In one study the DNA methylomes of 237 neonates were investigated and found to contain 1,423 regions that were highly variable between individuals, in contrast to a different set of DNA regions in which methylation was rather similar (Teh et al., 2014). The researchers estimated that 25 per cent of the differences in the variable regions could be explained by the underlying differences in DNA sequence (the 'genotype'), whereas 75 per cent of the variation was best explained by the interaction of genotype with different in utero environments, including maternal smoking, maternal depression, maternal body mass index, infant birth weight, gestational age and birth order.

An example illustrating the way in which a pregnant mother's choices have long-term effects on the epigenetic status of the offspring comes from ALSPAC. Using a cohort of 800 mother-offspring pairs, it was shown that maternal smoking during pregnancy influenced newborn DNA methylation in genes involved in fundamental developmental processes (Richmond et al., 2015). The methylation status in the offspring of smoking mothers still remained different compared to controls when measured at the age of seven and seventeen years, although in some genes, but not others, the altered methylation status was shown to be reversible by this stage. A comparison of paternal and maternal smoking and offspring methylation showed consistently stronger maternal associations, providing further evidence for causal intrauterine mechanisms. This study represents but one in a large collection of publications in which prenatal smoke exposure has been associated with reduced birth weight, poor developmental and psychological outcomes and increased risk for diseases and behavioural disorders later in life (Knopik et al., 2012).

As always there is a danger that the media will exaggerate the risks of negative post-natal outcomes as a consequence of the status of the pregnant mother, forgetting the myriad other factors that impact on the status of the newborn: paternal sperm quality; genetic variation; social expectations and traditional practices; poverty and much more. The following media headlines on this topic have all been recorded: 'Mother's Diet During Pregnancy Alters Baby's DNA' (BBC), 'Grandma's Experiences Leave a Mark on Your Genes' (*Discover*) and 'Pregnant 9/11 Survivors Transmitted Trauma to Their Children' (*The Guardian*) (Richardson et al., 2014). The relatively modest impact of the great majority of prenatal environmental influences therefore needs to be highlighted: DICI is important because it points to the complex interweaving of multiple components during the developmental process; it subverts maternal stigmatisation, whilst not under-estimating the importance of maternal choice. The prenatal environment really does make a difference.

That the epigenetic differences are not intrinsically determined is made clear by the remarkable differences in epigenetic profiles between identical ('monozygotic') twins at birth, although the discordance is greater between fraternal ('dizygotic') twins as compared to monozygotic (Ollikainen et al., 2010; Gordon et al., 2012). Some of the differences are presumably due to competition for resources within the uterus. The placenta develops from a layer of cells known as the 'chorion' which emerges after one week of blastocyst development. Around two-thirds of all monozygotic twins are formed when the embryo is split into two between five and nine days after fertilisation (Shur, 2009). Since by that time the chorion has already formed, the twins end up sharing the same placenta. But around one-third of identical twins are formed when the embryo splits into two before five days post-fertilisation, before the chorion is formed, so these twins end up each having their own placenta, just like non-identical twins. Different twin pairs will therefore have different intrauterine experiences, such as differences in infection, blood supply and space to grow (Czyz et al., 2012). At birth, let alone in later life, there are many reasons why monozygotic twins are not truly identical (Bell and Spector, 2012), a point pursued further in Chapter 6.

By the third trimester the fetus can hear and feel. Development of the auditory system begins very early in gestation (McMahon et al., 2012). All the major structures of the ear required for hearing are in place by twenty-three to twenty-five weeks and by twenty-six weeks the fetus can perceive sound and react to it. From twenty-six to thirty weeks onwards the hair cells in the cochlea (the auditory portion of the inner ear) are fine-tuned for specific frequencies and can convert acoustic stimuli into neuronal signals that are sent to the brainstem. After thirty weeks the auditory system is now developed sufficiently to respond to music, as inferred by a change in fetal heart-rate, and to distinguish between different speech sounds, presumably representing the very beginnings of language acquisition (McMahon et al., 2012). This hearing is known to be rather precise, with the fetus able to distinguish both unsubtle changes such as pitch, and subtle changes in rhythmic qualities and fine phoneme distinctions (Mahmoudzadeh et al., 2013). Studies on preterm babies (gestational age in the range 28–32 weeks) have revealed that even during this period, when neurons are still migrating to their assigned locations and the six-layered neuronal organisation of the cortex is still under construction, with its inter-connections only just beginning, subtle differences in human speech syllables can already be discriminated, a finding that underlines the innate capacity of the brain to process auditory information before any significant opportunity for learning becomes available (Mahmoudzadeh et al., 2013). Shortly after birth, neonates are able to distinguish between the mother's voice (which they prefer) and a strange female voice, the mother's voice and the father's voice, and between the mother's language and a foreign language, showing preference for the mother speaking her native language (McMahon et al.,

2012). Unfortunately, however, earlier claims that playing Mozart or other uplifting music to the fetus might be beneficial for its subsequent postnatal IQ or general well-being have not been confirmed.

On the brink of birth, the fetal brain is now poised to gain most benefit from the environmental inputs that will shape its architecture and functionality over the crucial first few post-natal years.

Post-natal Development

The timing of birth has a significant impact on the future health of the child and the epigenetic status of its genome, the latter observation likely to be highly relevant to the former. According to World Health Organization statistics from 2012, more than one in ten babies – around 15 million in total – are born prematurely each year. The great majority are born between thirty-two and thirty-seven weeks of gestation, but 1.6 million are born between twenty-eight and thirty-two weeks and 780,000 are born 'extremely pre-term,' before twenty-eight weeks (Abbott, 2015). Cerebral palsy affects 1–2 per cent of babies born at term, 9 per cent of those born earlier than thirty-two weeks and 18 per cent of those born at twenty-six weeks. In general children born preterm, defined as a birth earlier than thirty-seven weeks of gestation, are four to five times more likely to develop brain and cardiovascular disorders compared with infants born at term. Preterm infants are also more likely to suffer from a wide range of other diseases in later life. In one study 1,555 DNA sites were found to differ in their methylation status between term and preterm babies, but by the age of eighteen these differences had largely resolved (Cruickshank et al., 2013). However, even at that adult age, the methylation status of ten DNA sites still remained abnormal in the preterm cohort, appearing to reflect a long-term epigenetic legacy of preterm birth.

At birth, the human newborn is bombarded with a whole new array of sights, sounds, smells, touch, language, people, dogs, cats and other fascinating environmental phenomena. The physical development of the infant brain is critically dependent on such exposure. The infant brain is not a miniature version of the adult brain but a continuously self-organising system that only self-assembles correctly if the right inputs are available at the right time (Fox et al., 2010). As already mentioned, the adult brain contains around 10^{11} neurons, but in total twice as many brain cells are generated in the developing fetal brain with half of these being pruned away during development. At birth many of the fetal brain cells are still neuroblasts, undifferentiated cells that have yet to develop into mature neurons, so their axons are not yet myelinated and their synaptic connections sparse. It is the experiences of the growing baby that help shape the synaptic architecture of its brain. This is reflected rather dramatically in the average weight of the human brain which triples during the first year of life from 300 g to 900 g. After that the weight increases more slowly to reach 90 per cent of its adult weight (1,400 g for men, 1,250 g for women) by the age of five. This increase is not mainly due to an increase in the number of cells, but rather reflects the growth in the number of synaptic connections, the size of dendrites and axon bundles and myelination. Environmental inputs regulate transcription factor expression via epigenetic mechanisms, and the transcription factors in turn switch on a complex array of proteins and signalling molecules that are used to generate new synaptic connections (Fox et al., 2010). The branching process continues up to the age of two in some brain areas and up to the age of twelve in other areas (Klingberg, 2013, p. 16).

The synaptic architecture of the brain develops according to the principle of 'use it or lose it' (Linden, 2007, p. 74). Without neuronal activation, brain cells die by a process known as 'programmed cell death'. So the 'pruning' of half the total number of neurons that are generated during development involves the failure of these cells to generate efficient synaptic connections in comparison with their neighbours. The neurons that continue to flourish are those which receive strong signals via their growing collection of synaptic connections, and amongst those signals are those that counteract the cell death programme. So the neuronal architecture develops by a process akin to natural selection in which the 'successful neurons' are selected for in comparison with the 'unsuccessful neurons'.

The role of the environment in providing the signals that help construct the development of the post-natal brain is well illustrated by the existence of sensitive periods at which sensory inputs are vital for the development of specific brain areas. In early infancy these affect sight, hearing and balance. As reviewers of the field comment: 'Each sensory and cognitive system reaches a unique sensitive period, and thus identical environmental conditions will result in very different cognitive and emotional experiences for a child, depending upon his or her age' (Fox et al., 2010, p. 35).

Face recognition and discrimination provide some of the earliest postnatal environmental inputs. Newborns prefer face-like stimuli to non-face stimuli and familiar over unfamiliar faces and attend more to attractive over unattractive faces (Pascalis et al., 2011). At three to four months of age, infants respond to gender information in human faces. Specifically, young infants display a visual preference toward female over male faces, if raised by females, but of male over female if raised by males (Pascalis et al., 2011). Early visual impairment due to cataracts leads to persistent deficits in facial recognition, even when the cataracts are removed in the first months of life (Fox et al., 2010). The window of opportunity for learning how to discriminate faces is a narrow one. For example, six-month-olds show recognition memory for different human faces as well as for different monkey faces. However, such recognition memory for individual monkey faces disappears by nine months of age unless experience with such faces is provided (Pascalis et al., 2011). In fact, monkeys selectively exposed to human faces can only discriminate human faces, not monkey faces, and monkeys selectively exposed to monkey faces can only discriminate monkey faces, not human faces (Sugita, 2008). Taken together, the data suggest that experiences affect infants' face processing by the age of three months and it is these experiences that sculpt their subsequent face-recognition abilities. This early competence is then refined further during child development and on into adulthood (Pascalis et al., 2011). As a reviewer of this field suggests: 'Although biological factors may initially play some role in biasing the newborn's visual system toward faces in their environment, the existing evidence overwhelmingly suggests the important role of experience in the development of face-processing expertise' (Pascalis et al., 2011, p. 673).

The development of the ability to perceive visual flow motion provides another example of a 'sensitive window' in brain development (Morrone, 2010). This develops during the first few months of life and appears to be mediated by a specific region of the cerebral cortex. Children born with congenital cataracts in both eyes have large permanent deficits in motion processing, whereas children who develop cataracts from six to twelve months have no such deficit (Morrone, 2010). But up to around three years of age there appears to be considerable 'plasticity' in the relevant brain region, such that, in people blind up to or beyond the age of three but who are able to hear normally, the sound input is sufficient for the cognitive development required for the perception of normal visual flow motion (Bedny et al., 2010). But the same brain region cannot be activated by auditory inputs in later life – the plasticity seems to be restricted to the first few years of life. And providing visual information is available in the early years of life, then the influence of auditory inputs on this brain region are blocked.

Overall, it is the construction of the synaptic architecture of the visual cortex at the back of the brain in response to visual inputs that has been most thoroughly investigated (Levelt and Hubener, 2012). If a baby is blind from birth, then the visual cortex does not develop normally and permanent visual impairment can occur. In fact even with normal development, the size of the primary visual cortex in adults varies considerably. There are a whole series of 'sensitive periods' during which visual inputs are required for the neuronal construction of the developing visual cortex, many of the details having been elucidated using animal models (Huberman et al., 2008; Espinosa and Stryker, 2012). In the human newborn, binocular vision is absent at birth and is switched on when a baby is four months old: visual experience is key for its onset (Birch, 2012). Visual acuity, the ability to discriminate between, for example, stripes of different sizes, is significantly reduced in newborns with cataracts in both eyes following their removal and optical correction of vision at various times during the first nine months of life (Lewis and Maurer, 2005). However, even after one hour of patterned visual input, there was an improvement in acuity, which improved with further visual input, but acuity still fell below the normal range by two years of age, and again when measured in the age range five to eighteen years. So a deficit in visual input during the critical first nine months appears to cause permanent damage to normal visual development. However, if cataracts were removed by ten days of age, then acuity was reduced less, and in some cases, normal acuity was eventually attained with increasing age.

Almost one-quarter of the brain is normally devoted to processing visual information. Blindness profoundly re-organises the visual system. Most strikingly, developmental blindness enables 'visual' circuits to participate in high-level cognitive functions, including language processing, increased memory function and auditory abilities greater than those of sighted persons (Morrone, 2010; Bedny and Saxe, 2012). Once again there appears to be a sensitive period for the acquisition of language processing ability in blind persons since it depends on blindness in childhood and does not occur in those who become blind in later life (Bedny et al., 2011). One particularly enterprising blind person who had both eyes removed at the age of three due to cancer managed to learn how to echolocate like a bat, gaining the ability to 'see' by bouncing 'tick-tick' sounds off nearby objects and then listening to the echoes (Editorial, 2015). Those who are deaf from birth likewise undergo marked changes in their auditory cortex (located on the sides of the brain). Brain imaging and post-mortem studies in adults who have been congenitally deaf from birth have revealed that neurons that would normally be restricted to the visual cortex (in the far back of the brain) can also be found in the auditory cortex (Linden, 2007, p. 71). In normal development a few neurons may stray over into the auditory cortex but they are gradually eliminated. But in the congenitally deaf, such neurons not only stay but also sprout new synaptic connections. It seems that if there is a lack of auditory inputs, then neighbouring sensory areas in the brain take advantage of the deficit, a process helped in this case by the withering away of the auditory neurons by disuse. Conversely, in those who are congenitally blind, much of the visual cortex responds strongly to auditory and tactile inputs rather than to visual stimuli (Fine, 2014). The sensory regions of the brain display plasticity – the ability to adapt in their development depending on what inputs are, or are not, being received. Eventually, however, this plasticity is lost as the brain matures.

The importance of sensitive periods of brain development can be seen most strikingly, and at the same time most sadly, in infants who have suffered severe emotional and physical deprivation. The largest and most informative recent studies on such cohorts come from Romanian orphans. Romania's orphan crisis began in 1965 when Nicolae Ceaușescu took over as leader (Hughes, 2013). Over the next twenty-four years of his communist leadership, Ceaușescu deliberately adopted a policy of increasing the orphan population, banning contraception and abortions, with the aim of creating a population loyal to, and dependent on, the State. By the time revolutionaries executed Ceaușescu and his wife by a firing squad on Christmas Day 1989, an estimated 170,000 orphans were living in more than 700 state orphanages. In 2000 Charles Nelson, a neuroscientist from Harvard University, launched the Bucharest Early Intervention Project in which they enrolled 136 institutionalised children ranging in age from six to thirty-one months old, of whom half were randomly placed in foster care and half remained in state orphanages, subsequently tracking the physical, psychological and neurological development in both groups (Nelson et al., 2013). A third group was also recruited who had never been institutionalised. Since virtually no foster care was available in Bucharest when the study started, the research team had to recruit a whole new cohort of foster homes. Since that time more than sixty papers have been published describing their findings. The results are clear. Providing children were placed in foster homes by the age of two, significant gains could then be measured in psychological development, IQ and motor skills as compared with the control group in the state orphanages. At thirty, forty and fifty-two months, the average IQ of the institutionalised group was in the low to middle 70s, whereas it was about 10 points higher for children in foster care and about 100, the standard average, for the group that had never been institutionalised (Nelson et al., 2013). At around the age of eight, the children who grew up in institutions were found to have less white matter in their brains (the myelinated axonal sheaths involved in connecting up neurons) compared to those in foster care. This may explain why electroencephalographic (EEG) brain measurements in eight-year-olds likewise revealed distinct traces in those institutionalised up to that age, whereas those placed in foster homes before the age of two displayed EEG traces indistinguishable from the children who had never been institutionalised (Sheridan et al., 2012).

Further comparative studies have been carried out by the British psychiatrist Michael Rutter in which the development of children adopted from Romanian orphanages into homes in the United Kingdom was compared with a cohort of Romanian adoptees without any instutionalised background and a further cohort of children adopted within the United Kingdom. All the children in this study were adopted before the age of three and a half. When examined at age eleven, the Romanian adoptees who had previously been institutionalised had significantly lower attainments in almost every field measured, with a threefold higher incidence of psychiatric symptoms, most notably in those who had been adopted after the age of six months (Klingberg, 2013, p. 90).

Childhood institutionalisation is associated with a different epigenetic profile. In children from institutions compared to children who lived with their biological parents, more than 800 differentially methylated genes were detected, of which 90 per cent were hypermethylated (Naumova et al., 2012). In a further study saliva DNA specimens were collected on ninety-six maltreated children removed from their parents due to abuse or neglect and ninety-six demographically matched control children between 2003 and 2010. The maltreated children had significantly different methylation sites at 2,868 CpG sites (Yang et al., 2013). Individuals who experienced childhood maltreatment and who then subsequently committed suicide in later life were reported to have different gene methylation profiles compared to a control group who had experienced no childhood maltreatment and who had died suddenly from other causes (Lutz and Turecki, 2014). Translating such epigenetic differences into a detailed account of relevant gene regulation is an immense challenge, being helped greatly by animal model studies (Lutz and Turecki, 2014), such as the rat study already described earlier in which epigenetic modulation of the glucocorticoid receptor gene correlates with deprivation of maternal care.

Once again, as with prenatal findings, but now with post-natal development even more dominated by environmental inputs, the various narratives that can be told at different explanatory levels are all complementary in nature and thoroughly integrated in the developing infant. Political decisions by a communist leader in Romania; failure of care for young infants; consequent deficits in the normal epigenetic regulation of gene expression; cataracts in both eyes; failure of normal synaptic connectivity in key areas of the brain; genetic variation that contributes to more or less resistance to stress – these and many other levels of understanding all provide complementary accounts and all are required and important to gain a full appreciation of the complexity of human development.

Adult Development

Development never ends but continues throughout adult life until it finally ceases in death. Genetic information is changing as mutations take place in different cell lineages, some leading to disease. Human decisions change the environment of the growing individual in ways that impact on the epigenetic regulation of gene expression. Diet, smoking, exercise, drug taking, alcohol use, stress, choices about where we live, the daily choice of lonely environments or sociable environments and many other factors epigenetically modify our genomes in ways that change not only us but potentially also our children and perhaps even our children's children – a sobering thought.

The human individual seen from a biological perspective is a dynamic ever adapting, ever changing organism. This is seen perhaps most strikingly in the brain. At one time it was thought that at birth we were endowed with a fixed number of neurons and a predetermined set of synaptic connections, and after that ageing entailed an inevitable loss of neurons without any possibility of renewal. The past few decades of brain research have completely transformed such a picture. The hippocampus provides a striking example. Its early development is similar to other brain areas: an early prenatal increase in the number of neurons, and then a post-natal proliferation of synaptic connections, reaching a maximum by the age of two, followed by a process of pruning completed by around the age of five (Klingberg, 2013, p. 38). This initial developmental process proceeds faster than in other brain areas. For example, the prefrontal cortex, so critical for human cognitive functions, continues to thicken up to the age of around twelve and the subsequent pruning then continues on into the early twenties.

The hippocampus is the one brain area where new neurons are now known to be produced throughout life, a process known as "adult neurogenesis" (Christian et al., 2014). In fact, active neurogenesis occurs in two discrete brain regions: a small zone of the lateral ventricle, which acts as a kind of neuronal generation centre from which newly birthed neurons migrate to other areas, and the dentate gyrus of the hippocampus. The hippocampus is involved in learning, memory and emotion. Adult humans add around 700 new neurons to their hippocampus each day, corresponding to an annual turnover of 1.75 per cent of the renewable hippocampal cell population, followed by a modest decline during ageing (Spalding et al., 2013). The new neurons connect up to those already present and appear to have special functions, distinct from those already present, that contribute to brain functions (Christian et al., 2014). For example, the new neurons are highly excitable and generate nerve impulses more readily than mature adult neurons. In animals and humans, neurogenesis can be boosted by exercise and has been related to the prevention of cognitive decline (van Praag, 2008; Yau et al., 2014). Animal studies have implicated adult neurogenesis in long-term memory. Intensive learning in animals impacts the survival of the newly formed neurons. An interesting theory suggests (with some experimental support) that newly formed neurons help record memories in a way that distinguishes them from earlier memories already laid down (Kheirbek and Hen, 2014). In humans, experience, mood, behavioural states and antidepressants all impact on the regulation of adult neurogenesis in the hippocampus (Deng et al., 2010). Stress decreases and antidepressant treatment increases the number of new neurons generated (Banasr and Duman, 2012). Rats and mice raised with a high level of maternal care (licking and grooming) have higher levels of neurogenesis and synaptic density in the hippocampus compared to rats raised with less care, and the high-care rats display enhanced spatial learning and memory. As reviewers of this field summarise: '[A]dverse early childhood experiences can, and do, generate chemical and physical changes in the brain that can persist a life-time. Injurious experiences are not forgotten but integrated into the architecture of the developing brain' (Hoffmann and Spengler, 2014, p. 65).

Learning and memory provide further striking examples of brain plasticity. Memory is a multi-stage process which initially requires consolidation of the memory in the hippocampus. Once this process is complete, memories are then downloaded to the cortex in a process known as "systems consolidation." Both types of process require regulation by epigenetic mechanisms. Central to long-term memory in the cortex is the potentiation of neuronal circuits by modulating the synaptic inputs that a neuron receives so that it becomes more likely to 'fire'. So at any given moment during the day, some of our synaptic connections are physically weakening, whereas others are becoming stronger. The goal of every teacher and lecturer should be to change their students' brains (for the good) by the end of class. It is then up to the student to reinforce those synaptic connections involved in memory by effective revision.

Traumatic memories are particularly long-lasting. Remarkably, in mice at least, long-term memories can be attenuated by pharmacological manipulation of the epigenetic regulation system (Graff et al., 2014). Recall that histone acetylation provides one of the main mechanisms for the epigenetic regulation of gene expression. Enzymes known as 'histone deacetylases' function to remove these acetyl 'marks' from the histones. Inhibition of this class of enzymes therefore results in increased histone acetylation, thereby modulating the expression of multiple genes. In this particular study using mice, a long-term remembered fear response was much reduced by the administration of histone acetylase inhibitors (Graff et al., 2014).

Some striking differences have been reported in various brain regions in cohorts of adults whose hobbies or professions entail that they engage in various tasks, much of the data being obtained by means of various brain imaging techniques. A group of London taxi drivers were found to have enlargement of the posterior regions of their hippocampi in comparison with a relevant control group, an increase which correlated with the amount of time they had spent driving taxis (Maguire et al., 2000), a finding confirmed in a further study which reported increased grey matter volume in the mid-posterior hippocampi, changes that were not observed in London bus drivers (Maguire et al., 2006). In a further study, brain imaging was carried out on twenty licensed taxi drivers and compared with the results from twenty non-drivers. The study was focused on various functionally related centres in the brain, finding significant differences in the taxi drivers that also correlated with the number of years that they had been driving (Wang et al., 2015). Learning how to juggle also appears to impact on brain anatomy. A group of people who had never juggled before were divided into two groups and one group was taught how to juggle, involving daily practice over a period of three months, before having a rest period of a further three months during which they did no juggling. Compared to the control group, the jugglers displayed expansion of grey matter in two distinct brain areas, as revealed by brain imaging, but this expansion was already declining following the three-month rest period, underlying the plasticity of brain structures in response to environmental inputs (Draganski et al., 2004). Training techniques to expand working memory (the type of memory that keeps relevant facts 'in mind' whilst we are trying to solve a problem) have been correlated with specific changes in brain anatomy in many studies (Klingberg, 2013, pp. 122–4), as have a range of meditation techniques (Kang et al., 2012; Kumar et al., 2013). Playing a musical instrument has likewise been associated with more rapid cortical thickness maturation within areas implicated in motor planning and coordination, visuospatial ability and emotion and impulse regulation (Hudziak et al., 2014), just one of many studies relating the playing of musical instruments with changes in brain anatomy.

Underlying adult brain plasticity are a plethora of epigenetic changes, the great majority of which are only gradually being elucidated. Puberty is a time when significant organisation of the brain occurs, with a peak and subsequent decline in cortical grey matter and a continual increase in white matter volume in both the frontal and parietal lobes (Morrison et al., 2014). These changes are regulated by a release of a gonadotropin-releasing hormone which in turn triggers a cascade of downstream hormonal changes. Pubertal brain re-organisation entails large-scale epigenetic modifications of the genome involving both CpG methylation and histone acetylation, with additional inputs from hundreds of micro-RNAs, which display sex-specific patterns of expression (Morrison et al., 2014). Times of rapid brain development are also times when the brain is particularly sensitive to environmental insults, and there is currently intense research interest in the possible relationship between epigenetic changes induced during puberty by various types of chronic stress, or alcohol or drug abuse, and the onset of neuropsychiatric disease (Nestler, 2011). Psychiatric disorders are characterised by many epigenetic changes in specific brain areas, and mood-altering drugs likewise result in multiple epigenetic changes, although establishing cause and effect in such studies is notoriously difficult (Fass et al., 2014).

In adulthood as in infancy, epigenetic differences between monozygotic (identical) as contrasted with dizygotic (non-identical) twins are of particular interest as they help show how much of the similarities between twin pairs is due to the underlying DNA sequence, with its CpG methylation sites, and how much is due to differing environmental inputs. In one study, the investigators focused on a set of genes (the 'major histocompatibility complex') that generally varies extensively between individuals and which is intimately involved in the body's immune responses to antigens (foreign invaders). It turned out that the CpG methylation status of this DNA region was markedly different between both types of middle-aged twin pairs, with the differences between the monozygotic twins not much less than between the dizygotic twins, suggesting that by this stage of life environmental differences were the dominant factor in the differing epigenetic profiles (Gervin et al., 2011). A range of other studies carried out on adult twin pairs, this time assessing the complete 'methylome' of an accessible tissue, such as blood lymphocytes (white blood cells), have likewise shown that the differences in epigenetic profiles after some decades of life are not that different between dizygotic as compared to monozygotic twins. Having said that, the opposite is the case for certain specific CpG sites which display a much higher similarity in methylation status in monozygotic as compared to dizygotic twin pairs (Bell and Spector, 2012). It therefore seems that the underlying DNA sequence exerts strong constraints on the likelihood that certain CpG sites are methylated, whereas the majority of CpG sites are more adaptable in their responses to environmental inputs. The transfer of epigenetic information through successive cell divisions is also imperfect and so a random process of 'epigenetic drift' can also lead to a divergence of epigenomes with the passage of time.

The overall marked differences in epigenetic status between genetically identical (or nearly identical) individuals may help explain the otherwise rather puzzling observation that having identical genomes does not seem to make much difference to the degree of variation between identical twins when it comes to longevity (age at death) in comparison with pairs of siblings. In a Spanish study involving 184 twin pairs, the difference in the age of death (for those aged over 40 years) between the monozygotic twin pairs was seven years on average, but such averages hide the fact that the age differences ranged from a couple of weeks to eighteen years (Hayakawa et al., 1992). In the case of the non-identical twins, the difference in age at time of death was nine years on average and the range was three to nineteen years. So the differences between mono- and dizygotic twins were slight. Other studies on large cohorts of twins support the conclusion that the genetic contribution to the variation in age of death is significant but rather modest (McGue et al., 1993). One group of Swedish researchers compared longevity in 3,656 identical and 6,849 non-identical (sex-matched) twin pairs and concluded that '[o]ver the total age range examined, a maximum of around one third of the variance in longevity is attributable to genetic factors, and almost all of the remaining variance is due to non-shared, individual specific environmental factors' (Ljungquist et al., 1998). This may at least be of some comfort to identical twin pairs in which one member of the pair has already died. It turns out that our destinies are only partially influenced by our genetic endowment.

By means of this brief survey of human development, it should now be clear why the dichotomous language of nature and nurture is completely inadequate as a way of understanding human identity. Instead, DICI provides a road map for considering how the many different narratives that together contribute to human identity provide a much richer and more nuanced account that more accurately reflects the findings of contemporary biology. As it happens, DICI is equally relevant in discussing the identity of any biological organism that has a developmental history, as will become clear from the next chapter. Once that animal survey is complete, we will then be in a stronger position to consider in more detail the relationship between genomic variation and human behaviour, leading on, in turn, to the question as to whether and how certain human behaviours may be determined.

CHAPTER 5

Is the Worm Determined?

Gene Variation and Behaviour in Animals

> *Without doubt the fourth winner of the Nobel prize this year is Caenohabditis elegans; it deserves all of the honour but, of course, it will not be able to share the monetary award.*
>
> Sydney Brenner, Nobel Lecture, December 2002

> *You've heard about some of these pet projects, they really don't make a whole lot of sense and sometimes these dollars go to projects that have little or nothing to do with the public good. Things like fruitfly research in Paris, France. I kid you not.*
>
> Sarah Pailin, 2008[1]

> *Male prairie voles with lots of vasopressin binding sites are monogamous while montane voles, with few, are promiscuous.... As neuroscience uncovers these and other mechanisms regulating choices and social behaviour, we cannot help but wonder whether anyone truly chooses anything.... I suggest that free will, as traditionally understood, needs modification.*
>
> (Churchland, 2006)

Spending a chapter describing animals of increasing size and complexity in a book focused on the relationship between human genetic variation and human behaviour requires some justification. Three particular reasons come to mind for this inclusion.

The first is an evolutionary reason. All genes have an evolutionary history and nearly all human genes are represented in the genomes of other animals, many of them in animals with which we last shared a common ancestor many millions of years ago. Very often, but not always, it turns out that such genes have retained their same, or similar, functions over the millennia, and their study in animals can therefore provide us with very important clues as to their probable functions in humans.

The second reason comes naturally from the first: the use of animal model systems allows the scientist to investigate the relationship between genomic variation and animal behaviour in great molecular detail. This includes the genetic manipulation of colonies of animals (especially mice) in such ways as to shed light on the role of specific genes, and their variant versions, on animal behaviour. True, those behavioural traits in animals never mimic human behavioural traits precisely, but in many cases they can shed light on the human traits and give insights as to the relevant places to look in the human genome.

The third reason has already been mentioned at the end of the previous chapter: apart from unicellular organisms, such as bacteria, which simply replicate by dividing into two, most multicellular animals undergo an initial developmental programme before attaining adulthood. They therefore provide powerful examples of DICI, except in this case the way in which environmental inputs are integrated with genomic inputs during development to influence a subsequent behaviour can often be investigated in exquisite molecular detail.

To head off a possible misapprehension that may arise as a result of even describing such studies, the eventual argument will not be that animal studies shed some immediately relevant light on questions surrounding genetic determinism in humans: after all, we cannot enter into the mind of any animal to know from an internalist perspective its experience of its own repertoire of behaviours. Given that even our nearest living primate cousins, such as the chimpanzee and bonobo, are not in a position to communicate to us their experiences, the question as to whether 'up-to-usness' is amongst their repertoire of experiences simply does not arise. Given that fact, the question of gathering data that will count for or against the validity of that experience does not arise either. This is not to say that the experience does not exist, especially in the higher primates, only that we are not in a position to know that it exists. Of course we can infer that animals have animal minds based on their behavioural repertoire in a way not dissimilar to the way in which we infer that other humans have minds, including an inference to planned intentionality, but unfortunately this does not thereby enable us to know their experience of being intentional. Nonetheless, animal studies do have some relevance for the discussion about human freedom in relation to genetic variation for two distinct reasons: first, animal studies can shed great light on the repertoire of variant behaviours in respect of a particular trait that do, or do not, co-vary with a particular variant genome; and, second, animal studies in their evolutionary context illustrate the emergence of novel properties that arise in systems of increasing complexity, a point, it will be argued, of great relevance to the discussion concerning human freedom and determinism.

The examples that follow are selected from a huge range of possibilities and are included partly because they provide particularly well-studied models, but mainly because they illustrate so well the various aspects of the DICI framework. The focus here is on the behavioural effects of genetic variation in wild-type populations rather than on the vast literature describing the behavioural outcomes of genetic manipulation of animals in the laboratory.

Caenorhabditis elegans and their Feeding Behaviours

Questions of mind and freedom certainly do not arise from the tiny albeit famous little nematode worm, *C. elegans*, first described as a species in 1900, which is found in soils and decomposing matter such as compost heaps and garden waste. The name is a blend of Greek (*caeno* – recent, *rhabditis* – rod-like) and Latin (*elegans* – elegant). *C. elegans* has a global range and has been isolated from every continent except Antarctica (Kiontke and Sudhaus, 2006). It is incredibly abundant, attaining millions of individuals per square metre, and is also very small, only around 1 mm in length, with just 959 somatic (non germ-line) cells. Conveniently for researchers, it is also transparent. By stacking together 200,000 sections of the worm, it has been possible to generate a complete three-dimensional structure showing, for example, how all the 5,000–8,000 synaptic connections of its 385 neurons (in the adult male) join together (Jarrell et al., 2012; Sammut et al., 2015). The chemical properties of the worm's neurons in terms of the neurotransmitters they use are similar to those found in humans (Burne et al., 2011). *C. elegans* also has the honour of being the very first organism to have had its genome completely sequenced (in 1998); its genome contains around 100 million base-pairs (compared to the human 3.2 billion base-pairs) containing just over 20,000 protein coding genes (Thomas, 2008), a number similar to that of the human genome. This illustrates an important point that has become very clear from the sequencing of the genomes of many organisms: differences in levels of complexity do not stem from a basic difference in the amount of DNA present nor from the gene number. The worm genes are organised into exons and introns just like human genes. It is also interesting to note that around 35 per cent of *C. elegans* genes have human homologues (meaning genes sufficiently similar to identify them as the same genes), and in some worm genes the homology (sequence identity) reaches 80 per cent (Kammenga et al., 2008). In fact, various human genes have been incorporated into C. *elegans* in place of the worm's homologous gene by genetic engineering techniques and shown to function quite normally in this novel context.

The development, genetics and behavioural repertoire of C. *elegans* illustrate DICI brilliantly (Schaffner, 2000) – indeed so well that investigations of the worm have led to the award of at least six Nobel Prizes to date. The worm goes through a complex developmental process, including embryogenesis, morphogenesis and growth to an adult. The worm is androdioecious, which means that it can exist either as a hermaphrodite (the vast majority, 99.5%) which has both sperm and eggs and therefore self-fertilises, or as a male (a minority pursuit, only 0.5%) (Barriere and Felix, 2005b). When the hermaphrodite version self-fertilises, it produces about 300 progeny, a gift to laboratory researchers in a hurry for results, especially as the progeny are all genetically identical because there is only one parent (Kammenga et al., 2008). Development takes place as follows: eggs are fertilised within the adult hermaphrodite and laid a few hours later by which time the embryos have grown to about forty cells in size. The eggs hatch and the animals then proceed through four larval stages, each stage ending with a moult. Males and hermaphrodites are identical until the fourth larval stage, but gross morphological differences become apparent in the mature adult male (Hobert, 2010). The time from egg laying to a fully developed adult that can lay its own eggs depends on the environmental temperature: at 15 °C, it takes 5.5 days, at 20 °C, 3.5 days and at a balmy 25 °C, only 2.5 days. The worms live in total about two to three weeks and can be grown in the laboratory with ease by providing them with an agar plate covered with bacteria on which they happily graze (about 10,000 worms per dish). During development each cell-type has a particular lineage and the developmental origin of each one of the 959 cells has been worked out in great detail; more is known about cell development in C. *elegans* than for any other organism. A cell lineage is a description of all the cell divisions that occur to generate a specific group of specialised (differentiated) cells. The males cannot produce their own progeny but can fertilise hermaphrodites, which is experimentally useful as this provides a way of introducing genetic diversity into an otherwise clonal population of hermaphrodites.

C. *elegans* has been used very extensively to investigate the relationship between genetics, operation of the neuronal system and behaviour. As one reviewer of this research field comments:

Despite the simplicity of its nervous system, *C. elegans* shows an interesting repertoire of well-defined behaviours. With their sensory systems exquisitely developed, worms are able to find food sources, detect and escape noxious substances, find mating partners, display social or solitary feeding, and locate optimal oxygen concentration and temperature. Furthermore, *C. elegans* show sensitivity and intoxication to different drugs such as alcohol, nicotine and certain antipsychotic drugs, display different forms of non-associative and associative learning, and present sleep-like states during the quiescence phase associated with the four molts. (Burne et al., 2011)

C. elegans behaviours can be divided into four principal categories – locomotion, sensation, survival and reproduction (Rankin, 2002) – and all the worm's behaviours are mediated at some point by its sensory neurons, divided into chemosensory (detecting chemical changes) and mechanosensory (detecting movement and locomotion). Two neurons are also thermosensory (detecting changes in temperature). Chemosensory neurons have highly ciliated processes exposed to the external environment (Bargmann, 2006) so there is plenty of scope for environmental inputs into the worm's repertoire of behaviours. In fact, the roughly thirty chemosensory neurons can detect and distinguish a huge array of different chemicals, a truly economical system that puts this very small number of neurons to good use. By comparison the fruit fly *Drosophila*, considered further below, requires 1,000 neurons to detect the same number of chemicals, whereas mice use 10^7 neurons for such a purpose (Jovelin et al., 2003).

The vast majority of investigations into *C. elegans* behaviours have been carried out in the laboratory by artificially changing the genome. Many thousands of genes have been isolated and characterised in this way to study their role in worm behaviours (Kammenga et al., 2008). All the categories of *C. elegans* behaviour previously described have had at least one gene connected to the behaviour; more than 130 genes involved in locomotion have been identified. Indeed, virtually all the worm's behaviours are polygenic, that is, they are influenced by a whole range of different genes. However, inducing artificial mutations in the laboratory to study the role of genes in worm behaviour, useful as it is, also has its limitations. For example, most laboratory experiments are carried out on the same genetic worm strain known as N1. This is useful for comparing results between different research groups but ignores the fact that in real life different strains of worm, all members of the same species but with inter-individual genetic variation, may behave differently depending on their precise set of genetic variants (Schafer, 2005). For this and other reasons there has

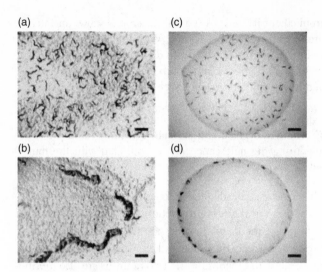

FIGURE 3 Solitary and social behaviour in *C. elegans*. (A) and (C): Solitary animals disperse evenly across a bacterial lawn and browse alone; (B) and (D): Social animals aggregate together into clumps and accumulate at the edge of the plate where the bacteria are most dense. (From de Bono and Bargmann, 1998.)

been increased research on different populations of *C. elegans* in the wild, especially since the pioneering studies of Hodgkin and Doniach (1997). Since that time hundreds of different strains of *C. elegans* have been identified, leading to some fascinating behavioural studies which in many ways are more relevant to our present theme, which is about how genetic variation between individuals contributes to the normal range of behavioural variation observed in wild-type populations.

By far the best studied example of a natural behavioural variation is the 'bordering/clumping' phenotype illustrated in Figure 3 (Barriere and Felix, 2005a). There is a natural variation in the tendency of different worm strains to clump together when feeding on an agar plate – also known as social feeding or aggregation behaviour. Worms that 'clump' also 'border', meaning that they form a ring around the edge of an agar plate (see figure), and they also move more rapidly on food than worm strains that graze in solitary fashion. Clumping behaviour only occurs when food is present, suggesting a chemotaxic response to an external stimulus (de Bono and Bargmann, 1998). Population density and environmental stress factors also modulate aggregation behaviour. One reason that the worms accumulate round the edge of the plate is that the amount of available oxygen is less there and the social feeders tend to migrate to the 'lower oxygen' regions of the plate (Gray et al., 2004). The clumping phenotype is more common than the solitary phenotype; the N2 strain is a notable exception, being solitary (de Bono and Bargmann, 1998).

Genetic sleuthing has identified the gene *npr-1* as a key player in controlling the difference between these two behavioural phenotypes (de Bono and Bargmann, 1998). The gene *npr-1* encodes a neuropeptide Y-related receptor (NPR-1), one of a family of receptors also found on the surface of neuronal cells in humans, that transmit signals inside the cell upon binding of the neuropeptide to the outer portion of the receptor (see Figure 4). NPR-1 is found also in other animals where it is involved in the regulation of behaviours such as food consumption, mood and anxiety (Hall, 2013b).

Remarkably, a single substitution of an amino acid at position 215 (indicated with an arrow in Figure 4) is sufficient to convert a solitary feeding behaviour into a clumping feeding behaviour (de Bono and Bargmann, 1998). Strains with an amino acid called phenylalanine at position 215 exhibit social behaviour, whereas those with the amino acid valine exhibit solitary behaviour. If the gene encoding phenylalanine at position 215 is incorporated into the normally solitary strain N2, then its progeny become social eaters. The reverse is also the case: if the gene encoding valine at position 215 is incorporated into social eaters, then they become solitary. So the presence of the genetic code for valine in the *npr-1* gene is both necessary and sufficient for solitary behaviour to occur.

How could such a tiny change at the genetic level, which in turn causes a change in a single amino acid, make such a dramatic difference to what is, after all, quite a complex behavioural difference? An important clue comes from the fact that the NPR-1 receptor which is encoded by the *npr-1* gene is located in sensory neurons found in various parts of the *C. elegans* nervous system, suggesting that it may be involved in regulating sensory inputs (de Bono and Bargmann, 1998). Furthermore, that single change in one amino acid in the receptor causes a change in the profile of the 'ligands' that bind to the receptor – a 'ligand' being a molecule arising from the environment surrounding the cell that requires a very precise conformation of the extracellular portion of the receptor to bind properly and transmit its signal to the inside of the cell (Rogers et al., 2003). Ligands that bind to the NPR-1 receptor in the solitary feeders give 'high inputs', whereas those that bind to the social feeder version trigger 'low inputs'. Even a single amino acid change in the intracellular portion of the receptor can be sufficient to change the binding properties of the extracellular portion. If ligand A then transmits a signal in place of ligand B, the response to environmental cues by the neuron can then be quite different.

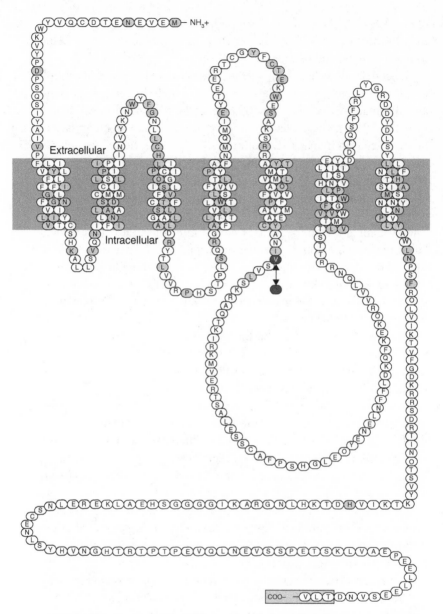

FIGURE 4 The structure of the NPR receptor. Each small circle represents an amino acid. The upper part of the structure, outside the plasma membrane that encompasses the cell, binds to neuropeptide ligands, whereas the long lower portion of the receptor transmits the signal inside the cell. The arrowed amino acid is the single substitution necessary to induce an alternative phenotype. (From de Bono and Bargmann, 1998.)

In the case of the NPR-1 receptor, the ligands that bind to the receptor form a particular 'family' of peptides; peptides are made up of a chain of amino acids, in this case eight to nine amino acids long, that make up the different ligands. Even such small peptides with slightly different amino acid sequences can adopt quite different conformations that then bind either better or worse to the extracellular region of the receptor.

Just as small differences in the amino acid sequence of the ligand can make big differences to the functioning of the NPR-1 receptor, so extensive studies have revealed just how complex the behavioural implications of a single amino acid change in the receptor itself may be (Boender et al., 2011). For example, like all animals, *C. elegans* has an innate immune system that defends it against attack by pathogens. There are certain bacteria that are poisonous for *C. elegans* to eat and it turns out that the NPR-1 receptor is important in defending the worms against such pathogens: those without the *npr-1* gene die faster when attacked by such bacteria (Reddy et al., 2009). Furthermore, the version of the receptor that has phenylalanine at position 215 (as in the social feeders) renders them more susceptible to attack than the solitary feeders that have valine at that position. But this affect appears not to be due to a direct action on the innate immune system, but rather arises from the fact that the worms are more susceptible to the pathogenic bacteria when they are clustered at the edge of the plate ('bordering'), and it is the lower oxygen level found there that attracts them to the edge. Social strains appear to link the lower oxygen level with the presence of a feeding group. When the oxygen level of the agar plate is forced up experimentally in the laboratory, then the social feeding disappears and the difference in susceptibility to being killed by bacterial attack also disappears (Reddy et al., 2009). The N2 version of the NPR-1 receptor, as found in solitary feeding worms, appears to 'protect' them against pathogens indirectly, by ensuring that the worms stay in areas of less oxygen and lower bacterial density.

Differential feeding habits and pathogen resistance are by no means the only *C. elegans* behaviours in which different versions of the *npr-1* gene have been implicated and include differences in fecundity and adult body size (Andersen et al., 2014). Again, as with pathogen resistance, these effects appear to be indirect, caused by avoidance of ambient oxygen concentrations, leading to a semi-starvation state as the worms aggregate to feed round the edge of the 'lawn' of bacteria. Such findings are highly relevant to interpreting human behavioural genetics: the behavioural phenotype is often 'connected' to a genotype indirectly by a long chain of mediating behaviours.

It should also be noted that none of the behavioural traits of the various strains of C. *elegans* described so far can be divorced from their development. It is as the worms develop following egg fertilisation that a particular behavioural repertoire emerges in which the variant information arising from variant genomes is integrated with cues arising from the micro- and macroenvironments. Development may take only 1 per cent of the time that it takes in a human context (providing the worms are kept at 25 °C), but the principle remains the same. Changing the cultivation conditions has long-term effects on subsequent worm behaviour. For example, if either social or solitary strains are cultivated at 1 per cent oxygen, then the adult worms in subsequent generations prefer living at oxygen levels between 0 and 7 per cent, whereas if they are cultivated at 21 per cent oxygen, then they subsequently prefer living at oxygen levels between 5 and 12 per cent (Cheung et al., 2005), effects presumably mediated by short-term epigenetic modification of specific C. *elegans* genes.

Other epigenetic changes have much longer-term transgenerational consequences. For example, the longevity of C. *elegans* worms is regulated epigenetically, partly via a protein complex called COMPASS which is involved in acetylating the histones that surround the DNA, an epigenetic modification described in Chapter 3. The worms do not use DNA methylation at all, unlike most other animals, and so histone acetylation represents a very important epigenetic mechanism. Mutations in the various genes that encode the proteins that make up the COMPASS complex result in worms that live much longer (around 30% longer than wild-type) (Greer et al., 2011). More surprisingly, the progeny of these long-lived mutants also live much longer down to the third generation, even though none of the progeny carry the COMPASS mutations (due to clever breeding techniques in the laboratory). By the fourth and fifth generations, the worms return to their original shorter lives once again. So there seems to be a transgenerational 'memory' such that the great-grand-offspring of the original mutant ancestors inherit a longer life and 'remember' the mutated COMPASS of their ancestors. Exactly how this works is yet to be sorted out, but an epigenetically mediated transgenerational process is clearly involved: indeed, mutant COMPASS-containing worms, or their long-lived progeny, showed differential expression of 759 different genes (Greer et al., 2011). Messing around with histone acetylation on a genome-wide scale has big effects.

Equally remarkable is the finding that continued exposure to certain odours can lead to populations of C. *elegans* that are sensitised to that particular odour for as many as forty generations, even though there has been no fresh exposure of the worms to the odour in question during the intervening generations (Remy, 2010). It has long been known that when *C. elegans* are exposed to an attractive odour (like benzaldehyde, each to his own) when present during the first larval stage of development, there is a life-long 'olfactory imprint' that enhances sexual performance and increases egg-laying rates in the adults when they are exposed to the same odour (Remy and Hobert, 2005). If the 'imprinting' of the same odour at the same larval stage is repeated over four generations, then the effect lasts for many generations even in the absence of the odour exposure altogether at the larval stage (Remy, 2010). Again it is presumed that this 'fixing' of a learned behavioural trait in the germ-line of an organism is due to epigenetic modification of some kind, although the molecular mechanisms involved likewise await further elucidation.

With this brief overview of *C. elegans* in place, we are now in a better position to ask some more general questions related to the DICI paradigm. Considering, for example, the role of the two versions of the *npr-1* gene in solitary versus social strains of *C. elegans*, it is clearly not the case that *npr-1* is a gene *for* social or solitary behaviour. Rather it is a gene that encodes the NPR-1 receptor that in turn is involved in multiple neurosensory pathways mediating a wide range of behaviours regulated by multiple sensory inputs. Could one then say that a particular variant of the *npr-1* gene is the *cause* of a particular behavioural trait? That does not sound right either, because there are hundreds if not thousands of 'causes' of a particular behavioural trait in the worm, involving many genes, epigenetic modifications, RNA molecules, proteins, neurons, hormones, pheromones, muscles, food, temperature, oxygen level, population density and much else besides: each cause requires a different kind of complementary narrative to do it justice. So it might be more accurate to say that it is a particular 'system state' that is causal for a particular behaviour. A somewhat similar point might be made in response to the question, So is the worm determined to display a particular behavioural trait? It all depends, yes, other things being equal, where all the other genomic, microenvironmental and macroenvironmental conditions listed previously comprise the 'other things being equal' list. But might it not then be possible to claim that expressing a variant *npr-1* gene encoding phenylalanine rather than valine at position 215 in the NPR-1 receptor causes *C. elegans* to engage in social rather than solitary feeding behaviour? That claim does seem more acceptable because now it is the *difference* between two behaviours for which the variant *npr-1* gene is claimed to be the cause. The variant gene is both 'necessary and sufficient' for the difference, which is distinct from the claim that the gene 'causes' a particular trait. The tricky question of causality will be further pursued in Chapter 11.

Foraging and Memory in *Drosophila* Fruit Flies

Drosophila melanogaster is the formal name for the fruit flies that swarm around rotting fruit in the orchard and, like *C. elegans*, is one of the classic models used for genetics research. Also like *C. elegans*, *Drosophila* are very easy to culture and genetically manipulate under laboratory conditions, the different strains being stored in hundreds of tubes where they live happily. Occasionally some escape and those, such as the present author, who have worked in departments in which a *Drosophila* research group is present become used to the occasional escapee fruit fly homing in on their lunchtime apple.

Drosophila fruit flies are about 3 mm in length, three times bigger than *C. elegans* and have around 300,000 neurons in their nervous system, 1,000-fold more than *C. elegans*(Kaun et al., 2012). The *Drosophila* genome was one of the first to be sequenced and contains 165 million base-pairs spread over four chromosomes, so in total size it is only 5 per cent the size of the human genome, albeit encoding around 15,500 genes of which approximately 13,000 encode proteins.[2] Of these, about 8,000 are homologous to human genes. Up to 75 per cent of human disease genes have related sequences in *Drosophila*, suggesting that flies can serve as an effective model to study the function of a wide array of genes involved in human disease.[3] As with *C. elegans*, *Drosophila* undergo a complex developmental programme prior to adulthood, a complete life-cycle taking around twelve days at room temperature. The *Drosophila* egg is about half a millimetre long. The female fruit fly lays between 750 and 1,500 eggs in her lifetime. It takes about one day after fertilisation for the embryo to develop and hatch into a worm-like larva. The larva eats and grows continuously, moulting one day, two days and four days after hatching. Two days later it then moults one more time to form an immobile pupa. Over the next four days, the body is completely remodelled to give the adult winged form, which then hatches from the pupal case and is fertile within about twelve hours. As with *C. elegans* the timing is highly temperature dependent. The timings here are for 25 °C; at 18 °C, development takes twice as long.[4]

One of the best studied behaviours in *Drosophila* is foraging. A natural gene polymorphism for foraging behaviour was first detected in *Drosophila* larvae in 1980 (Sokolowski, 1980). The two behavioural types are called 'rover' (forR) and 'sitter' (forS). In the presence of food, rover larvae have longer foraging path-lengths than sitter larvae, moving transiently from one food patch to the next and feeding briefly, while sitters tend to stay on one food patch and feed steadily (Sokolowski, 1980). In the absence of food, the behaviour of the two variant types is the same, indicating that the polymorphism is not a general locomotion defect, but specific to foraging behaviour (Yampolsky et al., 2012). The behavioural difference persists into adulthood, with rover adults more likely to leave a food source and seek a new one, and the sitters happy to stay where they are (Osborne et al., 1997). Roving is dominant to sitting, meaning that the roving behaviour is more likely to occur in wild-type populations, where the polymorphism is maintained at a frequency of around 70%:30% rover:sitter (Osborne et al., 1997).

The difference in behaviour between rovers and sitters is dependent on a single gene called *for* (foraging). This encodes for a protein kinase, an enzyme that attaches phosphate groups to other proteins, thereby modifying their properties (see Chapter 3). The *for* gene is primarily expressed in the neurons of the flies with no difference in its cellular localisation between rovers and sitters (Belay et al., 2007). Furthermore, the sequence of the *for* gene is identical between rovers and sitters. The key difference between the two behavioural types lies in gene expression: rovers have higher levels of *for* mRNA, and consequently higher kinase activity levels, than sitters, and it is this that contributes to the behavioural differences between the two types (Scheiner et al., 2004). The molecular details as to how one version of the *for* gene is expressed at higher levels in rovers than in sitters are still being worked out, but what is already clear is that multiple regulatory mechanisms are likely to be involved, including epigenetic modifications.

The *for* gene is also remarkably 'pleiotropic', meaning that it is involved in many different regulatory pathways in *Drosophila*. For example, the gene is involved in the responses of the different fly populations to sucrose (sugar), the sitters being much less responsive to sucrose than the rovers, correlating with their differences in foraging strategies (Scheiner et al., 2004). The rovers move further away from the sucrose source after feeding, whereas the sitters tend to hang around nearby. It has therefore been suggested that the difference in roving and sitting foraging behaviours is caused by different sensory responses to sucrose in these two types of *Drosophila*. But equally convincing is the suggestion that the differences in behaviour are primary, and these influence how the flies respond to food sources that include sucrose. This is the typical 'chicken-and-egg' scenario that crops up in behavioural genetics all the time: correlation does not equate with causation, so knowing what influences what is invariably problematic, even in a biological model as relatively simple as *Drosophila*.

The pleiotropic role of the *for* gene is underlined by a number of other examples. The sitters are more tolerant than rovers to higher temperatures, maintaining their ability to regulate their metabolism and move around as usual, whereas the rovers tend to wilt once the heat is on (Dawson-Scully et al., 2007). The *for* gene is also involved in learning and memory pathways (Kaun et al., 2007). Rover flies learn faster than sitter flies and also show better short–term memory formation and recall abilities, whereas sitter flies are better than rover flies at long-term memory, which may be more adaptive for their less active lifestyle (Mery et al., 2007). These various behaviours can be modulated according to particular environmental inputs. For example, differences in feeding behaviours between the rovers and the sitters only occur when food is abundant (Kaun et al., 2008). So the *for* gene provides a good example of the way in which gene-environment interactions occur, with varying outcomes depending on the situation.

The *for* gene also provides a good example of epistasis, which is the way in which different genes interact and collaborate to produce certain outcomes. A number of the interactions identified so far are connected with food and its metabolism. For example, investigation of food-deprived flies found significant evidence of epistasis between *for* and two 'positive regulators of insulin signalling' genes, *InR* and *dp110* (Kent et al., 2009). The *for* gene also interacts with a pathway regulated by a neurotransmitter called neuropeptide F, which is known to be important during both larval development and in the adult for food sensation (gustation) and particular feeding behaviours, including burrowing in agar and aversion to certain tastes, as observed in both *Drosophila* and *C. elegans* (Wu et al., 2003; Tan and Tang, 2006). There are many other examples, but the point is made: the *for* gene is but one component of a highly complex system in which gene products interact to mediate *Drosophila* behaviours. It is these complex interactions, involving multiple genes, that affect the overall 'fitness' of *Drosophila* populations in different environments, meaning the 'success' of a particular type in producing progeny. This in turn influences the ratios of rovers and sitters that are found in wild-type populations as they respond to changing environments and competition for limited food sources (Fitzpatrick et al., 2007). Epistasis is the norm, not the exception, in biological organisms.

So what causes different behaviours in *Drosophila*? The answer must surely be everything: the complete organism, which has developed, and is still developing in a particular way, in interaction with its complete environment and so we refer to the system taken as a whole. But, as with *C. elegans*, if we wish to ask the question as to whether the regulation of the *for* gene plays a causal role in *Drosophila* in the *differences* in behaviour between rovers and sitters, given a certain environment, then the answer is surely yes, providing it is noted that it is the regulation of the *for* gene that is making the difference (by mechanisms not yet fully elucidated), not variants in the sequence of the *for* gene itself (at least in its protein-coding portion).

Pair-bonding, Vasopressin and Oxytocin in Voles

It is not surprising to find a greatly increased repertoire of complex behaviours in animals with more complex nervous systems, albeit integrated with variant genomes in a way very similar to much smaller animals such as *C. elegans* and *Drosophila*. The example that follows illustrates the way in which the rather simplistic initial assignment of a particular gene to be the cause of a specified behaviour has been subverted by later findings that reveal a far more complex picture. The example comes from studies of vole pair-bonding.

Pair-bonding behaviour is the phenomenon of two individuals from the same species forming an exclusive long-term sociosexual bond, generally characterised by displays of affection and proximity, shared workload and feeding, rejection of outside individuals and exclusive or, more often, semi-exclusive sexual monogamy (McKaughan, 2012). Voles are the principal model species in the study of pair-bond behaviour due to the existence of natural variation in social structure between closely related species. Of the roughly sixty species of vole (*Microtus*), only a few species exhibit social monogamy while the remaining species are non-monogamous. In this respect voles are typical of mammals in general, only about 5 per cent of which display monogamous pair-bonding (Kleiman, 1977). The best studied vole species are the monogamous prairie vole (*M. ochrogaster*), along with pine voles (*M. pinetorum*), which are usually contrasted with the non-monogamous meadow vole (*M. pennsylvanicus*) and montane vole (*M. montanes*).

Two neuropeptides, known as 'vasopressin' and 'oxytocin', have been particularly implicated in mediating social bonding behaviours. Vasopressin is secreted by specialised neurons in the brain, and in larger quantities from the pituitary gland, and is involved in mediating a wide range of behaviours, including stress coping, territorial aggression, mate guarding, pair-bonding and paternal care. Vasopressin mediates its effects by binding to a specific receptor known as V1aR. While the structure and brain distribution of vasopressin are highly conserved among mammals, the behavioural effects of this peptide, and the neural distribution of V1aR, vary markedly across species (Donaldson and Young, 2013).

Vasopressin is known to facilitate pair-bond formation in monogamous voles both in the laboratory and in the wild; for example, it is released during and after mating in monogamous voles and its receptor density is increased in pair-bonded voles (Insel, 2010). Increasing vasopressin levels in the brain artificially by intracerebral injection promotes the formation of pair-bonds in both monogamous and non-monogamous species in partner preference tests (measured as time spent huddling with either a partner or a stranger after an extended period of co-habitation) (Young et al., 1999). The reverse is also true: if the interaction between vasopressin and its inhibitor is inhibited, then monogamous voles proceed to adopt a more promiscuous strategy (Barrett et al., 2013).

These and many other studies point to the regulation of vasopressin secretion and the regulation of vasopressin receptor gene expression as being important in distinguishing between the monogamous and non-monogamous voles. Indeed, earlier studies suggested that there was a regulatory sequence in the vole V1aR gene which acted like a 'switch' to control gene expression and, thereby, the actual amount of receptor present in key regions of the brain (Young et al., 1999; Hammock and Young, 2002). The switch was reported to be present in the monogamous prairie vole but not in the promiscuous montane vole. These findings, not surprisingly, attracted considerable press attention because it seemed that a relatively simple gene-expression switch could induce the transition from a monogamous to a non-monogamous animal behaviour, and vice versa – a dramatic result indeed. Alas, life with genetics is never that simple. For example, it was later found that across twenty-one species of vole worldwide, the proposed key regulatory V1aR gene switch was not associated with different social behaviours (Fink et al., 2006). This finding was followed by a whole string of papers suggesting that there was no significant correlation between regulation of the V1aR gene and pair-bonding, in wild populations of voles at least (Solomon et al., 2009). The key phrase in these studies appears to be 'free-living prairie voles' living in 'ecologically relevant conditions' (Mabry et al., 2011). Every single study using semi-natural or wild populations of prairie voles with variant V1aR gene switches found no correlation between variation in the sequence of the switch and the social strategy adopted by males (either pair-bonding or wandering) (Ophir et al., 2008; Solomon et al., 2009; Mabry et al., 2011). Meanwhile, laboratory-based studies have continued to insist that the key V1aR gene regulation switch *is* relevant in distinguishing between pair-bonding and non-monogamous behaviours in prairie voles (Castelli et al., 2011; Donaldson and Young, 2013). It is very likely that the differences in results between the laboratory bred vole populations compared to those out in the wild are explained by the fact that many of the normal selective pressures on social behavioural networks become disrupted in captivity when those same networks are no longer present.

If all this sounds very confusing, that is precisely the point. So often in animal behavioural studies, and even more in human studies, as we will consider later, correlation of a certain behavioural trait with the presence of a particular gene variant in animals displaying that trait is trumpeted as the gene *for* that particular trait. Behavioural differences in voles act as a warning that such a stance is naive.

Further studies have begun to unravel the complexities that characterise the role of various environmental inputs in influencing behavioural outcomes in the vole (Kelley et al., 2013, Wang et al., 2013). Ecological factors are clearly involved in vole monogamy versus vole 'wandering': population density, vegetation density, rainfall and temperature all affect mating system choice (Streatfeild et al., 2011). At low population densities and when resources are evenly spread, monogamy dominates, but when resources are more aggregated and populations larger, promiscuity becomes more common as home ranges start to overlap. Family and social factors could also be influencing social behaviour and mate choice. As a commentator on this research field reports: '[M]aternal nurturing, can induce physiological changes in the nervous system (e.g., modulating the expression of both peptides and receptors).... A vole's postnatal experience, including parent-offspring interactions, can have dramatic long term effects on partner preference and social bonding in adulthood' (McKaughan, 2012).

Vasopressin is one of a relatively small number of neuromodulators that influence social behaviour by shifting the balance of positive and negative affects elicited across a suite of connected brain regions (Phelps, 2010). The neuromodulator oxytocin appears equally important in this context, particularly in females, whereas the actions of vasopressin appear to be more dominant in males. In fact, oxytocin, vasopressin and the neurotransmitter dopamine are all released during sexual activity in all vole species that have been investigated. Oxytocin is also involved in the dilation of the uterine wall that leads to birth and in maternal nursing. In an analogous system to the vasopressin receptor, when an inhibitor of the oxytocin receptor is injected into a specific brain region of female prairie voles, it blocks the formation of partner preference and also of typical maternal behaviours, such as licking and grooming of the pups (Olazabal and Young, 2006). So activation of oxytocin receptors appears to be involved both in pair-bond formation and in maternal behaviours in voles.

Overall, it will be no surprise to hear that the vasopressin receptor gene has been promoted as the 'monogamy gene' (*New Scientist*, 2008).[5] The many other relevant media headlines include the following: 'The Gene Excuse: The Guy Can't Help It; He Was Born That Way' (*Pacific Sun*, 2004), and 'Why Men Cheat: Study Chalks Up Promiscuous Behavior to a Single Genetic Change' (*Science Now*, 2008).[6] ABC News reported, incoherently, that "[t]he monogamy gene is broken down into three parts – lust, romance and attachment and in some cases these three don't work together" (McKaughan, 2012). As part of the lead into 'The Cheatin' Gene: Researchers Find Men May Be Genetically Predisposed to Cheat' on the NBC Nightly News, anchor Brian Williams reported:

> Throughout history men have come up with all sorts of excuses for behaving badly. Now it appears they have a new one. It's in their genes, apparently. This is a line of research that started with rodents called voles. Now it's being applied to humans. (McKaughan, 2012)

During that programme an unidentified woman says, 'I would want to have my mate tested before I was married. And I am single and that would secure my marriage.' If such headlines in the popular press and on television appear somewhat extravagant, then they look positively restrained compared to the blogosphere, where we find this declaration of faith: 'So next time you get accused of adultery, next time you are unfaithful to your spouse, next time you have a big argument with your spouse over the remote control or the toilet seat – remember, it's not your fault. It's the vasopressin!'[7]

So what can we learn from the studies on genetics and different vole behaviours? In some ways the most important lessons for our present purposes belong more to the sociology of science and its communication to the public (McKaughan and Elliott, 2012). The transition from 'the discovery of an important genetic regulatory mechanism controlling monogamy' to 'oh dear the whole story is a lot more complex than we had first realised', all in the space of a decade, is fairly typical in the field of behavioural genetics, in humans just as much as in voles. The way in which anthropomorphic language can so quickly be applied to animal behaviours in inappropriate ways is also relevant. Articles in professional science journals reporting on different vole behaviours often talk of monogamy, promiscuity, fidelity and love, and indeed for convenience we have followed the articles' language likewise, at least in referring to monogamy, in describing some of the reported results. We can hardly blame the media for picking up on such language when it is precisely this language that is being used in the scientific literature. But when referring to monogamy in voles, scientists are using the word with a different meaning from its usage for humans. In the context of vole behaviour, biologists are talking about social monogamy, not about genetic exclusivity (sometimes referred to as 'genetic monogamy'). A 'monogamous social structure' is taken to include pair-bond formation in courting and mating relationships, frequent side-by-side contact ('huddling together'), care from both parents from birth through the early life of offspring, and even lifelong partnerships (Lim et al., 2004; Young and Hammock, 2007). But even among pair-bonded vole couples, genetic monogamy is rare. For example, among the so-called monogamous prairie voles, 45 per cent of males and 24 per cent of females adopt a 'wandering strategy' involving sex with individuals outside of their pair-bond (Nair and Young, 2006). Individuals in the human population who have sex outside of marriage, especially on such a regular basis, are unlikely to be referred to as 'monogamous'.

Despite the ambiguity of the data and the difficulty in extrapolating the application of animal studies into the human realm, philosophers also have not been shy at jumping on the 'vole band-wagon' to hold forth on the supposedly profound implications of the animal research. For example, the philosopher Patricia Churchland has made some interesting philosophical comments on vole bonding behaviours (Churchland, 2006). In terms of describing the genetics, her comments are exemplary, providing reasons for thinking that 'assumptions about "the gene for monogamy" oversimplify to the point of futility' (Churchland, 2012 p. 214, fn 71) and that oxytocin 'should not be dubbed the sociality/cognitive function molecule' since it, like our genes, 'is part of a complex, flexible, interactive network of genes, gene-neuron-neurochemical-environment interactions, and neuron-body interactions' (Churchland, 2012, p. 81). So far so good, but then Churchland also makes some big claims based on the earlier work on the role of vasopressin and oxytocin in vole behaviours, claiming that

> [m]ale prairie voles with lots of vasopressin binding sites are monogamous while montane voles, with few, are promiscuous. What determines the density of binding sites? Genes. Granting the effects of cultural complexity, something similar probably holds for humans too. As neuroscience uncovers these and other mechanisms regulating choices and social

behaviour, we cannot help but wonder whether anyone truly chooses anything.... I suggest that free will, as traditionally understood, needs modification. (Churchland, 2006)

Apart from the fact that the scientific statement claim made here is inaccurate (well of course the regulation of receptor genes is responsible for the density of receptors, but, as we have noted, the story on the relationship between vasopressin receptors and vole behaviour is by no means straightforward), the leap from voles to our understanding of human free-will does seem somewhat excessive. I suspect that ambiguous data derived from vole behaviours is simply not up to the herculean task of telling us anything useful about human free will. Over-enthusiastic extrapolations from the genetics of animal behaviours to questions concerning human morality and free will should be treated with suspicion.

None of these comments should be taken in any way as being sceptical of the roles of genomic variation, which in turn relate to different neuromodulators such as vasopressin and oxytocin; neurotransmitters such as dopamine; developmental variation dependent on different environmental inputs; a high level of maternal care of pups (or not); and a wide range of ecological inputs, all critical in influencing certain forms of vole behaviour. As previously emphasised when introducing the DICI paradigm, all animal phenotypes are 100 per cent genetic and 100 per cent environmental. This should not be taken to imply that the contributions of all genes, or of any kind of environment, to the differences in behaviours displayed by individuals in a species are of equal weight – far from it as clearly some variant factors within the overall complex system make more of a difference than others. But it seems very likely that the differential social behaviours of voles will ultimately be explained by hundreds of genetic variants and dozens of different neuronal signalling pathways, not by the switching on and off of a single gene.

The Genetics of Mammals Leaving Home

The genetics of mammals leaving the 'family home' is really important for the animal concerned, because it entails loss of social status and increased risk of injury, stress and mortality (21% of male macaques die within one year after leaving home). Furthermore, there is immense variation in such dispersal behaviours: some only involve males with their female sibs staying at home, or the precise opposite, or both males and females leave at the same time or at different times. The deficits involved in leaving home in terms of evolutionary fitness suggest that they must be balanced by other great gains (such as increased mate choice), and this in turn has led to an interest in the genetic variants that might contribute to dispersal behaviours. The genetic influence on the timing of dispersal has turned out to be quite high, for some species at least, and again is relevant to our aim of describing the behavioural consequences of natural genetic variation in the wild.

The pioneering work of Andrea Trefilov is of particular interest (Trefilov et al., 2000). This investigation involved the study of dispersal behaviours in a population of macaque monkeys living in the wild in Cayo Santiago, Puerto Rico. In rhesus macaques (*Macaca mulatta*) as well as in baboons, only males migrate, while females remain in their natal groups throughout their lives. But not all males migrate and the age at migration of those that do varies widely, with 87 per cent of males leaving within three and six years of birth. After leaving, males either join other social groups, form temporary small all-male groups or join other social groups on a temporary basis before returning to their original birth group ('nothing quite like mum's cooking'). The age at dispersal had previously been shown to be directly proportional to cerebrospinal concentrations of 5-hydroxyindoleacetic acid, which is a breakdown product of serotonin, a key brain neurotransmitter involved in the mediation of many different behaviours in mammals, with earlier dispersal being associated with low concentrations, and later dispersal with higher concentrations (Mehlman et al., 1995). In the study of Trefilov and colleagues, despite variations in the population size and density over a period of many years, around 37 per cent of the males left their natal group in any given year. The investigators decided to study the regulation of a particular gene involved in serotonin metabolism – the serotonin transporter gene which encodes the transporter protein responsible for taking up the neurotransmitter serotonin into neurons from the synaptic cleft between the neurons. This effectively brings to an end the actions of serotonin in transmitting the signal from one neuron to another. In macaques, as in humans, there are variable numbers of repeat DNA sequences in the regulatory region of the serotonin transporter gene, known as Variable Number Tandem Repeats (VNTR). There are two major versions of the VNTR, a short version (S), which reduces the amount of serotonin transporter protein that is produced, and a longer version (L), which increases the amount. So an increased amount of transporter will lower the amount of serotonin available for neurotransmission as the serotonin will be removed more rapidly from the synaptic clefts. Based on Mendelian genetics, it will be remembered that the gene can therefore exist in three different forms ('polymorphisms') so ranging from low serotonin transporter (SS), to intermediate (SL), to high (LL). Trefilov and colleagues found that the age when males disperse from their natal group correlates with their genotype with respect to the serotonin transporter gene. So SS homozygotes migrate from their natal group substantially earlier (57.1 ± 2.6 months) than LL homozygotes (71.5 ± 2.1 months), with LS heterozygotes intermediate (63.5 ± 1.5 months). This study was the first to detect an association between a genetic polymorphism and a behavioural trait in free-ranging primates. The data are consistent with the previous work showing a correlation between the breakdown product of serotonin (5-hydroxyindoleacetic acid) and the times of dispersal. So LL homozygotes, who are late dispersers, have higher expression of serotonin transporter and so an increased metabolism of serotonin as more of the neurotransmitter is taken up into the neurons and then broken down into its metabolic products. The rhesus males who exhibit less social competence and emigrate from their groups at a younger age have lower levels of serotonin transporter expression and so lower levels of serotonin metabolites.

So what are to make of this correlation between a polymorphism in the regulatory region of an important gene and variation in a complex behaviour in macaque monkeys? There could well be hundreds of other polymorphisms besides this one that correlate with the different dispersal behaviours described. On the other hand, the serotonin transporter gene is a known important player in behavioural studies in many other species, not least in humans, and it is entirely plausible that the serotonin pathway is involved in dispersal behaviours and, therefore, that genetic variation impinges on the generation of different behavioural dispersal traits. One of the problems with primate research of this kind is that replication studies rarely appear in the literature, and the present study has not been replicated at the time of writing. This in no way implies any criticism of the cited study. But experience from the kind of molecular genetic studies in humans that will be surveyed further in Chapter 7 suggests that replication studies of correlations between genetic variants and behavioural traits is extremely important in terms of establishing the trustworthiness of the correlation. This is because human populations, as macaque populations, vary considerably, and experimental conditions often vary between investigators working in different parts of the world, so replication always provides some welcome assurance. Some of the other challenges arising from behavioural molecular genetics will be further discussed in Chapter 7.

It is also worth noting that the same research group has carried out a further study on the same macaque population, only this time focusing on the timing of reproduction (Krawczak et al., 2005). The polymorphism in the serotonin transporter gene was found to be significantly associated with the timing of reproduction in the life histories of 580 males. Homozygotes were found to mate earlier than heterozygotes, and homozygotes had two peaks of activity, while heterozygotes had only one reproductively active peak. The authors suggest that this phenotype is mediated via the dispersal-age variation noted previously: because heterozygotes leave their natal group at an intermediate stage, they don't have time to mate with the natal group (as LL homozygotes do) and it takes longer to get settled with the new group than SS homozygotes who migrated earlier. However, according to the authors, the polymorphism explained only 7 per cent of the variance in reproductive timing, suggesting that many other variants are involved. Perhaps of greater significance in our present context is the reminder that it is often very hard to know what is the 'primary' behaviour that is being influenced by genetic variation, and what are the 'secondary behaviours' that arise from the primary behaviours. Very often gene variants are identified that correlate with a behavioural trait, and the assumption is then made that this is the prime behaviour (the 'endophenotype'), but further work then suggests that this is not the case at all, but rather the behaviour in question is a product of another behaviour, which in turn is a better candidate for being influenced by a particular variant gene (and so can now be referred to as the 'true endophenotype').

A good example of this is provided by a study of dispersal times in mice (Krackow and König, 2008). The onset of male antagonism towards other males (which occurs around two months of age in the majority of male mice) was reported to be positively associated with a polymorphic VNTR region in the mouse equivalent of the serotonin transporter gene. Antagonism towards rival males was considered by these authors to be an endophenotype for dispersal, as antagonism is a driver of migration in mice.

Variation in dispersal times has also been studied in a population of male red wolves (Sparkman et al., 2012). Among such cooperatively breeding species, delayed dispersal provides an opportunity for older siblings to help rear their young. Variation in age at dispersal of the wolves was found to be associated with a measure of genetic relatedness. Descendants from a particular founding wolf all tended to migrate at a similar age, which was significantly different from wolves descended from other founders. The investigators also found that age at dispersal was unrelated to litter size, litter sex ratio, natal home range or population density, all popular suggestions for environmental contributions to this trait. As the authors conclude: 'Identifying a multigenerational family resemblance in dispersal age is not conclusive evidence of a genetic basis for dispersal, as common environmental factors may be sufficient to create a misleading resemblance among closely related individuals.' Nevertheless the results are suggestive and important in at least apparently excluding some environmental factors from involvement in the trait.

So Can Animals Choose?

Animals clearly make choices; it is just how they experience those choices that is difficult (maybe impossible) for us to know. What is striking is the extent to which larger animals have a huge range of choices as they carry on their daily lives: the so-called pair-bonding monogamous prairie voles, who are not really monogamous in any human sense of the word, still have to choose whether to go wandering or whether to stay at home; and whatever predispositions mammals may have to leave home at a certain age, there is no doubt that there is a huge variation in this behavioural trait. Irrespective of whether or not a certain behaviour correlates with a genetic variant, the choice as to which day to leave home still has to be made.

Now it might well be objected that speaking of 'animal choice' is inappropriate because certainly animals do not choose in the way that humans choose, with their big frontal cortices providing powerful computing power enabling reflective decision-making in which a wide range of the likely consequences of one's actions can be reviewed. On the other hand, the larger animals that we have been considering in this chapter have plenty of neuronal computing power, surely sufficient to be aware of the likely consequences of at least some of their actions. Furthermore, these animals display clear signs of intentional behaviours. So even if we employ the language of 'choice' and 'intentionality' in a metaphorical sense, nevertheless the metaphors appear to refer to real behavioural equivalents in animals that have some connection with the meanings of those words in human discourse.

Our animal exemplars, both large and small, also illustrate just how entangled are all the many components that contribute to behavioural traits. Once again it is worth remembering that all the traits discussed in this chapter emerge as a result of a lengthy process of development in which genetic variation integrates with all the many environmental variations that characterise animal life in the wild. The behavioural influences of different genetic variants do not 'fall from the sky' as it were upon fully grown animals that then somehow behave differently: all the genetic differences are integrated with an emerging complex neuronal system that carries within it the tendencies that influence different behavioural traits. Some of those traits, especially those essential for survival, will be under tighter constraint, whereas with others there will be a much greater degree of flexibility, allowing adaptive strategies to be deployed according to circumstances.

Animal behavioural studies, of which this chapter represents the tip of a very large iceberg, be they on a worm with 302 neurons or on a vole with many millions of neurons, can be remarkably informative in seeing how multiple components in the genomic, micro- and macroenvironments integrate together to produce the observed behaviour. The underlying molecular mechanisms may be very similar, but investigating the role of genetic variation in differential human behaviours represents a challenge of a different order, as the next few chapters illustrate.

CHAPTER 6

Prisoners of the Genes?

Understanding Quantitative Behavioural Genetics

> *A century of familial studies of twins, siblings, parents and children,*
> *adoptees, and whole pedigrees has established beyond a shadow of a*
> *doubt that genes play a crucial role in the explanation of all human*
> *differences, from the medical to the normal, the biological to the*
> *behavioral.*
>
> Eric Turkheimer (2011, p. 227)

> *The sheer level of ignorance, distortion, and flawed reasoning that*
> *characterizes the 'anti-heritability' camp is unprecedented in science*
> *and philosophy of science.*
>
> Neven Sesardic (2005)

> *The biological worldview of postgenomics is characterized by extreme*
> *complexity, variability, multilevel reciprocal interactionism, and sto-*
> *chasticity as an inherent property of biological systems, which all*
> *contribute to what might be called the blurring of boundaries, in*
> *particular, the boundary between genes and environment.*
>
> Evan Charney (2012, p. 332)

> *Why is it that most geneticists do not understand that the phenotype*
> *is a developmental process?*
>
> Richard Lewontin, in a letter to Dobzhansky, May 1973[1]

The field of behavioural genetics tends to arouse great passions, especially when its findings and claims are used to justify educational or
other social views or policies. Some of its more controversial applications to traits such as intelligence, sexual orientation and religious practice will be surveyed in more detail in later chapters. For the present it is
worth pointing out that there are books that set out to rubbish the whole
field, sometimes written by those with a particular political axe to grind;

there are books written by professional practitioners, somewhat defensive at the perceived misunderstandings of the field by those who do not work within it; and then there are other books by non-specialists that passionately defend the field, again sometimes driven by political views of a somewhat dubious nature.

The situation is not helped by the way in which the field was brought into deep disrepute by some of the more horrendous events of the twentieth century as already described in Chapter 2, and as mentioned there, it was only from the 1960s onwards that behavioural genetics began to come back into the mainstream of scientific endeavour.

This chapter will take a middle-of-the-road position concerning the assumptions, practices and findings of this field. On one hand there is no doubt that the field has been, and continues to be, immensely valuable in the identification of genetic variation in the development of many traits of medical significance. This in turn has led to the search for the relevant genetic variants in the field known as 'molecular behavioural genetics' to be discussed further in the next chapter. There are also a number of traits within the normal spectrum of complex human abilities and behaviours for which the application of behavioural genetics has clearly demonstrated the relevance of genetic variation within specified populations. On the other hand, several of the assumptions underlying the field are somewhat shaky and difficult to justify, and beyond these technical points lies the over-interpretation of data regarding some traits in a way that often brings the field as a whole into disrepute, which is a pity. Plenty of examples of these somewhat sweeping generalisations will be found in the chapters that follow. In addition, the reporting of the findings of behavioural genetics can readily lend itself to the kind of dichotomous understanding of human personhood that does not fit well with the more integrated DICI approach that this book seeks to promote.

We will therefore pursue the topic in this chapter by first providing an introduction to the ways in which behavioural genetic studies are carried out. We will then move on to a critique of the field and its interpretations with a view to seeing how its outputs need to be interpreted with some caution.

The Assumptions and Practices of Behavioural Genetics

The aim of quantitative behavioural genetics is to attribute the proportions of variance with regard to a behavioural trait to genetic variation and to environmental influences in a specified population. The mathematical approaches utilised are known as 'biometrics'. The field draws heavily on the Fisher-Wright model (named after Ronald A. Fisher and Sewall Wright), which was established back in the 1920s and connects mathematically the continuous binomial distribution of genetic variants that impact on traits with standard familial Mendelian genetics. It is this model that led to the neo-Darwinian synthesis: the mathematical fusing of Mendelian genetics with population genetics. The model assumes that complex human traits are the result of a very large number of segregating genes, each of small effect. A 'trait' may refer to any behaviour that varies within a population: a medical trait that results in behavioural differences; personality; intelligence; sexual orientation; religious practice; number of hours per week watching television: you name it, behavioural geneticists have measured it. Whether the word 'trait' can be applied with an equivalent meaning across such a diverse range of human behaviours will be discussed later in this chapter. For the moment, we will simply accept its current usage as referring to any measurable behavioural variation.

The total amount of behavioural variability in a population is the *phenotypic variance* (Var(P)). According to quantitative genetics theory, and stating now its basic starting assumption more formally, the phenotype is a product of both genetic influences (G) and environmental influences (E) – consequently, phenotypic variance can be thought of as made up of two variance components:

$$Var(P) = Var(G) + Var(E).$$

The genetic component can be further broken down into two sub-components: *additive* genetic effects (A), and *dominance* genetic effects (D). Additive effects refer to the direct effect of a particular allele (gene variant) on a phenotype, for example, an allele for red flower colour in plants. If there are multiple loci, meaning multiple genes in the genome that impact on the trait in question, the effect at each locus 'adds up' to give the phenotype. Additive genetic effects represent the extent to which a trait breeds true through the generations: 'If a parent has one copy of a certain allele, say, A_1, then each offspring has a 50 per cent chance of receiving an A_1 allele. If an offspring receives an A_1 allele, then its additive effect will contribute to the phenotype to exactly the same extent as it did to the parent's phenotype. That is, it will lead to increased parent-offspring resemblance on the phenotype, irrespective of other alleles at that locus or at other loci' (Plomin et al., 2008). Dominance effects occur when one allele of a gene interacts with the other allele at that locus. Consequently, the phenotype is not just a result of adding up allele contributions, but depends on their interactions. Because each person inherits one allele from each

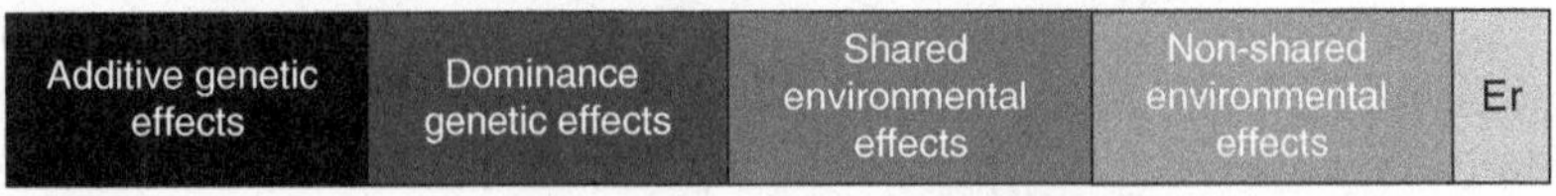

Additive genetic effects	Dominance genetic effects	Shared environmental effects	Non-shared environmental effects	Er

FIGURE 5 Components of phenotypic variance. Er = error, which is normally subsumed into the component 'non-shared environmental effects'.

parent, dominance effects are not passed on from parent to offspring, so are defined as being independent of additive effects. Incorporating these components into the model gives

$\mathrm{Var}(P) = \mathrm{Var}(A) + \mathrm{Var}(D) + \mathrm{Var}(E).$

The environmental component can be further broken down into two sub-components: *common* or *shared* environmental effects (C) and *unique* or *non-shared* environmental effects (E). The shared environment comprises events that make children raised in the same family more similar to each other and less similar to those who do not share it. Examples include socioeconomic status, nutrition or parenting style (Rijsdijk and Sham, 2002). Non-shared effects describe influences that are unique to an individual or have differential effects on them, such as smoking and drug taking, accidents or psychological trauma and which make individuals less similar to each other. Non-shared effects tend to affect just one family member. Within twin studies (on which more later) the non-shared effects may be defined as the uncorrelated variance between members of an identical twin pair. An error factor involved in measuring traits is also included in this component since it is assumed that measuring non-shared effects provides the least reliable data. Altogether these assumptions yield the model as shown in Figure 5.

The model can also be presented as

$\mathrm{Var}(P) = \mathrm{Var}(A) + \mathrm{Var}(D) + \mathrm{Var}(C) + \mathrm{Var}(E),$

which is also written as

$\mathrm{Var}\,(P) = a^2 + d^2 + c^2 + e^2.$

The various components are squared for statistical reasons. The reason that the various proportions of variance are written in lower-case letters is to make it easier to manipulate the components when deriving equations using this model.

The term 'heritability' was introduced into the quantitative genetic literature in 1940 in a paper on animal breeding to refer to the proportion of the variance of a trait that can be ascribed to genetic variation in a given population (Lush, 1940). Unfortunately, the word already had a clearly defined meaning in genetics, referring to the transmission of traits from parents to offspring. The word remains in common use with that meaning. But 'heritability' as used in quantitative genetics now has a second, quite different and more technical meaning: a statistical construct referring to a proportion of variance (Visscher et al., 2008). It is therefore unsurprising that the word 'heritability' is frequently misused in media reports of discoveries in genetics, and sometimes even biologists, who should know better, are prone to using the word with both its different meanings in the same text, without making clear that the two meanings are quite distinct.

Although the traditional derivation of Var(A), or a^2, refers to this value as representing additive genetic effects on the trait in question, this should not be thought of in causal terms: thinking more mathematically (the correct approach in this context), we can define Var(A) as the variance in phenotype that is associated with variance in genotype. In practice, Var(A) is generally written as h^2 rather than a^2 as it refers to the heritability as just defined in its technical sense within quantitative genetics. Written as h^2 it represents only 'narrow sense heritability', which refers to the additive effects of separate genetic variants on the trait in question. 'Broad sense heritability' is represented by H^2 and refers to both the additive and dominant genetic effects on a trait. In general, 'heritability', if unqualified, refers to narrow-sense heritability.

Statistically derived proportions of variance must, by definition, add up to 1. Heritability estimates are therefore always between 0 and 1, with 0 indicating that there is no genetic influence on phenotypic variation and 1 indicating that there is no environmental influence. For example, in many, but certainly not all, genetic disorders the disease will emerge if a certain genetic variation is present, irrespective of any environmental difference. It is also worth noting that heritability is often reported as a percentage, for example, in the statement that '55 per cent of intelligence is heritable'. This is the claim that 55 per cent of the phenotypic variance in intelligence in a given population is attributable to genetic variation, not that 55 per cent of an individual person's intelligence is due to their genetic endowment inherited from their parents.

Understanding the term 'heritability' in its technical sense is not always straightforward. Patrick Bateson provides this illustration to help us (Bateson and Gluckman, 2011, p.14). We notice that nearly all people have two legs. When they only have one leg, it is nearly always due to an accident, in other words, due to the environment. So the variation in leggedness in the population is 100 per cent environmental and 0 per cent heritable. That sounds odd because everyone knows that you need genes of a certain kind to develop legs. It is just that in practice the genomic system responsible for leg building during development nearly always encounters the same environmental context, and so the system builds two legs. Likewise, possession of a brain is heritable in the sense that it is inherited, but in biometrical terms the heritability of having a brain is zero because there is no variation in a population in this particular trait. So we note that heritability tells us nothing about the complex interplay of components described in Chapters 3 and 4 in the context of the DICI model that as a matter of fact lead to the development of traits. Heritability is a statistical construct referring to variation in a population, not about the interplay between genes and environment in any particular individual.

Imagine a population A in which all people can see. The heritability of sightedness in this population must therefore be zero since there is no variation in this population with respect to this trait. But now let us say that we introduce a few congenitally blind individuals into the population, now called population B. The trait of sightedness must now have a positive heritability value due to the variation. Does this mean that genes are more involved in the development of vision in population B as compared to population A? No!

If animals are cloned and fed on widely varying diets, then body size is 100 per cent environmental and the heritability of body size will be zero in this population. But if genetically variable animals of the same species are raised in identical environments, then the heritability of body size is likely to be 100 per cent.

Such reflections are important when interpreting the huge literature based on twin studies. One meta-analysis incorporated data published during the years 1958–2012 on the variation of 17,804 human traits based on more than 14 million different twin pairs drawn from thirty-nine different countries (Polderman et al., 2015).[2] The traits ranged from psychiatric and cognitive traits, and specific medical conditions, to social values, religion and spirituality. Virtually all twin studies within this fifty-year period were included. Across all traits the heritability was 49 per cent, and the authors report that for 69 per cent of traits the observed twin correlations are consistent with 'additive genetic variation'. The data on this cohort are inconsistent with 'substantial influences from shared environment or non-additive genetic variation'. For roughly one-third of traits, a simple additive genetic model did not sufficiently describe the population variance and there was a major influence of the shared environment in the observed variance in particular traits. The precise interpretation of such claims will be discussed further. For the moment it is worth noting that when Turkheimer described his so-called 'three laws of behavior genetics' in 2000, the first law was that '[a]ll human behavioral traits are heritable' (Turkheimer, 2000). The Polderman meta-analysis (Polderman et al., 2015) provides a striking confirmation of this contention, but does not at all entail what first impressions might suggest, as will hopefully become clearer in what follows. As Turkheimer himself has written (the 'distinctions' here referring to those made between genes and environment): 'the universality of heritability is best interpreted as a *reductio ad absurdum* of these very distinctions – as a way of observing that the endeavor of figuring out how genetic or environmental a trait is, let alone declaring that it is exclusively one or the other, is pointless' (Turkheimer et al., 2014).

This conclusion is illustrated by the observation, for example, that a high heritability value does not necessarily predict high concordance of a trait between identical twins. So we note, for example, that if one twin in a pair of female identical twins develops rheumatoid arthritis, which has a heritability in the range 53-65 per cent, then only 15 per cent of their twin sisters also develop the disease (MacGregor et al., 2000). Heritabilities of up to 80 per cent have been reported for schizophrenia but the pair-wise concordance level for identical twins is 41–48 per cent (Sullivan et al., 2003). A high heritability value likewise does not necessarily imply a high level of genetic influence (Visscher et al., 2008). Heritability values can also change in a population over time due, for example, to changing environments or to greater inbreeding or the influx of other genetic variants due to outbreeding.

The traditional methodology for a heritability analysis, which accounts for the vast majority of studies, is a family-based design. Family studies rely on the fact that relatives share both genetic and environmental influences on behavioural traits; the same influences are assumed to affect the trait in the relatives under study. The genome is assumed to be identical throughout life. The level of shared influence affects the proportion of the variance between relatives that can be assigned to either genetic or environmental effects. Relatives living in the same environment are assumed to share 100 per cent of common environmental effects, while those living apart are assumed to share no common effects. By definition, no relatives share unique environmental effects. Comparisons between trait variation in identical as compared to non-identical twins remain central to such heritability analysis. In practice, dizygotic (non-identical) twins, or any pair of full siblings, share 37–62 per cent of their genes (Visscher et al., 2006).[3]

A component of the variation is due to the different amounts of exchange of chromosomal material ('crossing over') that occurs in the parents during the generation of the sex cells, the sperm and the ova. And although identical (monozygotic) twins are assumed to be genetically identical, the reality is more complicated than that, as will be further discussed later in this chapter.

Behavioural genetics uses three main types of family-based design to calculate heritability. The first approach involves the study of monozygotic twins separated at birth and raised in different homes. However, this strategy is not that common, given that the number of twins separated at birth is happily rather small. We shall return later to the meaning of the phrase 'separated at birth'. The second more common approach is to compare monozygotic twins who have been raised together with dizygotic, that is non-identical twins, who have also been raised together. There are about 11 million identical twins in the world and more than seventeen twin registries each containing more than a million twins (van Dongen et al., 2012), so there is plenty of scope for such studies. The third approach involves adoption study designs which work in one of two ways: either genetically related siblings living apart from each other are compared, thus providing an estimate of genetic effects, or genetically unrelated adoptee-siblings living together are compared, thus providing an estimate of shared environmental effects. The adoption design can also be extended to compare adopted children with their biological and non-biological parents. Generally, an adoption study compares multiple sets of relatives.

To calculate the heritability, different approaches are required depending on the familial protocol being utilised. For twin studies, a basic approach is provided by Falconer's formula:[4]

$h^2 + c^2 + e^2 = 1.$

Although no longer used due to statistically more powerful approaches, the formula is helpful in understanding how a narrow-sense heritability value is generated. From Figure 5 it will be remembered that h^2 refers to the heritability, c^2 to the shared environmental effects and e^2 to the non-shared environmental effects. Using this formula it can be shown that the heritability can be estimated using the classic twin-design by multiplying the difference between the correlation of the variance between monozygotic twins and the correlation of the variance between dizygotic twins by two.[5] As already emphasised, heritability is therefore a statistical construct referring to variance within a population.

Falconer's formula provides a good rough estimate of narrow-sense heritability but can only be used for analysing a single behavioural trait and cannot compare multiple co-varying traits to detect an underlying common genetic or environmental influence. It does not provide confidence intervals for heritability estimates and cannot use data from adoption or family studies, only from twin studies. Furthermore, it is only accurate if there are no dominance effects for the behavioural trait being studied, which is unlikely to be the case. The formula is therefore not now used in formal heritability analysis, but understanding it is useful because it provides a good introduction to the range of assumptions that underlie the field of behavioural genetics.

The standard statistical methodology in current use in behavioural genetic studies is known as model-fitting, or structural equation modelling. Structural equation modelling allows complex models of genetic and environmental interaction on multiple traits to be broken down into their individual variance components. The basic principle behind model-fitting is the creation of theoretical models of how variance components interact with each other and the phenotype, where the interaction is measured using variances and co-variances (Plomin et al., 2008). A computer is then used to generate the *parameter estimates* (h^2, c^2, etc.) for each component being modelled that result in the closest fit between model and observed data. The iterative process of achieving the best fit is called 'optimisation', and produces a 'goodness-of-fit' statistic between observed and predicted co-variance matrices (Rijsdijk and Sham, 2002). The overall variance is said to be 'decomposed' into the various proportions of variance using such methods; despite its biological overtones, in this context the 'decomposition' involved is very far from the world of biology. The details of these various statistical approaches can be accessed in any standard text on behavioural genetics (Plomin et al., 2013a). For our present context it is only important to know that different types of comparisons between relatives require different statistical models. The ACE model, for example, is used when it is thought that the trait in question can be ascribed to the genetic variation (A), the shared environmental effects (C) and the non-shared environmental effects (E), without considering other factors, such as dominance (D) effects. If dominance genetic effects are thought to be important, then an ADE model is used. In practice, various statistical models are applied to the same sets of data until the 'best fit' is obtained. The data can also be represented visually as 'path diagrams' in which the different components of variance are mapped out to illustrate how they are thought to contribute to the trait in question.

As far as a 'quick and easy' way of mentally envisaging the derivation of a heritability value is concerned, Falconer's equation remains useful. If the correlation between identical twin pairs with respect to a particular trait is, on average, higher than for non-identical twins (in other words, the identical twins are more similar to each other than the non-identical twins), then by definition a positive heritability value must appear, based on this equation. And since the proportions of variance have to add up to 1, if one value goes up, such as the heritability, then another value, such as the variance attributable to a non-shared environment, must go down. None of the values have any units: they are statistical constructs.

What about Adding Up Gene Effects?

Although the original mathematics was based on the assumption that the effects of genetic variants on traits are additive, practitioners in the field are well aware that this is often not really the case (Plomin et al., 2008). So in practice in the statistical modelling, extra terms are employed for non-additive genetic effects, including interactions between the two different alleles at a single position in the genome (the effects of one allele may be 'dominant' over the other); epistasis – the interactions amongst alleles located at different places in the genome; and the interaction between alleles and the environment, often referred to as gene by environment interaction, or G x E. The G x E acronym may be defined as differential sensitivity to environmental risk as a function of genetic variability. Although it may seem counter-intuitive, most forms of G x E interaction increase identical twin similarity more than non-identical twin similarity, thereby increasing the genetic proportion of variance, and so the heritability value, in standard biometric approaches (Burt, 2011). G x E interactions include a genetic predisposition that then shapes the environment in such a way that it impacts in a different way on the individual had there been no predisposition in the first place. For example, a child with athletic ability will be picked out for coaching that might not be available to other children.

As far as epistasis is concerned, in practice it is the narrow sense heritability which is often measured in behavioural genetic studies because the extent of epistasis relevant for the trait being studied is unknown. But epistasis certainly does occur in the development of traits as emphasised in earlier chapters when presenting the DICI paradigm. One variant gene may have a synergistic rather than an additive effect in trait development, whereas another may have an inhibitory effect. For example, there is a rare gene mutation present in up to 0.5 per cent of Icelandic and Scandinavian populations that protects against the development of Alzheimer's disease. Compared with their countrymen who lack the mutant gene, Icelanders with even a single copy of the variant gene are more than five times more likely to reach the age of eighty-five without Alzheimer's and have a 50 per cent better chance of reaching their eighty-fifth birthday (Callaway, 2012). Without prior knowledge of that variant gene, it might readily have been assumed that epistasis was unimportant in the aetiology of Alzheimer's disease. This example is more the rule than the exception: the ability of mutations to cause disease depends to a significant extent on their genomic context (Jordan et al., 2015). In reality, epistasis is the way that the genome operates in the development of all traits and it is unlikely indeed that the variations observed in behavioural traits are somehow excluded from this basic biological insight. Having said that, it is the epistasis that contributes to variation in a population with reference to a particular trait that matters in this context, not the epistasis that contributes to everyone being the same (remember leggedness and brain possession).

As far as G x E is concerned, this was simply ignored in traditional calculations of heritability, but measuring G x E effects has now become a major emphasis in contemporary behavioural genetics, and some examples will be provided in the following chapter. For the moment, it is worth noting that in identical twins raised apart, for example, for whom G is assumed to be the same, epigenetic and other differences that affect gene expression in a way that affects the trait being measured will tend to accumulate statistically in the proportion of variance defined by E (the environment), even though the mechanistic differences are operating at the genomic level. The situation is further complicated by the observation, supported by a considerable body of data, suggesting that certain gene variants act as 'risk factors' contributing to vulnerability upon exposure to certain environmental events. So as well as the environment impacting the genome, the genome also affects the extent and outcome of that impact. Individuals differ by genotype in the extent to which they are affected, either positively or negatively, by exposure to certain environments.

Some Examples from Medical Genetics

The most fruitful applications of human quantitative behavioural genetics have been in the field of medicine. There are many pathological conditions associated with behavioural changes that lie outside the normal range in human populations. The use of twin studies, the most commonly used approach in this field, has been enormously useful in identifying genetic contributions to conditions that were previously thought to be due to environmental factors alone. A good example is provided by the autism spectrum disorders. Autism diagnoses rose to 1 in 68 US children (1.2 million in total) in 2010, up from 1 in 110 in 2006 (Weintraub, 2011). In one study carried out in South Korea on a sample of 55,000 people, a startlingly high incidence of 1 in 38 has been reported (Kim et al., 2011). Children living in a 900-square-kilometre area centred on West Hollywood in Los Angeles are four times more likely to be diagnosed with autism than are those living elsewhere in California (King and Bearman, 2011). There is a higher level of autism amongst cohorts good at maths as compared to the humanities (Baron-Cohen et al., 2007). Around four times as many males suffer from autism as females. Indeed it has been suggested that autism represents an extreme form of the male brain (Baron-Cohen, 2010), a theory that has received some empirical support (Baron-Cohen, 2005; Hall et al., 2013) and an equivalent number of detractors (Bejerot et al., 2012; Krahn and Fenton, 2012). Nearly half of the children diagnosed with autism have average or above average intelligence. It is unclear whether more intelligent children are developing the condition or whether they are being diagnosed at a higher rate than in the past (and the idea that vaccination is a cause of autism has been thoroughly debunked). The overall increase in autism compared with previous decades is the subject of intensive speculation and investigation; the increase does appear to be real and only partly explained by an increased efficiency of diagnosis (Weintraub, 2011).

Autism spectrum disorders are characterised by impairments in social interaction and communication of varying degrees of severity (hence the word 'spectrum') and by restricted, repetitive and stereotyped behaviour and interests. Severe forms of autism are also characterised by seizures and intellectual disability. Most of the features of autism are manifest during the first few years of life, during that critical period described in Chapter 3 when a huge increase is taking place in the formation of synaptic connections between neurons. Most current theories of autism revolve around the idea of the development of abnormal synaptic connectivity during early infancy.

Right up to the 1990s autism was still being blamed on birth trauma, infections, poor parenting and even child abuse. Psychologists suggested that unemotional 'refrigerator mothers' caused their infants to set up self-defence mechanisms which led to autism (Mahler et al., 1959). But it was noted that the relative risk of a child being diagnosed with autism is increased at least twenty-five-fold over the population prevalence in families in which a sibling is affected, suggesting a significant familial influence (Jorde et al., 1991). Later work using hundreds of twin pairs has since been used to derive the heritability of autism, generating values in the range 56–95 per cent (Colvert et al., 2015). In a further study carried out on a cohort of 192 twin pairs in California, heritability values in the range 14–67 per cent were reported for autism spectrum disorders and in the range 30–80 per cent for shared environmental factors (Hallmayer et al., 2011).

Overall, these and other findings point to both important genetic and environmental components in explaining the variance in a population with regard to the trait characterised by the term 'autism spectrum disorders'. As the weight of evidence has grown demonstrating the importance of genetic variation, a huge burden of unnecessary guilt also has been lifted from the shoulders of the parents of autistic children who had previously been blaming themselves for their child's condition. The positive pastoral impact of genetics on families in which a child is suffering from a particular disorder can be very considerable.

In addition to autism, it is also easy to forget how recently it was that psychoanalysts would speak of the 'schizophrenogenic mother' as if there was something about the mother's personality that induced the onset of schizophrenia in her children. It was not until the adoption study of Heston published in 1966 that it was clearly shown that this inference is quite false (Heston, 1966). The risk for developing schizophrenia increases with genetic relatedness: the general risk in the overall world population is 1 per cent; if there is a second-degree relative who already has schizophrenia (sharing 25% of their genes in common), then the risk rises to 4 per cent; for a first-degree relative (sharing 50% of their genes in common) the risk is 9 per cent; between dizygotic twins, it is 17 per cent; and between monozygotic twins, the risk rises to 48 per cent, in other words, the concordance is around 48 per cent. Expressing this another way: if one member of a monozygotic twin pair develops schizophrenia, then there is a 50:50 chance that their co-twin will also develop the syndrome (Gottesman and Wolfgram, 1991). If both parents are schizophrenics, then the risk of any one of their offspring developing schizophrenia is 46 per cent. Amongst monozygotic twins reared apart in which one member of the pair developed schizophrenia, there was a 64 per cent chance that the co-twin would also suffer from the disease, although these data were derived from only fourteen twin pairs (Gottesman and Wolfgram, 1991). A meta-review of twin studies in relation to schizophrenia (meaning the averaging of the results from many independent investigations) suggested that the heritability of schizophrenia is 81 per cent (range 73–90%), with 11 per cent of the variance being attributable to shared environmental influences (Sullivan et al., 2003). In a large population study carried out in Sweden, a heritability value for schizophrenia of 64 per cent was obtained, with shared environmental effects counting for 4.5 per cent of the variance (Lichtenstein et al., 2009). What is clear from such studies is that environmental variation plays an important role in explaining the variance in a population with respect to the trait of schizophrenia. The molecular genetics of both autism and schizophrenia will be considered in the following chapter.

An Example of a Complex Human Trait Assessed by Behavioural Genetics

Shifting now to the normal range of human behaviours, the measurement of personality provides a good example as to how quantitative behavioural genetics approaches such investigations. Research has focused on five broad dimensions of personality called the Five-Factor Model (FFM) (Goldberg, 1990). These include the best studied traits such as neuroticism, involving moodiness, anxiety and irritability, and extraversion, which involves traits such as sociability, impulsiveness and liveliness (Plomin et al., 2008, p. 239). The five traits included in the FFM lead to OCEAN: openness to experience (culture); conscientiousness (conformity, will to achieve); extraversion; agreeableness (likability, friendliness) and neuroticism (Plomin et al., 2008, p. 240). Extensive twin studies carried out in five different countries and involving 24,000 pairs of twins have reported heritability values of about 50 per cent for extraversion and 40 per cent for neuroticism (Loehlin, 1992). Overall heritability values in the range of 30–50 per cent are typical for measured personality traits based on both twin and adoption studies (Jang et al., 1998; Plomin et al., 2008). Children's personalities are more similar to their biological parents than to their adoptive parents. The environmental component of the variance is accounted for almost entirely by non-shared environmental effects (Loehlin, 1992). This of course does not mean that a shared environment is not vitally important in the development of personality in the life of an individual child in a specific home: all that we have reviewed about the DICI paradigm suggests that this must be the case. The point in the present context is that the common home environment does not seem to account for the variance in personality traits observed amongst siblings. Instead, around 30–50 per cent of the variance in a particular personality trait appears to be attributable to genetic variance and around 50–70 per cent to non-shared environmental influences outside the home, such as different schooling experiences, different peer group pressures, random life events and so forth. Furthermore, unlike the development of children's cognitive abilities, which nearly all parents deem to be a 'good thing', personality traits have both positive and negative aspects, and only very stringent parenting would seek to change basic personality traits, unless at the extreme they proved disruptive to the social life of the family. This might also contribute to the lack of variance attributable to shared common environments (Turkheimer et al., 2014).

It is also of interest that the extent of the contribution of genetic variation to the measured variance in traits such as neuroticism and extraversion remains remarkably stable over the lifetime of individuals, especially from the age of around twenty onwards (Turkheimer et al., 2014). By contrast measurements of the non-shared environment contribution to variance is less stable. After reviewing the extensive results on this point, Turkheimer invites us to

> [c]onsider a hypothetical pair of identical twins whose personalities develop throughout the lifespan. In the usual absence of shared environmental effects, their pair-average score on personality traits is highly, even perfectly, stable relative to the average of other pairs, largely as an expression of their genetic endowment. At the same time, within-pair differences between twins show environmental influences that come and go, systematically but temporarily, from childhood through late middle age. At any given point in time, something might happen to make one member of the pair more extraverted or more neurotic; then, as time goes by, the within-pair difference decays, and the twins return to their genetically influenced mean. Finally, in old age, new differential processes appear to be established that make within-pair differences much less stable over time. (Turkheimer et al., 2014, pp. 531–2)

Extensive tests on personality have been carried out using twins raised apart. (The precise meaning of 'twins raised apart' will be discussed later in this chapter.) Some of the main studies were carried out before the FFM system for measuring personality was introduced, so direct comparisons are not always easy in such studies. Nevertheless, the main take-home message remains the same as for the twin-study results previously summarised. One well-known study involved the use of the Minnesota Study of Twins Reared Apart (MISTRA), a project initiated in 1979 that has made very significant contributions to the field of behavioural genetics (Segal, 2012). A 1988 study utilised 44 monozygotic twin pairs raised apart, 27 dizygotic twin-pairs raised apart, 217 monozygotic twin pairs raised together and 114 dizygotic twin pairs raised together (Segal, 2012, p. 99). Based on eleven different psychological inventories used for the self-reporting of personality, the heritability of the various personality traits ranged from 39 to 58 per cent, very similar to the later twin studies described earlier based on the FFM system of measurement. In addition the results suggested that 36–56 per cent of the variance in personality traits could be attributed to the non-shared environment, whereas only 0–19 per cent was attributable to the shared environment, depending on the trait in question.

In contrast to the consistent data arising from twin studies, the data on personality from adoption studies has been more ambiguous. In studies in which the parents assess the personalities of their young children using cohorts where the child has been adopted and is then assessed by the adopting parents, little or no evidence for any genetic contribution to the variation in personality traits has been obtained (Plomin et al., 1991; Schmitz, 1994). One contributing factor to this puzzling outcome is what is known as the 'contrast effect', an effect which applies to twin studies as much as it does to adopted children. It has been shown that parents often exaggerate the differences between their children (biological or adopted), for example, reporting that one is very active in contrast to their inactive sibling, even though relative to other children of similar age they might not be that much different (Saudino et al., 1995, 2004). However, when the personalities of children are assessed by observers, then there is a much closer match between the results obtained from twin and from adoption studies (Plomin et al., 2008, p. 244).

A Methodological Critique of Behavioural Genetics

The many assumptions involved in the practice of behavioural genetics have been extensively critiqued. This section will summarise some of the methodological issues. The wider question as to, what does it all mean? – especially in light of the DICI paradigm that has been the focus of the book thus far – will be tackled in the following section.

Classifying Twins as Monozygotic or Dizygotic

Twin studies depend on being able to accurately identify monozygotic and dizygotic twins. There are three principal methods. The first depends on questionnaires filled in by parents regarding their small children, or questionnaires given to older twin pairs. Traditionally, questionnaires have been the primary method used in large-scale twin studies. Questionnaires assign zygosity on the basis of physical and behavioural attributes of twins, and sometimes on blood typing. Twins are typically asked questions about hair structure, eye colour and whether they get mistaken for each other. This method has an overall accuracy rate in the range 90–95 per cent (Rietveld et al., 2000; Chen et al., 2010).

A rough way to assign zygosity prenatally is chorionicity. Monochorionic twins who share the same placenta are assumed to be monozygotic, while dichorionic twins who have different placentas are assumed to be either dizygotic or monozygotic. However, there are problems with this methodology. As already described in Chapter 4, approximately one-third of monozygotic twins are dichorionic (Shur, 2009), and furthermore, chorionicity can make significant differences to future twin phenotypic differences (Nikkels et al., 2008). Simple mechanical factors like chord entanglement and more complex issues like transfusion syndromes are only found with monochorionic pregnancies (Machin, 2004). Additionally, monochorionic dizygotic twins occasionally develop through chimerism in which two fertilised embryos spontaneously form a joined placenta and display chimeric blood and placental cells but separate fetal tissue (Machin, 2009).

As costs decrease, genotyping is becoming the preferred method of zygosity determination, generally carried out with four to ten variable number tandem repeat (VNTR) markers. It may be remembered that such VNTR regions have already been encountered in Chapter 5 when discussing the regulation of the serotonin transporter gene. Any discordancy in VNTR markers is put down to dizygosity (Machin, 2009). Overall, VNTR genotyping has an accuracy rate of > 99 per cent, which is certainly good enough for any twin studies involving reasonably large cohorts (Chen et al., 2010).

In conclusion, inaccurate zygosity determination is not a significant problem for twin studies, especially when genotyping has been carried out and large twin cohorts are involved. However, older studies utilising small numbers of twin pairs should be viewed judiciously in this respect: a 10 per cent inaccuracy in zygosity determination in a small sample could easily skew the results.

Are Identical Twins Truly Identical?

Whether it be the classic twin study protocol of measuring similarities and differences between monozygotic and dizygotic twin pairs raised in different families, or whether studying twins raised apart, the underlying assumption used in all statistical analyses is that monozygotic twins are genetically identical, both in terms of additive as well as dominant effects. We now know that this is not necessarily the case (Shur, 2009). More than 400 studies have been published on identical twin pairs discordant for a large number of different physiological and psychiatric conditions (Zwijnenburg et al., 2010). This discordance arises for a whole range of different genetic reasons. The most commonly reported cause is 'chromosomal mosaicism' in which the chromosomal karyotype of each twin may develop differently (Machin, 2009). The karyotype refers to the overall structure of the twenty-three paired chromosomes that contain all the nuclear DNA molecules encoding the human genome. Another cause of discordance between monozygotic twin pairs is novel mutations in the nuclear or mitochondrial DNA. Once the embryo splitting occurs that leads to twin development, there is potential for DNA mutations to occur differently in either twin embryo. Examples of monozygotic twin pairs have been reported in which one twin but not the other suffers from genetic diseases such as cystic fibrosis, Huntington's disease and sickle cell anaemia (Machin, 2009). Copy number variations (CNVs) can also vary considerably between monozygotic twins (Zwijnenburg et al., 2010), referring to insertions or deletions of small segments of DNA, typically a few base-pairs in length.

The marked epigenetic differences that exist between monozygotic twins at birth have already been mentioned in Chapter 4, although in general the discordance appears to be greater between dizygotic twins as compared to monozygotic (Ollikainen et al., 2010; Bell et al., 2012; Gordon et al., 2012). Such differences are presumably due to their different intra-uterine experiences, such as differences in infection and blood supply, and to stochastic events. In one study investigating epigenetic differences in forty monozygotic twin pairs, ranging in age from three to seventy-four years, closely similar profiles were noted in early life, but in older twins a significant discordance rate of 35 per cent was reported for all the epigenetic parameters measured (Fraga et al., 2005). Those twins who had spent less of their adult lives together and/or who had different medical histories were those who showed the greatest differences in epigenetic profiles. Differences in smoking habits, diet, exercise and life events have all been proposed to be involved in such epigenetic differences. Overall the investigators concluded that approximately a third of adult monozygotic twins are discordant for the epigenome. Epigenetic differences imply that a gene might be permanently switched off from birth in one twin and not the other, in which case the person might as well not have the gene at all, because it is not expressed. A more likely scenario, however, is that a gene might be partially switched off in one twin and fully expressed in the other twin. Either way, from a functional perspective the identical twins will not have identical genomes.

In terms of the biometric assessment of G and E, if monozygotic twins are more similar with respect to a particular trait because of their epigenetic profiles, then this will increase the proportion of the apparent genetic contribution. To the extent that monozygotic twins differ, the proportion of variation will be counted as environmental. In reality, epigenetic regulation does not fit so readily into either category and so ignoring it can readily lead to inaccurate interpretations. In this context it is worth highlighting that the heritability values estimated for many animal behavioural studies are generally much lower than for human studies. For example, in a study on the variation in aggression in fruit flies, raised experimentally in apparently identical environments, a heritability value of only 10 per cent was estimated for this trait (Edwards et al., 2006), despite the fact that there is plenty of good evidence that genetic variation does indeed play an important role in the differential aggression noted in different lines of *Drosophila*(Edwards and Mackay, 2009; Zwarts et al., 2012). So how can it be that measures of human aggression give heritability values in the region of 50 per cent (Rushton et al., 1986)? Common sense might suggest that environmental inputs, not to speak of human choice, might make more of a difference in humans than in fruit flies. Common sense, in this case, is most likely correct. In *Drosophila* it is clear that both epistatic (Zwarts et al., 2012) as well as epigenetic affects (Xu et al., 2014) have extensive impacts on behavioural traits. Depending on the experimental approach being used, it is much easier to dissect out such effects in fruit flies than it is in humans. In the human context it is very likely that heritability values are overestimated due to the lack of information concerning epigenetic regulation, epistatic interactions and many other factors that are likely to distort accurate estimates of h^2.

In addition to these important ways in which monozygotic twins may differ genetically, it should also not be forgotten, as mentioned in Chapter 3, that the brains of monozygotic twins are also different due to the transposon insertion events into the DNA of the neurons of their hippocampus, events occurring throughout life which will be different for each twin (Singer et al., 2010). Furthermore, that chapter also described the many ways in which genetic information can be edited within the cell. For this and the other reasons summarised previously, the basic assumption of the biometric approach to twin studies, that the genomes of monozygotic twins are stable throughout life, is therefore incorrect.

Surveying this rather impressive collection of observations that, taken together, demonstrate that monozygotic twins are never identical at the biological level, one might then be tempted to become rather gloomy about the prospects of relying on twin studies to derive useful information in behavioural genetics. Indeed some investigators have adopted precisely this view, with one reviewer of the field suggesting that '[t]he evolving concept that no two people are genetically identical shakes the validity of a century of calculations' (Shur, 2009). This point may be valid for studies on small twin cohorts, in which a few non-identical monozygotic twin pairs might significantly bias the results, and/or in which earlier less sophisticated statistical analyses were employed. On the other hand, for large twin cohorts the proportion that are not identical at the genomic level in a way that makes a difference to the trait under study is likely to be rather small and statistically 'swamped' by the proportion of truly identical twins, although a much larger sample of complete twin genomic (and epigenomic) sequences than presently available will be necessary to confirm this point. Perhaps of greater concern, as already highlighted, is the impact of epigenetic differences between monozygotic twins that cause differential gene expression with an impact on the development of a trait, caused by environmental factors. Sophisticated statistical approaches have been developed to accommodate the finding that the genetic correlation between monozygotic twin pairs can no longer be assumed to carry a value of one but is rather subject specific (Molenaar et al., 2012). Furthermore, some recent DNA-based statistical methods, explained in the next chapter, that do not require twins or any assumptions about them have reached conclusions very similar to those from the classical twin studies (Turkheimer, 2011). In any event, for the moment it still remains the case that, taken overall, and despite the many provisos, monozygotic twins are more similar to each other at the genetic level than dizygotic twins, and therefore their comparison remains a useful method for carrying out behavioural genetic studies, providing the data are assessed with a considerable degree of caution (and sometimes scepticism) (Zwijnenburg et al., 2010).

What Does the Equal Environments Assumption Mean?

The concept of a 'shared' or 'equal' environment (EEA) (Var C) has previously been introduced. Identical or non-identical twins who share the same home upbringing are assumed to share an equal environment. This assumption has been the target of criticism ever since the field of population genetics began to be established in the 1920s (Joseph, 2004). In reality, differences in twin correlations might be at least partly explained by differing treatment effects from parents, teachers, peers and so on. Violations of the EEA cause artificially high heritability estimates, because they inflate the correlation between monozygotic compared with dizygotic twins and attribute the whole of the correlation to genetic influence (Kendler and Gardner, 1998). There are two versions of the EEA. The original version states that each member of a twin-pair is equally exposed to all environmental influences, regardless of zygosity. This formulation predicts that monozygotic twins do not ever experience more similar environments than same-sex dizygotic twins. However, several studies have demonstrated the increased environmental similarity of monozygotic compared to dizygotic twins. They are more likely to be confused for one another (ego confusion), be similarly dressed, socialise with the same group, share the same room and report a greater level of emotional connection with their co-twin (Joseph, 2004, Beckwith and Morris, 2008). It should be noted, however, that some of the evidence cited for these conclusions is now forty years old and there appears to be few recent studies investigating these claims more systematically. But other life events such as divorce are clearly experienced differently by different sibs, and if the influence of divorce on identical twins is more shared than its influence on fraternal twins, then the equal environments assumption again no longer holds. It also appears uncertain whether parents treat twins more similarly if they think they are identical rather than fraternal. This question has been investigated (Loehlin and Nichols, 1976), but the parents who responded in this study reported treating twins equally, regardless of zygosity, 98 per cent of the time, suggesting that the response may be retrospectively biased (Joseph, 2004). Twins' perceptions of how similarly they are treated seem very different from their parents' perception for some reason (Plomin, 2011). Parents think that they treat them similarly; the twins themselves often think not. Similarly, it is uncertain whether parents, or twins themselves, treat twins differently based on their perceived zygosity, where it differs from their true zygosity (Richardson and Norgate, 2005). The waters are further muddied by twin studies suggesting that negative life events (as reported by twins and their families) are themselves genetically influenced, correlating with personality traits, presumably because the genetic predispositions of individuals influence them to engage in behaviours that lead to undesirable life events (Billig et al., 1996). So genomic variation has at least something to do with the environments that individuals seek out, those environments in turn impacting on the development of certain traits in the individual, so the sum total of the influences become a very tangled web. It does indeed sound like an infinite regress: everything affects everything else.

These and other factors may help explain the results from a meta-analysis carried out (Turkheimer and Waldron, 2000) of forty-three different studies, which revealed that although non-shared environmental factors were apparently responsible for 50 per cent of the variation of various behavioural traits, identified factors that convincingly made a difference to the trait only accounted for 2 per cent of the total variation. For the purposes of this study the authors distinguished between an 'objective' sense in which the environment is not shared – that is, where the environmental effect can be verifiably measured to be experienced by only one member of a twin pair in a family – and an 'effectively non-shared' sense in that something makes siblings different, irrespective as to whether the 'something' is experienced by one or both. The study highlights the difficulty of defining the slippery term 'environment', an issue that will be discussed further, and in finding out what environmental effects might indeed be important in influencing identical twins (or any two sibs for that matter) to be different.

Given the likely invalidity of the original EEA assumption formulation, defenders of the twin method (Kendler and Gardner, 1998; Mitchell et al., 2007) have employed two counter-arguments. First, they now predominantly use a trait-relevant definition, which states that environmental influences that impact on the particular trait under study are equally shared between twins, regardless of zygosity. Under this formulation, it does not matter if monozygotic twins are more similarly treated in terms of dress, diet, rearing style and so forth, if those influences do not have any causal connection to the trait being investigated. Unfortunately, this raises a problem for testing the EEA as, for the vast majority of traits, environmental influences that do or do not have an impact are unknown, and the number of potential influences could be very large. So a certain element of circularity can creep in to assigning values to shared and non-shared environments, a process which has often been described more as an art than a science. Moreover, if identical twins live together, eat together, share the same room, have more common friends, experience greater levels of identity confusion and so forth, they are much more likely to be exposed to 'trait-relevant' environmental influences, known or unknown, as compared to non-identical twins. The trait-relevant definition therefore fails to address the key flaw in the EEA, because in reality monozygotic twins do have more similar environments than dizygotic twins for relevant influences.

Second, it can be argued that the EEA, in either its traditional or trait-relevant formulation, is not violated by the greater similarity of identical twins, because the similarity is a result, not a cause, of genetic similarity. This is known as the 'twins create their own environment' theory (Joseph, 2004) and argues that parents and society react to the greater genetic similarity of identical twins, as manifested phenotypically, by creating a more shared environment. Under this model, the existence of a more shared environment is evidence for a greater genetic similarity. However true that may be, as already noted, the reality is that in any given twin study, everything and everybody is influencing everything and everybody else all the time and genetic influence is present in all the players.

The shared environment assumption is clearly vital in assigning proportions of variance to different categories in population studies, since if one proportion of variance changes, then by definition another proportion, such as the non-shared environment, must also change (and vice versa). It is therefore also worth highlighting the fact that in studies of identical twins 'separated at birth', and therefore moved to non-shared environments, the real situation is not what one might infer from the language being used. In any event, by the time the twins are born they have already shared nine months of a fetal environment which, as already noted, is not identical for them, but still highly similar, during which time there has been massive neuronal growth and the common storage of events experienced in utero. Furthermore, separation of twins for adoption in reality rarely occurs immediately after birth. For example, in the Minnesota registry of twins raised apart, the average time together before separation is five months, but this ranges up to four years; the time apart until first reunion ranges from 0.5 to 65 years, and the average total time spent together before study is over two years (Bouchard et al., 1990). In another much used twin registry of Swedish twins 'raised apart', in total, 52 per cent of the twins were separated after the age of one, and 18 per cent not until after the age of five (Pedersen et al., 1988). These results are representative of all studies of twins reared apart (Kamin and Goldberger, 2002). So in practice even so-called separated twins live together first for some time before separation, and it is worth remembering the huge increase in brain synaptic density which occurs over the first two years of life – both twins responding to experiences shared in common during the first year of life.

Many studies have been carried out to test the equal environment assumption and it has been defended robustly by twin-study practitioners

(Loehlin and Nichols, 1976; Kendler and Gardner, 1998; Mitchell et al., 2007). An equivalent number of fierce critiques of the assumption have also been published (Joseph, 2004; Richardson and Norgate, 2005; Beckwith and Morris, 2008; Charney, 2012). And so the debate continues.

An Interpretative Critique of Behavioural Genetics

The more interpretative and methodological critiques of the field of behavioural genetics in reality overlap somewhat: interpretation is often difficult to disentangle from the technique itself. It is more a question of emphasis. Much of the interpretation revolves around the meaning of words. As Evelyn Fox Keller has aptly remarked: 'T[]he language of behavioral genetics is hopelessly polysemic' (Keller, 2010a, p. xii).

The notion of gene-environment interaction provides a good example of such polysemy. Within the DICI framework, the first and second I's for 'Integrated' and 'Interactionism', respectively, refer to all the myriad ways in which the genome interacts with, and integrates with, all the other events of cell biology, and in turn with the micro- and macroenvironments discussed in Chapter 4. It is a mechanistic and developmental concept because it refers to all the physical interactions that result in an adult organism that displays certain traits. So it refers to the *presence* of myriad physical events. In biometrics the meaning is quite different. The starting assumption until shown otherwise, going right back to the pioneering work of Ronald Fisher, is that the effects of different genes on a trait are additive. If the statistics indicate that this assumption is wrong, that there is an absence of linearity between the genetic variation and the variation in the trait under study, then this absence of linearity shows that there must be interaction between genes and environment. So the biometric definition of interaction started out as a concept arising from the statistical analysis of variance: if the proportions of variance of genes and environment failed to account for the total variance, then interaction became the remaining proportion of variance. So in Fisher's understanding, interaction was an annoying breakdown in additivity, an *absence* of summation, between the proportions of variance attributable to genes and environment. But the solution for the statistician like Fisher was simple: place the data on a non-linear scale, such as a logarithmic scale, and the interfering interaction would then go away.

These two concepts of interaction are so different that it might be thought that 'never the twain shall meet'. Indeed, this has tended to be the traditional stance of developmental biologists in their own robust

interactions with the biometricians over the past century. Hit the
nail on the head back in the early 1930s when he commented, in a
passage already quoted in Chapter 2, that there is a third class of variability which 'arises from the combination of a particular hereditary
constitution with a particular kind of environment' (Hogben, 1932). And
when the controversy instigated by Arthur Jensen on his publication on
variation in IQ and on the subject of race blew up in 1969 (Jensen, 1969),
Jensen was dismissive of Lewontin's critique that centred on 'genotype-
environment interaction' (Lewontin, 1975), claiming that because his
statistical analysis of variance had revealed no such interactions (meaning an absence of linearity), therefore interactions of the kind being
suggested were unimportant (Jensen, 1970). But in retrospect, and on
a more positive note, there seems no reason why all investigators interested in variation in behavioural traits could not gain useful information
from the two very different understandings of 'interaction'. In any experimental strategy it is important to recognise what the strategy is setting
out to explain. The biometricians have always had the aim of explaining
the statistical proportions of variance in a given population that contribute to the variation in a particular trait. The leading practitioners of biometric approaches have never claimed that their intent is to explain the
physical, mechanistic interactions that occur during the development of
an individual organism that leads to the emergence of a particular behavioural trait. So it seems a little unfair for blaming them (as some do) for
not extracting from their data an explanation for something which they
had never set out to explain (and anyway could not, given their methodology). If that is the case, then do we have to settle for two 'incommen-
surable' paradigms, to use the language of Thomas Kuhn, which operate
independently without any hope of useful communication between the
two experimental approaches? It is not clear why that needs to be the
case. If biometric approaches reveal significant G x E interactions based
on statistical data, then this is certainly a useful insight when it comes
to the investigation of molecular genetics and the search for the physical
interactions between gene variants and other molecular events in relation to environmental inputs that might be involved in mediating different behaviours. Several examples illustrating this point are provided in
the following chapter. Having said that, an absence of biometric G x E
interactions with reference to variation in a particular trait should not be
interpreted as meaning that there is not a massive amount of such interaction during the actual physical development of that trait in a given
organism, as there clearly is, but the biometric data might be indicating

that G x E interactions are less important for the variance of that trait in a given population. In the reverse direction, developmental biologists and molecular geneticists can help the biometricians by providing the most recent data that might challenge some of the assumptions on which a biometric approach might be based. We have already surveyed a considerable amount of data demonstrating the necessity of such a challenge, and to be fair on the biometricians, some serious statistical efforts have been made to address at least some of these concerns (Krueger et al., 2008; Partridge, 2011; Molenaar et al., 2012).

As far as heritability is concerned, the fact that it exists, in other words the value of h^2 is more than zero and less than one, is probably its most valuable contribution. Providing the value has statistical power (a point of relevance to some of the examples cited in following chapters), it is useful as a way of flagging up the probable importance of genetic variation in the development of a certain trait although, as noted earlier, there are many reasons why falsely positive heritability values might be generated. But in any event the actual value, a unit-less proportion of variance, is not that interesting. One reason for that is because the value depends on the particular population and the particular environment in which the trait is being measured. If a population has rather limited genetic variation, perhaps because it is located on an island in which the population has tended to breed with other islanders for many centuries, then the heritability may be relatively small, whereas if a genetically more diverse population was placed in the identical environment, then the heritability value might well be much higher. Likewise if two populations had identical levels of genetic variation but were placed into two different environments, then again the heritability values pertaining to a particular trait might well be different. A specific form of this complication has been dubbed 'bioecological interaction' (Pennington et al., 2009). The idea here is that when there is an optimal environment for the development of a particular trait, then any variation observed in a population with respect to that trait must be due to genetic variation, thereby increasing the apparent heritability of that trait. The analogy usually made is with a population of plants. Genetically variant plant seeds are sown in either a nutritionally poor or a nutritionally rich environment. The deprived soil results in a field of generally short plants. By contrast the plants in the good soil are able to fully express their potential for height, so any differences will be due to genetic differences, thereby amplifying the heritability (Burt, 2011).

So, unlike animal and plant populations, where it is possible to control for genetic and environmental differences, in human populations the heritability values have no inherent theoretical value (Crusio, 2012). A higher heritability value does not necessarily mean that genomic contributions to the development of a trait are more important than for lower values, though it might do. And a higher heritability value certainly does not entail that the trait in question is more or less amenable to change. It all depends. In a sense, what heritability is measuring is not a characteristic of the trait at all, but a measurement of environmental variability relevant to a particular population (Moore, 2013a). The conclusions of a very experienced practitioner in the field summarises the situation well: 'Heritability is greater than zero for all individual differences, and takes a determinate value for none of them. Figuring out how "genetic" traits are, either in absolute terms or relative to each other, is a lost cause: Everything is genetic to some extent and nothing is completely so. There is little more to be said' (Turkheimer and Harden, 2014). This discussion by no means exhausts the nuances surrounding the measurements of heritability, and good reviews are available for those who wish to delve further (Keller, 2010b, Johnson et al., 2011, Moore, 2013b).

A further contribution to the polysemic ensemble of confusing words that litter the field of behavioural genetics is the word 'environment', which has already been used previously multiple times with various nuances of meaning. The problematic biometric assumption of 'equal environments' has already been discussed. Coming from the perspective of developmental biology, the DICI framework has flagged up the importance of both micro- and macroenvironments. Hanging over all these usages of the 'environment' word is the wider question as to where does 'it' start and where does 'it' end? The very reification of the word entailed in rendering it an 'it' only exacerbates the problem. There is no general 'theory of the environment' as there is with genetics that will help in resolving this issue. As Earl Hunt remarks: 'Without a theory of environmental action it is hard to know where to begin' (Hunt, 2011, p. 244). Zoologically, genes and the macroenvironment do not interact, but living organisms certainly do, the environment here being understood to start where the skin or membrane of the animal ends and the air or water starts. The interaction with the environment arises from the behaviour of the animal and from the myriad environmental inputs into this intact living organism – via sight, hearing, temperature, gravity, food supply, social interactions, weather and on and on the list can go. The interface with the environment in this zoological sense is not static because the permeability of the skin or membrane are constantly changing. By contrast, within the DICI framework the microenvironment of the genome starts as close to the primary nucleotide sequence of DNA as molecules can get, in other words, very close indeed. The skin may be a very long way away, relative to the size of the genome in question, depending on which tissue is under consideration. From the biometric perspective, the environment is something different again: what contributes statistically to a proportion of variance in a population in respect to a particular trait. The cynic might suggest that the 'E' term in the proportion of variance statistics stands for 'Everything Else', everything impinging on trait variation that does not appear to be attributable to genetic variation. So the term 'environment' is a slippery one, with many different meanings according to context, and the only solution to the problem is to be alert as to how the word is being used in these different contexts.

Unfortunately the language of quantitative population genetics does not help in subverting the dichotomous view of human personhood that, we have suggested, is inconsistent with all that we know from a biological perspective about how personhood actually develops. It is not that practitioners in the field set out to promote such a dichotomous perspective – far from it, many make strenuous attempts to highlight the importance of G x E interactions – but the fact remains that once proportions of variance, labelled 'genetic' and 'environmental' are given numerical values, complete with pie charts or histograms illustrating those values, then it is virtually inevitable that such numbers will be understood by many not in the field as referring to actual 'essences', each of which contributes that numerical value of influence to the final individual product. It has to be said that not all working in the field are sufficiently careful about their language in this respect when presenting their results, leading to a confusing mingling of 'heritability' in its original sense of inheritance with 'heritability' carrying its technical biometric meaning, in the process generating a quite fallacious picture of the human person being somehow calved up into spheres of influence, one sphere ruled by the genes, the other sphere ruled by the environment. Hopefully by this stage it can be seen that such a picture is quite untenable.

It should also be clear by now that the field of quantitative behavioural genetics is not going to help us with questions of free will and determinism. So perhaps the field of behavioural molecular genetics might be of greater help in this regard. The next chapter addresses that question.

CHAPTER 7

Behavioural Molecules?

Understanding Molecular Behavioural Genetics

> A *wise man proportions his belief to the evidence.*
> David Hume (Hume et al., 1975, *Sect X, Pt I*)

> *The anticipated yield from genome-wide association studies gives much reason to be optimistic about the future vitality of behavior genetics.*
> Matt McGue (2010)

> *The fruitless search for genes in psychiatry and psychology: time to re-examine a paradigm.*
> Jay Joseph (2012)

At the close of the twentieth century, many were under the impression that the sequencing of the human genome would soon lead on to the rapid identification of genetic variants that contribute to differences in human behavioural traits. This hope has, so far, been realised only to a very limited extent. The partial reason is the sheer complexity of biological organisms in general and of humans in particular, strikingly highlighted by the advent of genomics, coupled to the many other insights arising from molecular biology reviewed in Chapter 3. In this chapter we will first provide an overview of the extent of genetic variation in human populations and explain the three main approaches used in molecular genetics, after which examples will be provided illustrating how these approaches work in practice.

Human Genetic Variation

One of the many impacts arising from human genome sequencing has been to reveal just how varied human populations are at the genetic level, much more than previously imagined. In fact it is now thought that single nucleotide polymorphisms (SNPs, pronounced 'snips'), those single nucleotide changes that litter our genomes, occur in 1 in every 300 nucleotide base-pairs on average. This means that there are roughly ten million SNPs in the DNA of any given individual, mostly, though not entirely, located outside the 1.5 per cent of the genome that encodes proteins.[1] Taking the human population as a whole, there are thought to be around 40 million SNPs in total, though this number is consistently being refined as more human genomes are sequenced. It is estimated that about 3–5 per cent of these SNPs make functional differences. There are also small DNA segments inserted or deleted, abbreviated as indels; there are copy number variations (CNVs) which are random duplications of small segments of DNA affecting as much as 12 per cent of the genome; and more rarely there may be larger re-arrangements of chromosomal segments. Until now more than 150 million different indels and other structural variants have come to light in the human population.[2] More than 99.9 per cent of variants between individuals consist of indels and short variants, but structural variants affect more DNA nucleotide bases because they are longer (Genomes Project, 2015). A typical genome differs from the 'reference genome' at up to five million locations involving around 0.5% (or more) of the genome in terms of actual number of base-pairs (Genomes Project, 2015).

Usually a genomic sequence is derived from only one of each chromosome pair, so it was quite a surprise when the first sequence appeared in 2007 of both chromosomes sequenced separately, revealing that the chromosome pairs in a single individual vary from each other by around 0.5 per cent (Levy et al., 2007). Each chromosome in a pair, one from the mother, one from the father, is a mosaic composed of a different family history, converging in a single individual to contribute towards the generation of a unique human being. New mutations are constantly being added to the human gene pool. It has been estimated, based on whole genome sequencing, that a newborn contains on average sixty new mutations not found in either parent. The exact number varies depending on the age of the father. Whereas a mother contributes on average fifteen new mutations, regardless of her age, a twenty-year-old father contributes an average twenty-five mutations to his child, whereas a forty-year-old father contributes around sixty-five new mutations (Kondrashov, 2012). So if you want to be a father, best to get on with it sooner rather than later. Having said that, the vast majority of the new mutations are without any functional consequences, as far as we know, but some clearly do have significant effects. In addition, it is of interest that, on average, every person carries mutations that inactivate at least one copy of 200 or so genes and both copies of around 20 genes (Callaway, 2014).

As if this inter-individual genetic variation was not enough to cope with, the differences in genome sequence and even structure between different cells of the same individual also need to be highlighted. Time was when each cell of the body was thought to contain an identical genome. We now know that is not the case. Chromosomal aneuploidy exists in many cells of the human brain. 'Aneuploidy' refers to sets of chromosomes that are different from the normal set – twenty-three pairs in human cells. Conservative estimates place the percentage of aneuploid neurons in the normal adult human brain at an astonishing 10 per cent (Iourov et al., 2009), although other estimates based on single cell sequencing have generated somewhat lower estimates (Knouse et al., 2014). Altogether there could be around 10 billion neurons in the human brain that have different sets of chromosomes as compared to the normal twenty-three pairs, and likewise different sets in 100–500 billion glial cells. It is hard to believe that at least some of this chromosomal diversity does not affect brain function, although finding ways of addressing that question experimentally is going to be difficult.

Variation between cells also exists at the level of the genomic sequence. It is now technically possible to sequence the genome of a single cell (Liang et al., 2014). All the data available so far suggests that there is considerable inter-cell genetic variation, even in cells from the same tissue in the same individual, although much of the developmental work on this new technology has been carried out using cell lines, and more definitive answers will be available once the methodologies have been further improved (Macaulay et al., 2015). Genetic variation of this kind should not be surprising: every time a cell replicates, there is the possibility of new mutations being introduced into the genome of the progeny cells. Also of significant interest is the finding that the epigenomic mutation rate, referring here to the rate at which methylation changes are detected in single cells, appears to be 100 times higher than the primary DNA sequence mutation rate (Gravina et al., 2015).

Now the question is, how much of all this variation makes any difference to the variation in behavioural traits that exists between individuals?

The Three Main Approaches of Molecular Genetics

The three main approaches will be described briefly and then some examples will follow to see how these work out in practice.

Linkage Analysis

This was the earliest approach to try and identify genetic variants that might be involved in the variance observed in behavioural traits. It is based on the fact, mentioned in the previous chapter, that during the formation of the haploid sex cells (meaning that the cells contain only one of the two chromosomes found in the parent's germ-line diploid cells), crossing over occurs between the two chromosomes. If genes are located well apart on the same chromosome, then they are unlikely to be involved in the same DNA segment swapping event, so they will tend to be inherited by the offspring in a typical Mendelian way, by random assortment, or in 'equilibrium'. However, if two genes are close together on the chromosome, then they are more likely to be swapped together during the crossing over process – in other words, they are 'linked' together. Rather confusingly, this phenomenon is known as 'linkage disequilibrium', a phrase trying to capture the idea that the independent assortment of a particular genetic variant does not happen. In practice, the approach is applied to family studies in which the inheritance of a trait is tracked through several generations and strongly correlates with the presence of a particular segment of a chromosome. Once that segment has been identified, there is then the added challenge of identifying the actual gene within that segment that may be influencing the trait. However, the approach is only of use in detecting alleles that contribute around 10 per cent or more of the total variance (Schaffner, 2006), as in a disease like cystic fibrosis. In behavioural rather than medical genetics it is now clear that the putative contribution of any one gene to the variance is far smaller than that, so the earlier linkage studies in this context, which often led to misleading results, will not be discussed further here.

Candidate Gene Studies

These experimental approaches, also called 'association studies', are typically carried out between families, rather than within families, as in the case of linkage analysis. The approach is simple: a gene is identified that is thought to be important in some key aspect of brain function, such as aggression, and then alleles of this gene are measured in a random population of individuals. If there is a significant correlation between the presence of the allele and the trait being measured, then the gene may be cited as contributing to the development of the trait.

This experimental strategy has been very popular in the study of human behavioural genetics, as will be illustrated further. However, there are a number of problems with this approach, two of which will be mentioned here, the others later. The first is generally known as the 'chopstick problem', the phrase arising from a well-cited paper by Eric Lander and Nicholas Schork:

> Suppose that a would-be geneticist set out to study the 'trait' of ability to eat with chopsticks in the San Francisco population by performing an association study with the HLA complex [a set of immune response genes that frequently vary between ethnic groups]. This would-be geneticist suspects that immune response genes, that he believes are involved in the autoimmune disease multiple sclerosis, may also affect manual dexterity in normals. The allele HLA-A1 would turn out to be positively associated with ability to use chopsticks – not because immunological determinants play any role in manual dexterity, but simply because the allele HLA-A1 is more common among Asians than Caucasians. (Lander and Schork, 1994, p. 2041)

In practice there are additional controls and tests that can be used to exclude such spurious interpretations, although they are not invariably used in such studies.

The second problem is rather different and has been dubbed the 'winner's curse' (Xiao and Boehnke, 2009), a theory first described by auctioneers. The winning bids in auctions are likely to over-value the real value of the item so the bids in this winning category are biased upwards. In genetic association studies, the first report of an association plays the role of the 'winning bid', since the attention is placed on the association that displays statistical significance, resulting in an 'ascertainment bias' whereby only those samples are ascertained that display evidence for such significance. If the evidence for association between an allele and a behavioural trait is pegged too high, then replication studies are likely to fail, as they may not be able to demonstrate this high degree of association, even though there might indeed be an association, albeit with a correlation factor lower than that originally reported. A review of the winner's curse phenomenon suggests that it is indeed a reality in association studies (Ioannidis et al., 2001). Eric Turkheimer expresses the situation robustly:

> Faced with thousands of potential alleles, thousands of potential outcomes, intense pressure to publish, and a stringent peer-review system that prefers positive results, even a well-intentioned and honest community of scientists will produce effect sizes that are severely biased upward. The winner's curse is not exclusive to genomics; it is rampant in the behavioral sciences generally. (Turkheimer and Harden, 2014, p. 177)

Once again, sophisticated statistical methods are available which aim to correct for ascertainment bias (Xiao and Boehnke, 2009), but as Turkheimer also reports: 'It has been extremely difficult to screen reports of association studies for what is known as the "winner's curse"' (Turkheimer and Harden, 2014, p. 177). It should also be noted that the ascertainment bias spills over into the media reports of new results, since the first report of an association with a behavioural trait is often reported with breathless enthusiasm, but then the various failures to replicate that often appear in the literature thereafter are generally ignored because negative results are not deemed to be newsworthy.

Genome-wide Association Studies

The advent of genome sequencing has led to a popular and extensively used method for screening DNA for significant associations with particular traits known as a genome-wide association study (GWAS). This approach capitalises on the millions of SNPs that are present in the human genome. It is now routine to present at least a million DNA SNPs on a dense microarray to a sample of DNA taken from an individual in such a way that the person can be 'genotyped' for all those SNPs. So the SNPs act as 'flags' to mark different segments of the genome which vary slightly between individuals. If a particular SNP keeps on associating with a particular trait at levels above chance, then the inference is made that there should be a variant allele nearby, either in a protein-coding or gene regulatory region, that contributes to the development of the trait in question. Most commonly it is the nearest protein-encoding gene to the SNP that is identified as being relevant, though this is not invariably the case (Smemo et al., 2014).

An interesting point to note about GWAS is that it is conducted atheoretically, unlike candidate gene studies in which a hypothesis is in mind entailing the possibility that a particular gene is involved in a particular trait. In other words, it is a fishing exercise looking for any variant gene that might be involved in a particular trait. Furthermore, the statistical standards for significance ($p < 10^{-8}$) are set extremely high to exclude random associations, though even then the overlap in terms of 'SNP hits' between different replication studies can be discouragingly low. The GWAS approach, like any association study, is also not immune from the chopstick problem nor from the problem of the winner's curse, and the data have therefore to be assessed accordingly.

To overcome some of these challenges, a method known by the clever acronym GCTA (genome-wide complex trait analysis) has been introduced: 'the first new human quantitative genetic technique in a century' (Plomin and Simpson, 2013, Yang et al., 2013). In brief, the method depends on the fact that DNA similarity between pairs of unrelated individuals varies by +/- 2 per cent. Unlike GWAS, there is no attempt to use this method to identify specific SNPs. Instead of using genetic similarity from groups differing by a known degree of genetic similarity, as in twin studies, GCTA uses genetic similarity (statistically greater than +/- 2%) for each pair of unrelated individuals based on that pair's overall similarity across hundreds of thousands of SNPs; each pair's genetic similarity is then used to predict their phenotypic similarity (Plomin and Simpson, 2013). In other words, GCTA relates SNP similarities to phenotypic similarities between individuals. In this way, a sample of 6,000 individuals, for example, provides eight million pair-by-pair comparisons, providing considerable statistical power. Applications of this method are of particular use when interpretations of twin-based data remain ambiguous, as illustrated in the chapter that follows. The approach has been used with some success in the attempt to identify genetic variants that correlate with personality traits (Power and Pluess, 2015).

The GWAS approach has resulted in some important discoveries in the domain of medical genetics, as illustrated in the next section, but has not lived up to initial expectations in the field of behavioural genetics. It is informative to describe the GWAS approach to identifying SNPs that associate with variation in height to see how and why it has been so difficult to obtain meaningful results from the field of behavioural genetics. The hunt for meaningful alleles to help explain the 80 per cent heritability of height that exists in different populations has become the 'poster child' of the GWAS approach. The reasons are not hard to discern. Measuring height is easy and non-invasive and therefore it is feasible to carry out GWAS on very large populations.[3] One GWAS investigation of 253,288 individuals of European descent revealed that if the level of statistical significance was set at a very stringent level ($p < 5 \times 10^{-8}$), then 697 variant alleles were identified that contributed around one fifth to the heritability (Wood et al., 2014). However, if the p value was relaxed down to 5×10^{-3}, then nearly 10,000 alleles could be identified that contributed to around 30 per cent of the heritability. And if the results of this study were pooled with others, then sufficient SNP variation has been identified which could explain about 50 per cent of the heritability. Keeping to the more stringent collection of 697 variants, the authors (Wood et al., 2014) demonstrated that many of the genes identified (because the SNP was either within or close to the relevant gene) had biological relevance to the construction of height. But it is clear that the average contribution of each variant gene, considered individually, is tiny, around 0.001 per cent or less. The large number of variant alleles per se should not be at all surprising: with increased or decreased height, everything has to change – the size of organs, the expansion of the skin, the construction of the neuronal system and so forth.

What is more surprising is the so-called missing heritability that is noted even in a study investigating a quarter of a million people. In fact, the percentage of 50 per cent contribution to variation in height measured by pooling several large studies is at the upper limit of what has been achieved so far in the GWAS approach to any traits, be they medical or behavioural. Typically the relevant SNPs identified explain less than 10 per cent of the variation in a given population with respect to a particular trait. This has led to an extensive discussion in the literature as to what accounts for the missing heritability. There are many theories, not necessarily mutually exclusive, and the reasons are likely to be different in various studies. In studies of physical traits like height, it may well be that if several million people were included in a study, for example, then the statistical power would become so great that the final set of SNPs contributing to the 'missing' 50 per cent of the variation would be identified. Another explanation is that the heritability value is heavily inflated due to the reasons discussed in the previous chapter, such as epistasis and/or G x E interactions, so in reality there is nothing missing to explain, so-called phantom heritability (Kaprio, 2012; Zuk, 2012). Other explanations include the presence of rare SNP variants that contribute to variation in the trait under study but which are unlikely to be included in the SNP arrays routinely used for investigation. And as already highlighted, there is a huge amount of variation in the genome that is not included in the category of SNPs at all, but rather DNA insertions, deletions, duplications of segments and so forth.

Medical Genetics

Medical genetics is not the main focus of this book, but there is of course a range of syndromes associated with abnormal mental and behavioural conditions, and these overlap with the field of behavioural genetics. There are at least 5,000 medical conditions described in the literature that display familial patterns of classical Mendelian inheritance and are best explained by the presence of a single defective gene. It is here that we come closest to a stringent form of genetic determinism in which there is no sense of 'up-to-usness' in the development of the syndrome: in many cases if the mutant gene is present, then the development of the genetic disease will be virtually inevitable. Hurler syndrome, for example, is a severe lysosomal storage disease which develops when there are particular mutations in a gene which encodes the enzyme α-iduronidase; both copies of the gene located in the chromosome 4 pair need to contain the severest mutations for the disease to develop. If only one copy is mutated, then the other 'good' gene can make sufficient enzyme to prevent the disease. The baby affected with Hurler syndrome often looks normal at birth, but further development then virtually ceases from the age of two onwards with progressive deterioration of the central nervous system, mental decline and usually death before the age of ten. Although there is as yet no cure, enzyme replacement therapy and stem cell transplantation can bring some slowing of disease progression, and there are attempts in progress to tackle the disease by gene therapy. But generally the child is, sadly, 'constrained to follow one particular future', at least unless responding to the latest therapy. But when gene therapy becomes successful, as it surely will one day, then the outlook for these sufferers will be transformed, and the spectre of genetic determinism, at least as it arises from a dysfunctional α-iduronidase gene, will disappear.

In other well-known genetic diseases, such as cystic fibrosis, the situation is different. There are more than 1,500 different mutations in the gene that cause cystic fibrosis (Bobadilla et al., 2002), but the consequences of having two genes with the same mutation, or having two genes with different mutations, can be quite different, leading to headaches for clinicians who wish to predict the likely outcomes for patients carrying the rarer and therefore less studied mutations. Cystic fibrosis represents a range of disease types rather than a single disease. Furthermore, there is now a range of therapies that improve the quality of life and life expectancy of cystic fibrosis patients. So the possible outcomes for patients are probabilistic rather than deterministic, and such is the case for many other genetic diseases due to mutations in single genes. In practice, the vast majority of diseases in which genetic variation is involved are polygenic in nature and therefore cannot readily be tracked through families by the rules of standard Mendelian inheritance. Having said that, as we have just noted, even point mutations in single genes can result in complex traits that vary considerably in their clinical manifestations, so at the clinical level the distinction between monogenic and polygenic syndromes is often not as clear-cut as might at first be imagined.

Two medical syndromes with high positive heritability and with distinctive behavioural consequences have already been introduced in the previous chapter: schizophrenia,[4] perhaps better called 'the schizophrenias' as there is quite a range of clinical categories, and autism. In both cases GWAS approaches have led to some very significant findings. In one major study involving nearly 37,000 cases of schizophrenia and more than 113,000 controls, 128 different SNPs were reported that associated with the disease at greater than chance levels, identifying 108 different relevant locations in the genome (Consortium, 2014). Of these, 75 per cent of the locations included protein-encoding genes, many thought to be biologically relevant to the aetiology of the disease. For example, the dopamine receptor D_2 (DRD2) was identified, the target for all successful antipsychotic drugs. A sizeable cohort of the other associated genes have protein products that are also involved in synaptic neurotransmission. More of a surprise was the finding that a number of genetic variants involved in the immune system are also associated with schizophrenia, pointing to a possible immunological involvement in the aetiology of the disease. A further paper identified variants in a gene called C4, which encodes a complement factor (a part of the immune system) as being particularly significant (Sekar et al., 2016). The complement pathway, of which C4 is a part, is involved in the 'synaptic pruning' mentioned in Chapter 3 whereby the brain's synaptic architecture continues to be sculpted right into early adulthood. Altogether up to a half of the variance in a population can now be explained by the common SNPs that are generally displayed on the chips used for analysis. But this is based on additive approaches to the assessment of heritability, and a thorough examination of epistasis, somewhat neglected in the schizophrenia field, might well change that assessment. In addition, it is now becoming clear that rare copy number variations are also involved, many of them de novo, meaning that they represent new mutations that have arisen in a particular schizophrenic individual (Fromer et al., 2014). A further study revealed a considerable burden of rare disruptive mutations distributed across many genes associated with schizophrenics but less so with controls (Purcell et al., 2014). Increased retrotranspositional insertion events, of the kind described in Chapter 3, have also been described in the brains of schizophrenics, many of them located within or close to genes thought to be important in the development of the disease (Bundo et al., 2014). The idea here is that environmental and/or genetic factors might cause the increased number of insertions into DNA that in turn could lead to the faulty regulation of certain key genes or to the instability of DNA in that region, in turn leading to further mutations. In some studies, overlap was reported between the variant genes that contribute to the development of schizophrenia and those that contribute to autism and to individuals with intellectual disability.

Environmental risk factors for schizophrenia are well established, such as prenatal infection (Brown, 2006) and obstetric complications at birth (Lewis and Murray, 1987). Complex interactions with immune responses have also been described (Khandaker, 2015). Yet how environmental factors integrate with the genomic variation in the development of schizophrenia remains quite unknown. One common suggestion is that there is an accumulation of risk factors coming from both genes and the environment, and then just one or a few more mutations coordinate to take the system 'over the edge'. The appropriate analogy from the world of physics would be a 'phase transition' in which a series of very small changes occur in a material until the critical point comes at which one further change causes the material to acquire a quite different state. This raises the provocative question as to whether the development of schizophrenia might be almost entirely due to genetic variation. If that were the case, the reason that monozygotic twins are only 45 per cent concordant for the disease might be due to the fact that in one twin, but not the other, one or a few rare mutations have occurred against a background of high genetic risk, shared by both twins, and those few 'extra' changes are then sufficient to trigger the phase transition leading to development of the disease.

Like the schizophrenias, the autism spectrum disorders have also been the subject of intense investigation using GWAS and other approaches. A summary of six independent GWAS studies revealed 200 SNPs that appear to selectively associate with autism, each one contributing a tiny percentage to the overall heritability, many of them identifying genes that encode proteins that build integrated biological pathways of possible relevance to the aetiology of autism (Poelmans et al., 2013). De novo mutations have also received particular attention in the autism field. In one study, genomic sequencing was carried out in around 2,500 families in which only one of two sibs was on the autistic spectrum (Iossifov et al., 2014). The total burden of novel mutations discovered in the autistic cohort was significant for up to 45 per cent of those with autism. The rate of novel mutations likely to disrupt gene function was nearly twofold higher in the autistic cohort compared to their unaffected sibs. A total of 391 novel gene-disrupting mutations in a total of 353 target genes were identified in the autistic children but not in their non-autistic sibs, and of these, 27 target genes were recurrent. At the time of writing, more than 700 DNA regions have been implicated in autism, although only a small sub-set of these have been replicated (Geschwind and State, 2015).[5] There are also rare cases of autistic spectrum disorders that appear to be caused by a single gene disorder (Zoghbi and Bear, 2012). For example, Fragile X Syndrome is so called because the single mutant gene that causes the syndrome is located on the X chromosome and around one third of the carriers of this mutation develop an autism spectrum disorder (Ebert and Greenberg, 2013). As expected, differences in epigenetic regulation are also involved in the aetiology of autism as shown by analysing the methylome in identical twins either concordant or discordant for autism (Wong et al., 2013).

As with schizophrenia, the genetic analysis of autism taken overall has led to the identification of many hundreds of mutant genes that appear to contribute to disease development, many of them elucidating the biological pathways that regulate synaptic connectivity between neurons. There is every hope that eventually the delineation of these pathways will lead to novel pharmaceutical interventions that will either prevent disease development or at least ameliorate the symptoms once they develop. But as we move now to consider the molecular genetics of behavioural traits within the normal range, we find ourselves in a very different kind of territory when compared with pathological conditions such as schizophrenia and autism.

Aggression and the Molecular Genetics of Candidate Genes

There have been very few GWAS studies of aggression, most likely due to the difficulty in specifying a specific measurable trait, but perhaps also due to the domination of candidate gene studies in the field.[6] Why look for hundreds more gene variants when the candidate genes being studied already (such as the one described later in this chapter) are enough to be getting on with? In one GWAS study, 8,747 Americans, both male and female, average age fifty-four, were genotyped in relation to the traits of angry temperament and angry reaction, the former relating more to intrinsic temperament, the latter to induced responses, as assessed by self-reporting questionnaires (Mick et al., 2014). The authors reported a genetic susceptibility for the unprovoked 'angry temperament' variation in scores, but not for the provoked 'angry reaction' variation. Genetic association with each anger phenotype was assessed using DNA arrays containing 677,643 SNPs. A total of thirty-eight DNA regions were found to be associated with the variation in angry temperament, some of which contained genes with protein products that, arguably, might be involved in brain development and responses. However, ironically none of the thirty-eight regions identified contained the candidate genes that have been the subject of intensive investigation as possible contributors to the various traits found under the general label of 'aggression'.

There have been a huge number of candidate gene studies relating to aggression. In one meta-analysis Vassos and colleagues identified 185 candidate gene studies, published in the period 1992–2011, covering thirty-one genes relating to anger, hostility, aggression, violence and criminality (Vassos et al., 2014). In ninety-two of the papers reviewed (50% of the total), a statistically significant result was reported for an association between a variant gene and a trait included within the 'aggressive phenotype'. However, of the thirty-one genes reported to have a significant association, only eleven had been studied in at least three samples, the quality criteria set by the authors. When these eleven were meta-analysed, not one significant association was reported. As the authors report: 'Our study provides evidence that the candidate gene approach has not succeeded in identifying genes associated with these outcomes. This is consistent with recent observations in the field that candidate gene studies of human characteristics and complex diseases at large have failed to produce consistent and clinically useful findings' (Vassos et al., 2014).

So, might the results be more positive if a candidate gene was investigated in the context of G x E interactions? Studies of this kind on the gene encoding the enzyme monoamine oxidase A (MAOA) provide some informative examples of this approach. There are good biological reasons why MAOA might be thought to be involved in aggression. The gene encoding the enzyme is expressed throughout the frontal and temporal lobes of the brain and metabolises the key neurotransmitters serotonin, dopamine and noradrenaline – that is, once they have been taken up into the neuron from the synaptic cleft – thereby overall reducing their actions at synaptic junctions. Numerous animal studies point to an important role for MAOA in modulating animal behaviour: for example, deletion of the MAOA gene from mice led to a colony displaying increased aggression (Cases et al., 1995). Of further interest is the fact that the MAOA gene is X-linked, meaning that males are 'hemizygous' for the gene – they only have a single copy. Expression of the gene is regulated by variable number tandem repeats (VNTR) of the kind already encountered when discussing the varying dispersal times of macaque monkeys in Chapter 5. The number of repeats of small DNA sequences in the regulatory region of the gene changes the mRNA expression levels of the gene, at least under laboratory conditions. The genes containing 2-, 3- or 5-repeats are generally characterised as low-activity, meaning low expression of the gene (MAOA-L), while 3.5- and 4-repeats are high-activity (MAOA-H) (Kuepper et al., 2013). The 3- and 4-repeats are the most common variants in all populations (Cerasa et al., 2010). There is however a fly in the ointment in this apparently clear-cut classification. Normally, it is assumed that if a gene is transcribed more vigorously into mRNA, then there will be more protein, but this is not invariably the case as translation of the mRNA can be inhibited by various factors. Some investigators have assessed in vivo levels of MAOA enzyme and reported no differences in protein levels or activity for L and H carriers (Cirulli and Goldstein, 2007; Fowler et al., 2007), and similar conclusions have been drawn from measurements of MAOA levels in post-mortem brain tissue. The assumption in the research field, possibly a trifle optimistic, is that the dynamic epigenetic regulation of the gene in vivo (Shumay et al., 2012) and/or differences between the MAO-L and MAO-H forms during brain development (Aslund et al., 2011) may explain the behavioural differences in the carriers of the two variants, differences described later in this chapter. Indeed, differences in brain structures have been described between carriers of the MAOA-L variant as compared to the MAOA-H variant (Buckholtz and Meyer-Lindenberg, 2008; Cerasa et al., 2010), although given the large number of other gene variants that likely characterise such carriers, it is invariably difficult to assess the causal role of the gene being investigated based on such findings, at least in the human context.

Two landmark papers launched the current interest in MAOA as a possible player in the behavioural genetics of aggression in humans. The first arose from a chance encounter in 1978 between a geneticist Han Brunner from the University Hospital of Nijmegen in the Netherlands and a lady who came by his office seeking help for the outbursts of aggression that characterised many but not all of the males in her extended family (Morell, 1993). Fifteen years later Brunner described the family tree of this family (geneticists often need a lot of patience) and the pattern of inheritance (Brunner et al., 1993b). This was followed up soon after by a description of the mutation involved – a very rare mutation that causes premature termination of transcription and thereby complete deletion of the MAOA enzyme in affected males (Brunner et al., 1993a). As expected, lack of the enzyme led to a major disruption in the metabolism of key neurotransmitters. The fourteen males having the mutation, spread over four generations, were characterised by mild intellectual disability, together with outbursts of anger and aggression including arson, rape and exhibitionism. For example, one affected male raped his sister at the age of twenty-three years. Aggressive behaviour tended to cluster in periods of one to three days during which the affected male would sleep very little and experience frequent 'night terrors' (Brunner et al., 1993b). Female behaviours were found to be within the normal range. This pattern of inheritance is typical of X-chromosome-linked disorders. Assuming that females carry the mutation in only one of their two X chromosomes, so that the normal gene on the second chromosome is sufficient to generate enough MAOA enzyme, females then become carriers of the mutation without displaying any phenotypic evidence of its presence. By contrast, the XY chromosomes of males entail that there is no 'back-up' gene and so the mutation of the gene on the single X chromosome causes complete deficiency of the MAOA enzyme. On average, 50 per cent of males will be characterised by the deficiency because there is a 50:50 chance of inheriting the X chromosome carrying the defective gene rather than the mother's other X chromosome. In the present case, measurement of MAOA activity in cells from the females showed that they were within the normal range, whereas virtually no activity was detectable in the affected males (Brunner et al., 1993a).

Unfortunately, little information is available from the study of this family concerning the relationship between early childhood experiences and the development of aggressive episodes. The abnormal behaviour was documented in males living in four different families in different parts of the country, so environmental differences must have been considerable. The aggression varied markedly in severity and, over time, even within a single family.

For twenty years it was thought that this was the only extended family in the world with this particular mutation. But in 2014 males in three generations of a French family were found to display autism spectrum disorders, attention deficits and aggressive behaviour (Piton et al., 2014). Further analysis showed that there was a mutation in the MAOA gene in the affected males, different from the one in the Dutch family, but sufficient to reduce MAO enzyme activity by around 80 per cent, a reduction associated with a much lower expression of MAOA protein than normal. Unlike the reports on the Dutch family, in this case much more was known about the varying environmental conditions of the affected males. The two maternal uncles of the index case displayed the most severe behavioural traits and in both cases they had suffered familial neglect, maltreatment and sexual abuse during childhood with parental psychiatric problems and substance abuse, leading to their being placed in care at the age of five and seven years. Some amelioration of symptoms was obtained by the use of psychotropic drugs. Affected males in their mid-thirties had developmental ages of typical two- to five-year-olds. No symptoms were reported for the females in the family, again typical of X-linked inheritance.

Such family studies will be revisited in Chapter 10 in the context of criminal responsibility. For the moment we will turn to the second landmark paper that has led to intensive study of trait differences between the carriers of the 'low' and 'high' activity versions of the MAOA gene (Caspi et al., 2002). This paper was based on the Dunedin study in which around 1,000 children born in Dunedin, New Zealand, have been intensively studied every two to three years since their births in 1972–3. Out of this cohort, 442 males aged in their mid-twenties were genotyped for the high or low activity forms of the MAOA gene, of whom 63 per cent had the high form and 37 per cent, the low form. The study aimed to correlate the extent of early childhood mistreatment with the high or low activity forms of MAOA. Between the ages of three and eleven years, 8 per cent of the study children experienced 'severe' maltreatment, 28 per cent experienced 'probable' maltreatment and 64 per cent experienced no maltreatment. Within the cohort with no evidence of childhood maltreatment, there was a slightly smaller incidence of antisocial behaviour amongst those with the low MAOA activity gene, as Figure 6 illustrates, although this difference was not statistically significant. By contrast, there was a higher reporting of anti-social behaviour amongst those who had experienced probable childhood maltreatment if they also carried the low activity MAOA gene and a much higher incidence of antisocial behaviour amongst the severe maltreatment group (Figure 6). Although individuals from this latter cohort accounted for only 12 per cent of the total male birth cohort, they accounted for 44 per cent of the cohort's violent convictions by the time they reached the age of twenty-six. This study therefore concluded that having a higher activity version of the MAOA gene somehow protected some children, at least, from the longer-term consequences of maltreatment. However, the data do not suggest that the MAOA gene has any effect on the incidence of antisocial behaviour per se, providing children are raised in a non-abusive context. In fact, if anything, having the low activity version correlates with lower levels of antisocial behavior, in the absence of childhood maltreatment, a trend confirmed by others in other studies in which statistically significant differences were noted. For example, in one study the high activity MAOA version correlated with higher impulsivity and aggression (Manuck et al., 2000), and in a meta-analysis of results up to 2006, the higher activity MAOA gene was found to significantly correlate with higher levels of mental problems amongst boys aged seven (Kim-Cohen et al., 2006). A further study based on 593 young people from the Estonian Children Personality, Behaviour and Health Study found that the low activity MAOA allele correlated with the attainment of higher education by the age of twenty-five (Kiive et al., 2014). In contrast to such results, a meta-analysis that pooled results from thirty different studies suggested that the low-activity allele contributed to a 'modest effect' on raising the incidence of antisocial behaviour

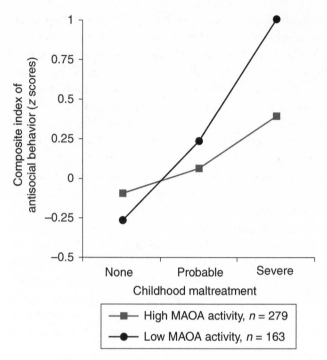

FIGURE 6 The correlations in a cohort of males between mean values on a composite scale of antisocial behaviour, severity of childhood maltreatment and carrying the high activity and low activity forms of the MAOA gene. (Caspi et al., 2002)

in young male adults (Ficks and Waldman, 2014). The ethical significance of these findings will be discussed in later chapters.

As always, replication is the key in such studies, and many attempts have been made to replicate the G x E findings of the 2002 study (Caspi et al., 2002) with somewhat mixed results. A meta-analysis carried out by pooling the results from five different studies concluded that the association between early familial adversity and later mental health problems was significantly stronger in the low-activity MAOA versus the high-activity MAOA groups (Kim-Cohen et al., 2006), a conclusion confirmed in a larger meta-analysis pooling results from eight separate studies (Taylor and Kim-Cohen, 2007). More recent studies have produced conflicting results. On one hand, a further major meta-analysis published in 2014 included data from twenty different studies involving 11,064 male subjects, once again supporting a moderate association between early life adversities with later aggressive and antisocial outcomes in the low-activity MAOA-expressing cohorts (Byrd and Manuck, 2014). Furthermore, a study of 622 violent Finnish offenders found a significant correlation between their low-activity MAOA genotype and their history of violence, though in this case no correlation between the MAOA genotype and childhood maltreatment was noted, unlike the original 2002 Caspi study (Tiihonen et al., 2015). In addition, a study based on 3,356 white men and 960 black men drawn from the US-based National Longitudinal Study of Adolescent Health found no significant relationship between childhood maltreatment and later antisocial behaviour in the low compared to high MAOA groups, using methods very comparable to the original 2002 Caspi paper (Haberstick et al., 2014). The authors comment that'[t]he increased sample sizes in our study afforded enough statistical power to detect an effect size, if present, as small as 0.001. This suggests that previous replications in smaller samples could be false-positives and underscores the potential difficulty of detecting gene–environment interactions involving common genetic variants.' There have also been other papers reporting a failure to replicate the Caspi study, or even an opposite effect (for example, Huizinga et al., 2006 and Prichard et al., 2008).

What can we conclude from this rather bewildering array of contradictory data? It should be emphasised that there can be numerous reasons for a failure to replicate: differing methods of assessing childhood maltreatment; differing methods for quantifying later antisocial or aggressive behaviour; populations with different proportions of the different alleles; differences in sample size and so on. There is also the depressing conclusion that the original observation may be simply mistaken. A more likely interpretation is that the variant MAOA alleles represent a tiny percentage of all the hundreds of genetic variants that contribute to the population variation in the diverse range of phenotypes subsumed under the overall heading of 'aggression' and that identifying such small contributions is always going to be experimentally problematic. The MAO gene story is rather typical of the history of investigation of other candidate genes that are thought to influence complex behavioural traits: the initial positive finding of an association is greeted with acclaim by both the academic community and, often in a distorted or exaggerated form, by the media (on which, see later in this chapter); extensive attempts are then made to replicate the original finding, some positive, some negative in outcome; increasingly large cohorts are then studied to increase the statistical power of the study; very often the original finding eventually fades from view through lack of positive replication. Whether the high/low activity MAOA allele story has yet reached this final stage is too early to say – the jury is still out. But this account provides a textbook example of the general trend, one repeated in the investigation of numerous other candidate genes. Ironically, despite the huge numbers of studies on G x E interactions, involving many thousands of individuals, it is currently the two family studies described previously in which MAOA gene mutations reduce the enzyme activity to a much reduced or zero level (Brunner et al., 1993a; Piton et al., 2014) – families in which the genotype and phenotype can be tracked over several generations – that provide the most convincing evidence for an involvement of MAOA in aggressive human phenotypes.

It is now time to step back from this tsunami of somewhat contradictory data (of which only a small part is cited here) and ask ourselves the question: what does all this mean? Some salient points are worth bearing in mind. First, as mentioned earlier, despite the designation of 'high activity' MAOA-H and 'low activity' MAOA-L alleles, there are no data showing that these variants actually lead to different neuronal enzyme activities in vivo and, indeed, as already cited, some evidence that the in vivo enzyme activities in the brain are comparable. In this context it should be noted that in those rare families in which male individuals have either no or much reduced MAOA activity (Brunner et al., 1993a; Piton et al., 2014), the hemizygous females appear to have MAOA enzyme activities within the normal range, despite having only a single functional copy of the MAOA gene. For a long time MAOA inhibitors have also been used in the treatment of depression, but such inhibition does not result in aggressive behavior. The assumption is, therefore, that during brain development, differences in regulation of the variant MAOA alleles lead to different synaptic architectures, an idea supported by the apparent anatomical differences in the brain reported between the allelic sub-groups, and it is these differing architectures that then modulate aggressive responses. Scientists tend to believe data more when there is a clearly delineated idea as to what mechanisms are involved in explaining the results. In the present context, there are some data and plenty of theories, but little mechanistic understanding as to how the data may be explained.

Second, whether this (presumed) change in synaptic architecture is directly or only indirectly involved in aggressive behaviour remains quite unclear. Indeed there are some suggestions that aggression is downstream of more immediate proximal effects. For example, it has been suggested that MAOA has a more proximal effect on 'negative urgency', which refers to the tendency to react impulsively to experiences of negative affect. Negative urgency is one of four facets of impulsivity that also include lack of perseverance, lack of premeditation and sensation seeking (Chester et al., 2015). The lower functioning MAOA genotype was found to correlate in this study with measurements of this putative trait of negative urgency.

Third, the low activity MAOA allele is extremely common in all human populations and is carried by 35–40 per cent of western white populations and 77 per cent of the Han Chinese population (Lu et al., 2002). The vast majority of these populations are not especially aggressive, though if they were maltreated as children, then some cross-cultural reports, but not others, suggest that such individuals are more likely to express aggressive behavior patterns in adulthood. But in any event, even if the data on this point are believable, the results would have no relevance for the majority of the population.

Fourth, some commentary is necessary on the way in which the media, together with ethical and political commentators, have been handling these results. The original 2002 Caspi paper made a big media impact and by 2004 the MAOA-L allele was already being dubbed the 'Warrior Gene', this in the respected journal *Science*, of all places, by a journalist who was commenting on the correlation between MAOA alleles and an apparent increased level of aggression in a small group of macaque monkeys (Gibbons, 2004). The terminology unfortunately stuck, and to this day the ridiculous phrase is still wheeled out in the media when referring to the MAOA gene. The situation was not helped when in 2006 two New Zealand geneticists from Wellington's Victoria University reported during a talk at an international meeting that the low activity allele of the MAOA gene was found in 60 per cent of Maori people, data based on a sample of just seventeen Maori individuals (Merriman and Cameron, 2007), an estimate later adjusted to 56 per cent (Lea and Chambers, 2007). Based on these data they concluded that 'positive selection of MAOA associated with risk taking and aggressive behaviour has occurred during the Polynesian migrations' and termed the low-activity variant the 'warrior allele'. In further publicity surrounding this announcement, the MAOA allele was then linked to antisocial behavior in the Maori people. The *Christchurch Press* of 9 August 2006 reported that Maori men were genetically predisposed to 'violence, criminal acts, and risky behavior', whilst the Wellington-based *Dominion Post* opined that the monoamine oxidase gene 'goes a long way to explaining some of the problems Maori have. Obviously, this means they are going to be more aggressive and violent and more likely to get involved in risk-taking behaviour like gambling.'[7] And all this based on the fact that Maori individuals have a somewhat higher proportion of individuals carrying the low activity MAOA gene compared to some other populations. Despite the negative criticism that faced the initial announcement, the same authors continued to insist that the higher proportion of the low activity MAOA allele was connected to the fact that the Maori are well known as 'fearless warriors', whilst dissociating themselves from the idea that the allele was in any way linked to increased criminality amongst the Maori people (Lea and Chambers, 2007). Quite how this fits with the 77 per cent of the Han Chinese who express the low activity MAOA allele, was not made clear. This whole sorry story has been critically analysed by genetic bioethicists ever since, for example, by Perbal (2013). Scientists need to take great care over the public presentation of their results, particularly seeking to avoid the pitfall of extrapolating wildly from scanty data to broad societal conclusions.

The New Zealand story about the Maori is not the only racial profil- ing account to emerge. The background to this research is the finding that the 2-repeat allele of the MAOA gene has even lower activity when measured in the laboratory than the 3-repeat allele and is associated with higher levels of self-reported serious and violent delinquency compared to groups that do not carry this allele (Guo et al., 2008). A further study reported that this 2-repeat allele of the MAOA gene was found to be higher in African Americans (5.2%) than in Caucasian males (0.1%), and a comparison within a cohort of around 170 male African Americans suggested that the incidence of the rare 2-repeat allele correlated with a higher lifetime risk of being in prison when compared with a group of African Americans who did not carry this allele (Beaver et al., 2013). It should be noted that both these studies conflict with those other studies analysed using a meta-analysis, already mentioned previously, in which antisocial behaviours were associated with the higher activity MAOA alleles in cohorts who had been normally treated as children (Byrd and Manuck, 2014). In our present context, it is interesting to note that the 2-repeat allele became quickly dubbed in the media 'the extreme war- rior gene' and began to take on racial overtones. When a disproportion- ally high number of males of an ethnic group are found to carry a rare genetic variant not present in other ethnic groups that then correlates with some behavioural trait or other, then one can be fairly sure that racial stereotyping will soon appear. For example, the author Nicholas Wade has suggested that the relative prevalence of the lowest activity MAOA allele amongst black African Americans (albeit present in only 5.1% of this population) might 'explain' the higher level of criminal- ity observed in this community (Wade, 2014). The alternative and more obvious alternative is that the presence of a rare allele might identify a group in which criminality is higher due to social deprivation and lack of employment opportunities. Once again the mantra that 'correlation does not entail causation' is an important one in this context, with the 'chopstick phenomenon' also being relevant.

Besides the racial stereotyping that arose from reports on the MAOA alleles, the earlier behavioural reports also triggered some highly alarmist and irrational speculation by bioethicists and other commentators. A lawyer, Jennifer Brooks-Crozier, proposed that screening at birth for the low activity MAOA-L allele should be made mandatory in the United States, identification of low-allele carriers being then subjected to intensive social intervention to prevent future crime (Brooks-Crozier, 2011), commenting that

> [t]he 'disorder' occurs frequently in the population, can be screened for, and is treatable. The threat posed by MAOA-low males is real and substantial and the means adopted by states to confront that threat – mandatory newborn screening for the MAOA-low genotype followed by intervention services – is likely to be effective. Thus a court is likely to find that the state interest in mandating newborn screening for the MAOA low genotype is compelling. (p. 551)

The author also commented that the prevalence of MAOA-low males 'is also a kind of epidemic, especially given the devastating tangible and intangible costs of violent crime' (p. 565). Unfortunately the paper failed to cite the numerous papers, some of which are previously discussed, that render the core data at least debatable, and certainly insufficiently robust to provide any justification for a screening programme that would, inevitably, label more than one-third of the US population as 'potential criminals'. Despite the huge number of contradictory findings during the period 2002–11, the author chose to base her costings of such a programme on the original rather dramatic data reported in the 2002 Caspi paper (based on a relatively small population in New Zealand), so that, not surprisingly, the cost-effectiveness of 'saving crimes' came out looking rather good. The author appeared to be unaware of the 'winner's curse'. At least the emphasis on social intervention in families where there is reported maltreatment of children is sound, but that happens anyway in all countries with developed social services, and genotyping at birth would add nothing to the effectiveness of such programmes.

The proposal to genotype an allele at birth which has such a disputed role in influencing antisocial behaviours sounds irrational, and so does the speculation (presented as such, not as a proposal) that embryos might be screened by preimplantation diagnosis to discard those embryos that carry the low activity MAOA allele (Moreno, 2003; Wasserman, 2004; Savulescu et al., 2006). Apart from the fact that up to 77 per cent of the embryos (in China, for example) used in such in vitro fertilization procedures might test positive for the low activity MAOA allele, in any case, if a body of data (cited previously) can be believed, then the low activity allele would provide some modest *protection* against future antisocial behaviour, so there would be a positive argument to choose that embryo, not to discard it. Only if the parents were planning on mistreating their child, once born, would there be a reason not to choose the embryos carrying the low activity MAOA allele. In other words, the speculation about preimplantation diagnosis is incoherent.

One broader observation may help put these somewhat tortuous discussions about 'genetic predispositions' to violence (on which more later in this chapter) into some kind of perspective. Out of 131 countries worldwide, an average of 93 per cent of the prisoners are male,[8] and the gene that identifies this population is the Sry gene found on the Y chromosome. So universal is the correlation between the Sry gene, without which males would not be male, and criminality, that we can safely say that no other genetic correlation will ever be found between a particular gene and criminality that surpasses this one. And yet we still hold nearly all males responsible for their criminal actions and put them in jail as soon as they are convicted. Furthermore, we note that most people who possess a Y chromosome go through life without committing a crime. So the Sry gene does not *cause* criminality, although clearly we cannot go to the opposite extreme and say that it has nothing to do with human behaviour at all. Genes and criminality will be discussed further in Chapter 10.

Predisposition or Differential Susceptibility?

It is commonplace to read that a particular gene variant entails that the carrier has a 'predisposition' for a certain outcome. The terminology seems quite acceptable where the gene in question operates in a way only partially dependent on the environment; for example, the mutant *BRCA1* gene has a direct lifetime effect on the risk for breast and ovarian cancer. However, the situation becomes more complicated when some complex behavioural phenotype is in mind. For example, it is simply not the case that the low activity MAOA allele entails a predisposition to aggression. Apart from the fact that the published data are contradictory, many of the results, as we have noted, entail a G x E interaction such that 'if A then B, but only if C pertains'. It is quite likely that all confirmed cases of the correlation between gene variants and behavioural phenotypes will turn out to have such G x E constraints. To capture this situation, Jay Belsky has introduced the idea of 'differential susceptibility' in an evolutionary context (Belsky,

1997), but the same point applies equally well in the context of behavioural genetics. James Tabery has helpfully defined a differential susceptibility as

> [t]he presence of a genetic difference between various groups that both *increases* and *decreases* the probability of individuals from one group, in comparison to individuals from the other group(s), developing a particular trait *depending on* the measured environmental condition. (Tabery, 2014, p. 179; author's italics)

Differential susceptibility clearly applies to the Caspi 2002 paper and those positive replications that have followed in its wake. The low activity MAOA-L allele is no warrior gene, nor does it render individuals susceptible to aggression, but if we believe the data, then it contributes to a susceptibility to develop antisocial behaviour under particular environmental conditions of maltreatment.

The 'language game' is important here and there are no perfect words to express the complex reality. 'Susceptibility' could sound like something that is done to a person, so might be more appropriate to use in the context of medical genetics, whereas a 'predisposition' could be used more for the normal variation of behaviours, highlighting the fact that whatever predispositions might be inherited, the individual remains capable of bringing those predispositions (if unwanted) under control. Such linguistic issues become particularly problematic when considering cases such as the very rare mutation found in the Dutch family, causing a total loss of MAOA activity. It seems reasonable to infer that these male individuals have a strong 'predisposition' to aggression and, likewise, that they are 'susceptible' to aggression: both words fit. Yet if more information concerning 'environmental provocations' becomes available on this or similar families, then this might give a clearer indication as to which word is more appropriate.

The Way Forward for Molecular Genetics

Space has only allowed for consideration of one main gene implicated in aggression, a gene which remains the subject of continued extensive investigation. Nonetheless, as already intimated, the history of the study of other genetic variants in relation to various behavioural traits has often followed a similar pattern. For example, the serotonin transporter (introduced in Chapter 5) is a protein involved in the uptake of neurotransmitters back into the neuron after their release into the synaptic cleft. As with MAOA, there is a range of variant versions of the gene that generate transport proteins with a range of different activities. In 2003 Caspi and colleagues published

a paper, once again based on the Dunedin cohort, reporting that some allelic forms of the gene endowed their possessors with protection from depression following a variable number of stressful life events (Caspi et al., 2003).
As with the MAOA studies, meta-analyses have either confirmed or failed to
confirm these initial findings in roughly equal proportions: a meta-analysis
of fourteen different studies involving more than 14,000 participants published in 2009 failed to replicate the original findings (Risch et al., 2009),
as did another meta-analysis published the same year involving five studies
(Munafo et al., 2009). But then, in 2011, a further meta-analysis was published, this time including data from fifty-four different studies (carried out
up until December 2009), involving 40,000 research participants, and a
positive result was obtained, agreeing with Caspi's original findings (Karg
et al., 2011), supported also by a further meta-analysis carried out in 2012
(van Ijzendoorn et al., 2012). Such a catalogue of 'warring meta-analyses'
can lead to confusion on the part of investigators and the wider public. It is
worth noting that in this case the authors of these various papers also chose
to include different groups of papers in their analyses. The earlier meta-
analyses that produced negative results were very restrictive in their choice,
focusing on papers that reproduced the conditions of the Caspi 2003 paper
very precisely, whereas the later study by Karg et al. (2011) was more inclusive, including studies, for example, that involved studying the effects of a
wide range of stressful events in relation to later depression and the serotonin transporter genotype. This does not resolve the issue of what the correct
answer might be, but at least helps us understand why meta-analyses do not
always agree with each other.

    Other candidate genes that have been the focus of intensive investiga-
tions in relation to human behaviour, include variants of the DRD4 gene
which encode receptors for the neurotransmitter dopamine that operate
with more or less efficiency (Bakermans-Kranenburg and van Ijzendoorn,
2011); the catechol-O-methyltransferase gene which encodes an enzyme
that degrades neurotransmitters such as dopamine (Caspi et al., 2005);
the variant genes that encode receptors for the neurotransmitter serotonin (Pavlov et al., 2012) and other genes as well – too many to list here. All
the examples listed are plausible candidate genes in that they are known
to modulate the actions and efficacy of key neurotransmitters involved
in the emotional life of humans. So it is a perfectly reasonable assumption that variants of these various genes will influence brain development
from fetal life onwards in such a way that personalities will be more or
less impetuous; more outgoing or more inward looking; more or less resistant to emotional stress; more or less resistant to the kinds of stress that

can trigger depression and so forth. The data, confusing as it may be, is perfectly in accord with DICI: variant genomes interacting with variant micro- and macroenvironments integrate in the development of people with varying behavioural traits and personalities. The problem with candidate gene studies is that there are likely to be hundreds of genetic variants, maybe thousands, that impinge on the development of complex human behavioural traits (Kim-Cohen et al., 2006). So attempting to demonstrate the effect of a single genetic variant is like looking for a needle in the proverbial haystack. Apart from those variants that contribute to clearly defined pathologies, the enormous collection of gene variants that are involved in the normal range of human behavioural traits are likely to contribute only a very small percentage effect to the overall heritability, an effect that is experimentally difficult to extract from a background of the other hundreds of variants that are likewise contributing to the same trait. An animal study provides a clue as to the complexity involved. Zwarts et al. (2011) bred a strain of hyper-aggressive *Drosophila* fruit flies, finding differences in the transcription levels of 4,038 genes in homozygous hyper-aggressive flies versus controls and 1,169 genes that were co-ordinately up- or down-regulated in all hyper-aggressive homozygous flies, with epistatic interactions for more than 800 genes. If that is the situation in flies, then one cannot be surprised if the situation is at least that complex in humans.

So in the context of human behavioural genetics, we are left with the slightly gloomy question: is it really worth it? What possible gain can there be for identifying the hundreds of variants involved, each of tiny effect? As Turkheimer rhetorically poses the question (in the context of genes and personality): 'Can one point to a field of science that has been successful by stringing together the multiple effects of such tiny associations?' (Turkheimer et al., 2014). In pathological conditions, the answer might well be that 'there could be great gain because we may find that the relevant gene variants all line up to contribute to one or a few well-specified pathways that are then candidates for pharmaceutical intervention'. This is relevant to genetic investigations of conditions such as schizophrenia or the autism spectrum disorders, but as far as the normal range of human behaviours is concerned, such goals are, with rare exceptions, thankfully not in mind. If some initial studies had uncovered gene variants with large and repeatable differential behavioural effects, then the question of pointlessness might not arise. But that is not the case, so far at least, and in the examples that follow in the next two chapters, that same question will continue to rear its head.

CHAPTER 8

Mensa, Mediocrity or Meritocracy?

The Genetics of Intelligence, Religiosity and Politics

> [I]ntelligence does push a lot of buttons. It's like waving a red flag to a bull.
>> Robert Plomin (Panofsky, 2014, p. 211)

> [W]hat is inherited is not this or that trait, but the manner in which the organism responds to the environment.
>> Theodosius Dobzhansky (Dunn and
>> Dobzhansky, 1952, p. 132)

> We all make mistakes, but intelligence enables us to do it on purpose.
>> Will Cuppy[1]

Attempts to measure the relationship between inheritance and intelligence go all the way back to Galton and, as outlined in Chapter 2, represent a decidedly murky history. That history has been told often before from various perspectives (Kevles, 1985; Paul, 1998; Carson, 2007; Gillette, 2011) and will not be reviewed further here. The way in which the behavioural genetics field became more fragmented and controversial following the 'Jensen affair' of 1969, followed by *The Bell Curve* twenty-five years later (Herrnstein & Murray, 1994), has also been well described from a sociological perspective (Panofsky, 2014). For those who wish to dismiss the whole field of behavioural genetics, intelligence testing has provided a tempting target (Gould, 1996; Murdoch, 2007; Joseph, 2013), with plenty of robust defences of such critiques from within the research community (Carroll, 1995; Bouchard, 2004; Gottfredson and Saklofske, 2009). In practice the professional field as a whole has, in general, been careful to dissociate itself in more recent years from its earlier focus on the heritability of IQ, particularly as such assessments were then used to influence national educational policies (White, 2006),[2] and its attention has turned to more general theories of cognitive ability. Our aim here is to provide a snapshot of the contemporary science of genetics in relation to intelligence testing, bearing in mind our overall goal of assessing the extent to which genetic variation between individuals does, or does not, differentially constrain or predict their biographical histories.

What Is Intelligence?

'Few constructs are as mysterious and controversial as human intelligence. One mystery is why, even though the concept has existed for centuries, there is still little consensus on exactly what it means for someone to be intelligent or for one person to be more intelligent than another' (Davidson and Kemp, 2011, p. 58). Indeed, all definitions continue to remain disputed to the present day. Whilst recognising this fact, the main focus on measuring intelligence in western academia involves assessing analytical skills such as verbal dexterity, logical-mathematical skills and problem-solving ability, precisely the set of skills that help in making good progress through western educational systems (Ang et al., 2011). In Western Europe and North America, intelligence is largely related to one's cognitive abilities. However, '[t]he overall picture is that intelligence is defined and perceived differently by people from different parts of the world, and that these differences are largely reflective of long-standing cultural traditions' (Niu and Brass, 2011, p. 640). This is highly relevant to questions concerning the measurement of intelligence and, therefore, to the genetics of intelligence. In turn, this is part of a wider discussion as to what counts as a 'trait' and whether a 'trait' is a local cultural construct or one that is genuinely universal. A ruler cannot be used to measure an invisible entity of unknown dimensions. In reality there are many different 'intelligences' – systems of abilities – only a few of which can be captured by the standard psychometric testing techniques discussed further in this chapter (Neisser, 1996).

IQ and Intelligence Testing

During the past century, hundreds of intelligence tests have been developed, both by theorists and by commercial testing companies. They are used, particularly in the United States, by schools and universities, military forces, governments and employers for a variety of reasons including clinical diagnosis, school admissions, performance assessment and testing job suitability. In the United Sates scores can have an enormous impact on life outcomes: a score can mean the difference in being offered a college place or a job. It can even mean the difference between life and death – whether a convicted criminal has a death sentence commuted. This situation arose from the *Atkins v Virginia* Supreme Court decision of 2002 in the United States which forbad the use of the death penalty in cases where the defendant was cognitively impaired, but left it up to state legislatures to decide exactly how such impairment is defined. This has led to psychometricians battling it out in courtrooms with different testing methods to prove that a defendant has a particular IQ score, either above or below a crucial cut-off point (previously 70 in Virginia and Florida, for example) (Murdoch, 2007).[3] Despite a steady barrage of critiques, including those that question the whole rationale of IQ testing, a large number of countries have embraced intelligence testing with almost equal enthusiasm to the United States, particularly in the developing world, where there is a pressing need to use limited public educational and employment resources effectively (Urbina, 2011). In the United Kingdom, IQ testing is largely past history, except in academic research studies and in the diagnosis of certain medical conditions entailing cognitive disability.

What do IQ tests measure? The history of the idea has been described elsewhere (Murdoch, 2007, Urbina, 2011). In brief, Lewis Terman, an academic at Stanford University, made substantial revisions in 1916 to the French Binet-Simon test (mentioned in Chapter 2) to adapt it for use in the different U.S. cultural context; it was published as the Stanford-Binet test. This was the first test to produce an IQ score. Terman borrowed the Binet-Simon methodology of a quotient, but multiplied each score by 100, as this meant that scores below the average still came out as positive and as non-decimal numbers, resulting in an 'intelligence quotient'. Therefore, a child who was able to answer the average number of questions for the child's age group, meaning that the mental age was identical to chronological age, would have an IQ of 100. In this way IQ became defined as Mental Age multiplied by 100 divided by the Chronological Age. Therefore, if a six-year-old completes tasks designated as 'eight-year-old level' – in other words, is more intelligent than the average – the child will have an IQ of $(8/6) \times 100 = 133$. 'In spite of the fact that the Stanford-Binet test was primarily suitable for children, this scale dominated the field of individual intelligence testing for the next few decades' (Urbina, 2011). A significant innovation was the 1939 publication of the Wechsler-Bellevue Intelligence Scale, the first test designed specifically for adults rather than being adapted from tests intended for children. Wechsler discarded the quotient system of calculating IQ scores and introduced deviation scores, a much more accurate methodology for adults. The rationale of linear ratio scores is only legitimate for children, as the relationship between mental age and chronological age breaks down at around age sixteen (Snyderman and Rothman, 1988). There is no difference in mental age between a twenty-five-year-old and a sixty-five-year-old (in other words, these groups can *on average* perform equally well on all subtests), rendering ratio scores completely meaningless for adults. The score an individual receives on the Wechsler-Bellevue test is compared with the scores of a standardisation group, who are the same age and nationality (and sometimes ethnicity or social class) as the test-taker. Almost every current intelligence test produces a deviation score. But in addition Wechsler, in a move that has caused more than seventy years' worth of misunderstanding and confusion, opted to retain the IQ terminology. He referred to a deviation score as an IQ score, even though it is not a quotient of anything, and standardised the scores so that the average was 100 with a standard deviation of 15. By definition 95 per cent of a given population has scores within two standard deviations of the mean, in other words between 70 and 130. Only a small number of test batteries still refer to the measurement score as an IQ score from historical habit, although some of these are the most used, ensuring the current preservation of the IQ terminology.

In today's intelligence testing there is a wide range of batteries of tests to choose from, some designed for clinical use, others for general adult use, others for children, and so forth. Each test battery is composed of between ten and twenty subtests, each of which contains multiple items (questions). Some items are timed, some are not. Each subtest measures a different facet of intelligence, such as visual-spatial reasoning or short-term memory, and subtests are often put together to yield scores for specific domains of intelligence, such as perceptual reasoning or comprehension. There is no agreed set of subtests that must be used in a battery. Instead, modern subtests are designed or reused based on the specific theoretical model being used by the test designers. The lack of agreement on a theory and structure of intelligence is the major reason for so many intelligence tests, although many share similar or overlapping components (Gottfredson and Saklofske, 2009).

IQ scores are often misinterpreted or misrepresented, so it is worth summarising the most common misunderstandings. First, a deviation IQ score is not a *fixed* or *absolute* score and, therefore, has no intrinsic or absolute meaning – it is entirely based on comparison with the standardisation group. So IQ values change around every ten years depending on the re-calibration of IQ based on the standardisation group so that the average IQ remains at 100, of particular relevance if you happen to be in a

U.S. state prison in which the IQ value is used to decide the application of the death penalty.[4] Second, the standardisation group must be representative of the test-taker, otherwise the result will make no sense. For example, a fifteen-year-old's IQ score would be much lower if calculated using a standardisation group of fifty-year-old college graduates than a standardisation group of other fifteen-year-olds. Third, an intelligence test score *by itself* describes only the reality of the test-taker's cognitive ability at the time the test was taken. Without more information, it cannot say anything about the degree to which the score is dependent on hereditary or environmental influences. Fourth, IQ scores cannot predict the potential future development of intelligence for a given individual. For example, someone whose score is affected by adverse environmental factors (such as childhood malnutrition) may receive a higher IQ score later in life if that individual's environmental conditions improve. Fifth, because of the re-standardisation of scores for historical consistency, IQ scores appear to be directly comparable. However, comparisons between scores obtained on different tests should be made with care. Each test battery is measuring a slightly different facet of cognitive ability, although it has been shown that test batteries correlate very highly with each other, due to their overlapping nature (Urbina, 2011).

Most investigators who are now working in the neurodevelopmental field think that when trying to understand differences between children, and why some children struggle at school, IQ is not a particularly helpful concept. Instead, most investigators tend to study particular cognitive skills such as working memory, executive control, attention, specific language skills and phonological skills. These provide a more mechanistic account of learning differences, in most cases are better predictors of specific aspects of learning, and there is a far better idea of the cognitive and neural processes underpinning these more specific skills, which ultimately therefore provide better targets for interventions.[5]

General Intelligence g

One of the many controversial aspects of intelligent testing has been, and continues to be, the measurement of so-called general intelligence known as g, now often referred to as 'general cognitive ability' (Plomin et al., 2013a; Bouchard, 2014). The concept has its roots in 'factor analysis', first described by Charles Spearman in 1904 to explain his observations of the differing abilities of pupils at a village school in Jersey (Mackintosh, 2011). Spearman noticed that pupils who did well in English also tended to do well in mathematics, while pupils who did badly in one subject tended to do badly in the others as well (Carroll, 1997). In other words, the test results were correlated with each other. Factor analysis then eventually became a technique to measure how much each intelligence test contributes to the general cognitive ability, g, which in turn involves a measure of the correlations between different tests. It turns out that all tests correlate with each other positively, although measures of spatial and verbal ability (for example) correlate with each other more highly than other measures, such as non-verbal memory tests (Plomin et al., 2013a). Some tests contribute more to g than others in a way that is related to the cognitive ability being assessed; for example, abstract reasoning provides a better assessment of the value of g than less complex cognitive processes such as simple sensory discriminations (Plomin et al., 2013a). Imagine a matrix containing the values of a wide range of psychometric tests measuring differing aspects of cognitive ability; now measure the correlations between the values in all the different tests – it is the highest set of correlations that contribute most to the value of g. The validity of g has been promoted on the grounds that its value has long-term stability after childhood, reportedly greater than the stability of any other behavioural trait (Deary, 2012). Furthermore, use of different intelligence tests on various populations, including twins reared apart, has generated identical or nearly identical values for g (Johnson et al., 2004; 2008). The value of g also positively correlates with many life outcomes, such as good health, occupational status and educational attainment, although on the face of it, the last factor is not very surprising, given that the western educational system depends to a considerable degree on precisely the cognitive abilities that psychometric testing measure and that psychometric methods are influential on the structure of the various exams necessary to gain entry to U.S. universities. Furthermore, the degree of motivation makes a big difference to test scores in young people (Duckworth, 2011), so low values may reflect a kind of self-fulfilling prophecy in which those with low motivation score poorly in psychometric tests and thereby fail to gain entry to the best universities. If differential motivation is influenced by differential genetic variation, for which there is some evidence (Scarr, 1966; Spector, 2012, p. 84), then perhaps the heritability of intelligence test results are more to do with motivation than with intelligence per se.

In practice, concepts such as IQ and g are now found embedded within hierarchical models in which different levels representing varying degrees of generality give a meaningful account of the variation in intelligence that exists between individuals. For example, Carroll (1993) has provided an influential scheme in which level 1 assesses the variation that exists among the various intelligence tests, level 2 relates to the broad domains of cognitive function – so-called group factors – and level 3 relates to g. Such hierarchies are often pictured as a pyramid with g at the top as the most 'fundamental' of these entities. Translating each level into daily life, the first level reflects the fact that some people are particularly good at certain narrow mental skills; the second level reflects the fact that people who score highly in one domain, such as verbal ability, are also good at other related tasks within that same domain; the third level, g, reflects the fact that people who are good at one type of mental task tend also to be good at another quite different mental task (Deary, 2012). So when a battery of tests is applied to a population, some of the variation that exists between individuals is shared by all tests, some is shared by tests that relate to a particular cognitive domain, and some relate to one specific test.

What exactly does g represent? No one really knows. It clearly represents a statistical construct: correlated abilities at obtaining certain scores on psychometric intelligence tests. Some theories point to a single main mechanism underlying g, such as speed of brain information processing, whereas other theories suggest a combination of different cognitive abilities. There are plenty of reported correlations between values of g and various aspects of brain anatomy (Deary, 2012), although cause-effect relationships in such human studies are notoriously difficult to establish. There has been some success in relating specific variant genes to inter-individual differences in anatomical features of the brain (Hibar et al., 2015). The variation in the theories reflects the current lack of knowledge in the neurosciences as to how measured differences in cognitive abilities translate into variation in brain mechanisms.

The Genetics of Intelligence

The heritability of the main products of psychometric testing, IQ and g, has been the initial focus over many decades into the attempts to investigate the role of genetic variation in intelligence. As already discussed, to what extent such parameters can be equated with 'intelligence', broadly defined, is open to question, but since the literature is dominated by the genetics of the fruits of psychometric testing, that will remain the focus here.

Estimates of the heritability of IQ have varied widely depending on the population under study and the precise methods employed. The typical value for the heritability of IQ is normally given as around 50 per cent (Plomin et al., 2013a). But some investigators have reported values as high as 80 per cent (Jensen, 2000), whereas others have reported values in the range 30–35 per cent (Cavalli-Sforza and Feldman, 1973; Devlin et al., 1997). One report described a meta-analysis of more than 200 studies that estimated the heritability of IQ, including twin, family and adoption studies, and involving more than 50,000 pairs of relatives, concluding that the heritability of IQ is 34 per cent (Devlin et al., 1997). All such studies, as their authors generally emphasise, highlight the fact that environmental factors contribute a large slice of the variation in IQ as measured in a given population. For example, in the study by Devlin, modelling suggested that the in utero environment was an important factor in contributing to the observed variation in the populations being analysed (Devlin et al., 1997). Given the close statistical links between IQ measurements and 'general intelligence', it is no surprise to find that heritability estimates for g are roughly in the same range as for IQ, although the precise values depend once more on which particular method is being used. For example, when the heritability of g was assessed using genome-wide complex trait analysis (GCTA), the method introduced in the previous chapter, in a population of more than 3,000 unrelated older individuals, values in the range of 40–50 per cent were reported (Davies et al., 2011). It is also of interest to note that the heritability of IQ and of g increases with age, a fact that partly helps explain the variation in heritability values reported in the literature.

As the discussion on heritability in the previous chapter has highlighted, the precise heritability values are of dubious significance (Visscher et al., 2008; Johnson et al., 2011). And as already emphasised, it is certainly not the case that high heritability, a population measurement concerned with variance, has anything necessarily to do with the development and extent of intelligence in individuals. But the fact that the heritability values are not zero does suggest that genetic variance is important in contributing to the variance that exists in populations with respect to IQ and g values generated via psychometric testing. This is, of course, of no surprise at all when considering such measurements within the DICI framework. Given the fact that the majority of protein-encoding genes in the genome are expressed at various stages of brain development, it would indeed be miraculous if genomic variation between individuals made no difference at all to a person's cognitive abilities. As a reminder, complex human behavioural traits, which include 'intelligence' (however that may be defined), are 100 per cent due to the environment and 100 per cent due to genomic information, integrated together to generate unique human individuals with traits that vary when measured in populations. Measurements of the heritability of IQ and g have value in emphasising this fact, but interpretations of the data that seek to extrapolate these population data to the development of intelligence (however defined) in individuals, make no sense at all. Truly it is here that we detect the 'mirage' in the relation between nature and nurture that Evelyn Fox Keller has so helpfully highlighted (Keller, 2010b).

The practical difficulties entailed in estimating precise proportionalities between 'genetic and environmental influences' that impinge upon variation in populations are well illustrated by the extensive adoption studies that have been carried out using psychometric testing. On the face of it, the data based on adoption studies look rather impressive (Plomin et al., 2013a). First-degree relatives (who are 50% genetically related) are moderately correlated for g (at about 0.45), but when siblings from the same genetic family are reared apart by adoption, the correlation for g between them and their genetic parents falls to 0.24, generating a heritability value of around 48 per cent. Measurements of g on monozygotic twins reared apart have yielded heritability values in the region of 72 per cent. But problems in adoption studies emerge once one starts drilling down further into the data (Joseph, 2010). One problem is the selective placement of adoptees. Adoption agencies often attempt to place children they perceive as 'bright' in similar homes, seeking to fit the adoptee's home into the social and educational background of their parents. In one interesting study, in large agreement with the findings of (Duyme et al., 1999), data were used from the U.S. National Collaborative Perinatal Project, which included a large national sample of American mothers, who were enrolled into the study during pregnancy ($n= 48,197$), and their children ($n= 59,397$), who were followed from birth until age seven (Turkheimer et al., 2003). The key finding was that, whereas children from high socioeconomic status had a measured heritability of 72 per cent with respect to IQ, those children who came from low socioeconomic families had a heritability close to zero – in other words, the environment explained nearly all of the IQ variation. However, most variables traditionally thought of as markers of environmental quality also reflect genetic variability (Plomin and Bergeman, 1991). Children reared in low socioeconomic households, therefore, may differ from more affluent children both environmentally and genetically (Elston and Gottesman, 1968), and the models employed in this particular study do not allow the various possibilities to be distinguished. This study therefore provides a good example of the complex interactions that are always in progress between the various environments and genetic variation and underlines the intrinsic incoherence entailed in the goal to quantify the different contributions as if they were somehow 'discrete entities', which of course they are not. In this context it is interesting to note that adoption itself from a low to high socioeconomic status family may raise the adopted child's IQ by as much as ten points (Schiff et al., 1982). In another study, adopted black children in the state of Minnesota showed closer IQ correlations to their white adopting parents than to their biological mothers (Scarr and Weinberg, 1976).

A further complication in adoption studies arises from so-called range restriction (Joseph, 2010). IQ adoption studies, as we have already noted, frequently assess the correlation of adoptees and their adoptive parents with respect to a trait such as IQ, but this correlation is distorted by the fact that adoptive parents represent a specially selected and therefore non-representative population. For example, there is the selection of families who wish to adopt a child by adoptive agencies, including the strict criteria used in allowing a family to adopt, plus the decision of the family to adopt in the first place plus the further decision to allow their children to be included in an adoption research study. As far as the adoptees are concerned, many come from single teenage mothers who may be poorly placed economically and educationally to raise their child well in the early months or years of life prior to adoption, whose offspring may therefore receive 'sub-optimal obstetric care' (Rutter, 2006). Furthermore, adoptees as a population are at a greater risk for being diagnosed with a psychiatric disorder as compared to the general population (Faraone et al., 1999, p. 43).

In a helpful illustration used by (Eysenck and Kamin, 1981), we may imagine the situation with boxers who have been organised into subdivisions according to weight. Given that fights can only take place between boxers of similar weight, the correlation between weight and boxing success is therefore necessarily low. In terms of the cohorts of adoptive parents, it is as if they are all in the heavyweight division, because they have been selected to represent a rather special category of families, and so there is a rather low parent-child IQ correlation observed in adoptive families. But in the unlikely event that poor families were selected to adopt more often, then the correlation would presumably be much higher. We might also erroneously infer from all this that there is no inherent relationship between weight and boxing success, but we would be quite wrong. As psychologists Richardson and Norgate observe, the assumption 'that the adoptive situation approximates a randomized-effects design' is not supported by the evidence (Richardson and Norgate, 2005). So even though adopted children may correlate more individually with their biological parents than with their adoptive parents, as a group they might in fact be more similar to their adoptive parents than to their biological parents (Schiff and Lewontin, 1986, p. 179). 'Thus focusing on parent-child correlations at the expense of evaluating group mean IQ differences, as behavioural geneticists frequently do, can paint a misleading picture of the potential roles of genetic and environmental influences on intelligence' (Joseph, 2010, p. 599).

Many other environmental variants have been found to correlate with differences in IQ. For example, a study led by Christopher Eppig at the University of New Mexico suggested that infections may be a critical factor affecting the maturation of the brain, thereby impacting on the results from intelligence testing (Eppig et al., 2010). Eppig assessed WHO statistics on the number of disease years caused by the twenty-eight most common infections using data from 192 countries in Africa, Asia and the Pacific region. Data on intelligence tests were available from 142 countries, allowing the investigators to estimate that 68 per cent of the differences in mean test scores between the countries could be attributed to their levels of childhood infection. The correlation was the strongest of all the variables tested, including educational level, malnourishment and gross domestic product. The usual proviso that correlation does not equate with causation is important in this context, but the findings represent but one striking example, out of dozens that could be cited, of correlations between IQ, g and environmental factors.

A further observation, known as the 'Flynn effect', has also led to much discussion concerning the relative impact of genetic variation and varying environments on IQ measurements. The observation is that the average IQ in different countries has been rising steadily over the past half-century or more since IQ measurements started to become available (Flynn, 2012). For example, American average IQ has been gaining at a rate of 0.3 points per year over the past half-century. The same tendency has been observed around the world. South Korean children gained at double the U.S. rate over the period 1989–2002 (Flynn, 2012). Children in urban Brazil (1930–2002), Estonia (1935–98]) and Spain (1970–99) made gains similar to the U.S. rate of increase. As always in IQ testing, the results are only meaningful if the same or highly similar tests are used. Between 1952 and 1982 young Dutch males gained 20 IQ points on a test of forty items selected from Raven's Progressive Matrices, one particular widely used intelligence test which measures what is termed 'fluid intelligence', the ability to solve non-verbal problems on the spot without a previously learned method for doing so (often contrasted with 'crystallised intelligence', the ability to apply acquired knowledge). This is of particular interest since the Raven's test has been designed to be largely independent of cultural differences. The gains in the Dutch study could not be dismissed on the grounds that the cohort being tested was maturing earlier than its forebears.

Many theories have been put forward to explain the Flynn effect (Hunt, 2011, pp. 261–70). What is quite clear is that the changes are taking place far too quickly for genetic variation to play a role: half a century is a very short period in relation to the timescale required for significant changes in population genetic diversification. James Flynn himself prefers to explain the changes by invoking the Industrial Revolution as an 'ultimate cause', with the intermediate causes being 'probably its social consequences, such as more formal schooling, more cognitively demanding jobs, cognitively challenging leisure, a better ratio of adults to children, richer interactions between parent and child' (Flynn, 2012, p. 15). Flynn argues that the great gains noted upon use of the Raven's test suggests that people have learnt how to tackle problems less concretely and more abstractly over the years. Other explanations, including methodological bias, have also been suggested (Dodge and Rutter, 2011, pp. 194–200). The main point in our present context is that there is nothing 'fixed' about the average values obtained from intelligence testing on populations – powerful environmental effects are at work. From a practical perspective, U.S. courts have taken the Flynn effect seriously in terms of assessing the IQ values of those facing the death penalty (Flynn, 2012).

The Molecular Genetics of Intelligence

Many candidate gene studies have been carried out in earlier years in an attempt to identify variant genes that correlate with variance in IQ and g as assessed by psychometric testing, but none of the identified genes have been sufficiently replicated to achieve general assent as to their relevance (Deary et al., 2009). An exception is the apolipoprotein gene which has three different alleles, confusingly called alleles 2, 3 and 4. The frequency of allele 4 is about 15 per cent in the general population and about 40 per cent in those with Alzheimer's disease. The results translate into a sixfold higher risk for late-onset Alzheimer's disease for individuals who have one or two of these alleles, in turn correlated with lower IQ (Plomin et al., 2013a, pp. 182–3). The incidence of homozygosity among stretches of DNA also correlates negatively with cognitive ability (Joshi et al., 2015). Arguably such results are more to do with medical conditions than with IQ variation within the general population. In any event, candidate gene studies have given way to some big studies using GWAS and GCTA. The initial GWAS studies had the statistical power to detect as little as 0.01 of the variance, but in practice the associations with the largest effect accounted for less than 0.005 of the variance (Davies et al., 2011; Kirkpatrick et al., 2014). However, a GWAS study on 126,559 individuals testing the association of SNPs with educational attainment did report three SNPs that together 'explained' about 2 per cent of the variation in attainment levels (Rietveld et al., 2013). Further studies have focused on the GCTA approach using many thousands of subjects (up to nearly 18,000), revealing that up to 66 per cent of the heritability of general intelligence, g, could be attributed to the associated SNP markers identified in aggregate, but in no case did any single SNP reach statistical significance (Plomin et al., 2013b; Benyamin et al., 2014; Plomin and Deary, 2015). So the conclusions so far are that thousands of genetic variants contribute to the variation in psychometric test scores and that any one gene variant is unlikely to make more than a 0.1 per cent contribution to the variance in the scores, not unlike the findings from the GWAS studies on height described in the previous chapter.

The authors of one of the GCTA investigations cited previously comment in the introduction to their paper that '[f]ew discoveries would have greater impact than identifying some of the genes responsible for the heritability of cognitive abilities' (Plomin et al., 2013b, p. 562). Really? The field does face a significant 'so what?' challenge, at least as it presently stands, given that it now seems clear that the contribution of any one genetic variant to the heritability of intelligence is likely to be extremely small. At the same time one cannot but help admire the perseverance of those in the field who will now no doubt take the next obvious steps: exomic sequencing in cohorts of high IQ individuals, followed by whole genome sequencing as the costs continue to decrease (Yong, 2013); focused studies on mathematical geniuses who lie on the autistic spectrum; epigenetic studies on high IQ cohorts and so forth. The present trend in both behavioural and medical genetics is to find cohorts at the extreme end of a spectrum: people who are *really* depressed, people with super-high IQ (above 130) and so forth. But all the evidence on IQ and g at present suggests that the outcomes will generate data-sets of increasing complexity in terms of their ability to 'explain' variation in psychometric test scores. An optimistic view predicts that eventually genomic and epigenomic profiles may emerge that correlate with various cognitive abilities. Such findings would fit perfectly well within a DICI framework in which multiple complementary levels are integrated during development to generate behavioural traits. But equally, the data make clear that enthusiastic transhumanists who think that genetic engineering might be used to generate a cohort of more intelligent humans are living in cloud-cuckoo-land. In this respect the complexity of the genome remains its best defence against ill-advised manipulation. So what useful broader purposes might such research serve? We already know that the best educational systems are those that allow children with different capabilities to achieve their optimal potential. We already know that people vary in their capabilities to achieve certain goals, be they academic, economic, sporting, working with their hands or whatever. It is not clear that the identification of thousands of different genetic variants that may contribute to the variation in these various capabilities will be that interesting, given that the possible contribution of each one will in any case be tiny. The field might perhaps direct its energies more usefully to animal studies in which there is more scope to integrate the molecular and behavioural aspects into a coherent account during development and to computer modelling which allows the manipulation of complex systems in silico.

The Heritability of Religiosity

One of the sociological characteristics of the behavioural genetics community is that it has been generous in giving out its methodologies to all and sundry (Panofsky, 2014). This has led to a huge number of studies on the heritability of virtually any human trait one can think of, as has already been noted. This has been a mixed blessing. On one hand, it has encouraged researchers in other disciplines to become interested in genetics, no bad thing. But on the other hand, it has resulted in a sizeable cohort of papers in the literature coming from research groups with little understanding of genetics or of the complexities entailed in interpreting the heritability data. With all due respect to the humanities, this is particularly noticeable in papers coming from research groups whose grasp of the science is perhaps less than it should be. In practice, the estimation of heritability is very straightforward, at least as far as the traditional twin-study format is concerned and readily taught to anyone with a minimal grasp of statistics. The problem lies more with the interpretation of the results so obtained, an interpretation that is non-trivial for the reasons already outlined. Furthermore, the literature then becomes littered with heritability values reported for this, that or the other trait, typically without any follow-up studies. Measurement of the heritability of religiosity provides a typical example of such a trend.

If intelligence is tricky to define, then words like 'religion', 'spirituality' and 'religiosity' are ten times worse. Within the social sciences, religiosity is treated as a multidimensional construct, incorporating numerous elements, or domains (Holdcroft, 2006). Although these are generally given new names with each new author, they include beliefs, values and attitudes, practices including participating in services, ceremonies or rituals, knowledge of the religion, religious experience, personal faith or devotion and denominational or institutional loyalty. Religiosity thus covers a range of social, cultural, cognitive and behavioural elements and includes both how a person feels about or experiences their religion and how they act and behave religiously. Domains do not necessarily align perfectly – a person can hold religious beliefs without engaging in religious practices.

Others have come to the definitional questions more from the perspective of evolutionary psychology, defining religiosity as 'the mental ability to be religious' (Voland, 2009, p. 21). This is a much narrower definition and redefines religiosity as a biological capacity, rather than as an actuality. Under Voland's model, the phenotypic expression of religious behaviours, both thoughts and action, is called 'religiousness': 'the individually varying psychic and behavioural manifestation of religiosity' (Voland, 2009, p. 10). This terminology has been widely used in behavioural genetic studies, which generally use religiousness to refer to public or social domains such as church attendance and religious practices. Religiousness is also used to describe the domain of religious salience, that is, the importance of religion in a person's life as assessed by their actions (Koenig et al., 2005; Bradshaw and Ellison, 2008). The words 'religiousness' and 'religiosity' have often been treated interchangeably.

'Spirituality' is also sometimes assessed in the behavioural genetics literature. 'As the terms religiousness and spirituality have evolved over time they have acquired much more specific connotations. Currently, religiousness is increasingly characterized as "narrow and institutional" and spirituality is increasingly characterized as "personal and subjective"' (Zinnbauer et al., 1997). Overall, spirituality is generally described in terms of connection to, or a feeling of oneness with, the wider world, other persons and the transcendent or supernatural (Voland, 2009). Its overlap with religious practice seems to be highly culture-dependent (Hyman and Handal, 2006; Del Rio and White, 2012).

Behavioural genetic studies have tended to assess only one or a few religiosity domains, which are rarely the same from study to study (D'Onofrio et al., 1999). The most common domains assessed are religious practices, church attendance and religious conservatism or fundamentalism, followed by personal devotion and religious attitudes. More recent studies have tended to assess a larger number of domains, acknowledging that all are relevant and related (Bradshaw and Ellison, 2008; Vance et al., 2010).

In eight different published studies investigating the heritability of religiosity in adolescents (aged 11–18) using the standard twin method, all were in agreement that the heritability is essentially zero or so small as to be unlikely to be of significance, especially in younger adolescents. This is reflected in the meta-analysis of Polderman and team who grouped sixty-three separate twin studies on religion and spirituality, reporting that for age range twelve to seventeen the concordance between monozygotic twins was only very slightly higher than that between dizygotic twins.[6] The actual traits measured included religious beliefs, religious affiliation and practices such as church attendance, religious fundamentalism, a nine-point scale measuring 'religiousness', a forty-five-point scale measuring religious values and 'spiritual involvement' and religious salience (Boomsma et al., 1999; D'Onofrio et al., 1999; Winter et al., 1999; Koenig et al., 2005, 2008; Eaves et al., 2008; Kendler and Myers, 2009; Button et al., 2011). In contrast to these studies, measurements of religiosity in adults (over age 18) have consistently reported positive heritability values in twelve different studies[7], of which a representative sample are cited here (Truett et al., 1992; D'Onofrio et al., 1999; Vance et al., 2010). Most of the traits measured in the adolescent cohorts were also included in the adult cohorts, and in addition, some studies included extra traits, such as spirituality, personal devotion and religious conservatism. Heritabilities are, on average, between 30 and 50 per cent but, for some, domains are much lower (16%) or much higher (64%). There is inconsistency within domains. For example, religious conservatism, the most thoroughly assessed domain, was found to be approximately 60 per cent heritable by Vance et al. (2010), but only 16 per cent heritable in males by Truett et al. (1992). Similarly, personal devotion was found to be 46 per cent heritable by Koenig et al. (2005), but only 29 per cent heritable by Kendler and Myers (2009). Environmental effects on variance are equally as large as genetic effects in most domains. The data are inconsistent on whether church attendance and other external or public religious practices are heritable: (Truett et al., 1992) found non-significant genetic effects in their study, but (Bradshaw and Ellison, 2008) estimated the heritability of church attendance at 32 per cent, which is broadly equivalent to other domains. No study found a positive heritability for affiliation with specific religions or denominations (Kendler et al., 1997; D'Onofrio et al., 1999). Spirituality does not appear to have greater heritability than religious practice or attitudes. Average heritabilities for spirituality domains include 23 per cent for spiritual well-being (Tsuang et al., 2002), 37–41 per cent for self-transcendence (Kirk et al., 1999), 65 per cent for religious transformation (Bradshaw and Ellison, 2008) and 51 per cent for general religiosity

(Vance et al., 2010). These heritabilities are not significantly higher than the heritabilities for scales of religious fundamentalism or church attendance (Kendler et al., 1997; D'Onofrio et al., 1999). All the twin study data are subject to the normal methodological limitations of twin studies as discussed in Chapter 6. One problem with several of these studies is small sample sizes – samples only rarely achieve more than 1,000 twin pairs, and twin studies need very large samples (more than 10,000) to achieve really good statistical power; in consequence, confidence intervals tend to be very wide and often include 0, which renders some of the heritability values reported somewhat suspect on statistical grounds.

How does one interpret the fact that heritability of these various religious traits in adolescence are zero or close to zero, whereas in adults the heritability values are at least generally positive, even if one does not take the precise values that seriously? The usual explanation provided is that shared environmental effects on religious practices and values should be considerably higher in childhood and adolescence than in adulthood, because childhood behaviour is under parental control. It would thus be expected that genetic effects would increase variation in early adulthood as adolescents gain more independence and are able to make their own behavioural choices (known technically as 'niche picking'). Well, maybe – although no coherent theories exist as to why genetic variation might relatively quickly start to exert measurable effects in adulthood on the variation in such complex behavioural traits which, as already noted, are themselves difficult to define clearly. Since the epigenetically regulated gene expression profile changes during development, perhaps this amplifies the role of genetic variation on the trait in adulthood? Since variance in personality traits, as discussed in Chapter 6, are well established as displaying positive heritability values, correlations between certain personality types and particular forms of spirituality or religious practice might perhaps go some way towards explaining the results (Kandler and Riemann, 2013).

In science it is usual to seek simpler explanations for data in place of more complicated or implausible explanations. In this context we need to remember that the three correlation components $h^2 + c^2 + e^2$ together have to add up to 1. So when the children leave home and go their separate ways such that the influence of the shared environment c^2 dramatically declines, ipso facto something has to change and the correlation with genetic variation, the heritability, now begins to loom large. As it happens, there is a simple explanation for the heritability difference between adolescence and adulthood that need not involve genetic variation at all, at least not in any direct sense. It should be remembered that all that is necessary to generate a positive heritability value is for the covariance between pairs of identical twins to be more correlated than between pairs of non-identical twins with respect to a particular measurable trait (like church attendance). It is quite possible that the growing disparity in scores between the different twin-types in adulthood is due to the closer links and emotional bonds that are known to be more of a characteristic of identical as contrasted with non-identical twins once twins leave home. No one is a more 'significant other' in the life of an identical twin than their co-twin. For example, an interesting retrospective study from Germany based on 133 monozygotic and 60 same-sex dizygotic older twin pairs reported data on the degree of closeness between the different twin pairs over time (Neyer, 2002). There was much greater frequency of contact, emotional closeness and level of mutual support throughout adulthood reported by the identical as compared with the non-identical twins. Other studies have likewise reported the greater degree of contact of monozygotic as contrasted with dizygotic twins in adulthood (Lykken et al., 1990; Rose et al., 1990; Posner et al., 1996). It is not difficult to imagine how this ongoing interaction and greater emotional attachment throughout life might contribute to the measured greater correlation in traits of religious attitudes and practices between monozygotic as compared to dizygotic twins in adulthood. So one possibility is that the heritability of the variation in religiosity, at whatever age, is actually zero, and genetic variation is irrelevant to differences in religiosity traits. Two observations, however, argue against this conclusion. The first is that when a small sample of adult twins 'reared apart' were measured for five different aspects of religious interests, attitudes and values, heritability values of around 50 per cent were determined (Waller et al., 1990), data consistent with other similar studies measuring various facets of intrinsic and extrinsic religiousness (Bouchard et al., 1999; Segal, 2012, pp. 255–6). On the face of it, given their separation, the 'greater contact' explanation could not apply in such contexts. Having said that, the numbers of each type of twin pair involved in such studies are small and the number of years of separation ranged from zero to sixty-nine (separation time was arbitrarily set at zero for twin pairs who were reared in different homes but who had periodic contact during childhood) (Waller et al., 1990). For these and reasons discussed previously, data obtained from such studies are not as easy to interpret as might first appear. The second observation that might count against the 'greater contact' argument is that there is no heritability for religious affiliation in adulthood, whereas the greater social contact of monozygotic twins in adulthood might be expected to generate more conformity in such commitments.

A further curious fact usually ignored in reports on the heritability of religious traits is that if atheism is treated as merely the low end of the religiosity scale, then its heritability value will be identical. For example, a scale assessing church attendance is, at the same time, a scale assessing non-church attendance. If identical adult co-twins correlate very highly for church attendance (and they do), it follows that they also correlate very highly for non-church attendance. The statement 'church attendance is x% heritable' could equally easily be written as 'non-church attendance is x% heritable'. This is equally true of any of the religiosity scales used by the various twin studies. However, in the same way that the church attendance domain measures only a small facet of religiosity, so non-church attendance measures only a small facet of atheism. There could be numerous social reasons for not attending religious services, undertaking religious practices or observing religious holidays in addition to atheistic beliefs. But in any event, if the suggested non-genetic 'simple explanation' for the heritability of adult religiosity is correct, as it may well be, then any discussion of the heritability of religiosity or of atheism is likely to be irrelevant anyway.

Believe it or not, there have been attempts to identify specific genetic variants that associate with religiosity, but as with other candidate gene studies, no replicable 'hits' have been identified, an unsurprising result given the discussion so far. In *The God Gene: How Faith is Hardwired into Our Genes* Hamer (2004) claimed to have identified a polymorphism of a gene (VMAT2) which encodes a generic monamine neurotransmitter transporter and that supposedly associates with the trait of 'self-transcendence'. However, the study was never published in the peer-reviewed literature, nor has it ever been taken seriously by others in the field. It is claims such as those made by *The God Gene* that often lead to hyped up accounts in the media but which, by the same token, unfortunately bring the field as a whole into disrepute.

Heritability of Politics

The literature on the heritability of political attitudes and commitments goes back more than forty years, providing one includes broad social traits such as 'conservatism' and 'radicalism'. More recent research has focused on the measurement of specific political traits using the standard twin methodology. As with 'religiosity', it is a challenge to know exactly what to measure, and different studies have homed in on different political traits (Hatemi et al., 2011; Hatemi and McDermott, 2012). A useful list of papers and a historical overview are provided by Hatemi (2012) and Plomin et al.

(2013a, pp. 284–5). Studies have included: identifying the heritability of voter turnout (Fowler et al., 2008); exploring common genetic influence on attitudes and voter choice using multivariate twin models (Hatemi et al., 2007); utilising extended kinships and assortative mating to identify special twin environments and passive gene–environment covariation (Hatemi et al., 2010); conducting longitudinal twin studies (from children to adulthood) on attitudes (Hatemi et al., 2009) and exploring the covariance between personality, partisanship and intensity of political comment (Verhulst et al., 2012). Twin studies have consistently reported positive heritability values in the range of 48–76 per cent for traits such as political participation (Fowler et al., 2008), 'political sophistication' (Arceneaux et al., 2012), 'political interest' (Klemmensen et al., 2012) and 'foreign policy preferences' (Cranmer and Dawes, 2012). By contrast, actual political party identification was reported to be due mostly to shared environmental influences (Hatemi et al., 2009). In this respect the findings are rather similar to those for religiosity for which twin studies revealed no heritability for religious or denominational affiliation.

How can one interpret such findings? No one believes that there are 'genes for political preferences' or 'genes for conservatism' or 'genes for foreign policy preferences', despite some media headlines to the contrary. The reification of such concepts is problematic (Charney, 2012). The questionnaires used generally focus on twenty-first century notions of American 'conservatism' or 'liberalism', including attitudes to 'pyjama parties, nudist camps, computer music and horoscopes' (Alford et al., 2005). But such ideas are highly culture and time dependent. Maintaining conservative attitudes does not even mean the same between the United Kingdom and the United States, let alone Japan or North Korea or Saudi Arabia. The problem comes in treating a 'political attitude' as if it were something akin to height, body mass index or Type 1 diabetes. But it is not. The prevalence of Type 1 diabetes is 45 per 100,000 in Finland and 1 per 100,000 in Venezuela. It is a trait with a clear universal definition and highly dependent on differences in the polygenic and environmental compositions of different populations. A political attitude is simply nothing like that: it can mean something quite different depending on era and cultural context. As Charney comments, this becomes even more problematic when trying to interpret the reports of positive heritabilities for mobile phone use (including amount of time spent texting) (Miller et al., 2012) and consumer preferences for soups and snacks, hybrid cars, science fiction movies and jazz (Simonson and Sela, 2011). An approach which 'explains everything' runs the risk of ending up explaining nothing.

On a more positive note, clues about the interpretation of the heritability of political engagement traits arise from the fact that they correlate with many other facets of human personality, such as prosocial personality and behaviour (Cesarini et al., 2008), 'obedience to traditional authority' (Ludeke et al., 2013), openness to experience (Xu et al., 2013) and the 'need to evaluate' (Bizer et al., 2004). People who trust others more, a heritable trait (Sturgis et al., 2010), are more likely to join political organisations and civic associations (Uslaner and Brown, 2005). One study reported, with respect to the heritability of two key aspects of political orientation, that a 'substantial proportion of this genetic variance can be accounted for by genetic variance in personality traits' (Kandler et al., 2012). Other attempts have been made to move beyond correlations to causation. One study suggests that rather than personality traits causing people to develop political attitudes, covariation between the two is largely due to common genetic influences (Verhulst et al., 2012). A further study measured personality and political traits ten years apart in a large cohort of adult Australian twins, concluding that personality changes were not causal for differences in political attitudes over this time period (Hatemi and Verhulst, 2015). However, disentangling behavioural causes and effects in any biological organism is tricky, let alone when it comes to complex human behavioural traits, and it is not even clear what it really means to suggest that one complex trait is 'causal' for another: the fact that one trait changes, on average, before another trait in a population does not necessarily establish a causal relationship. For the moment it is perhaps safer to accept the conclusions of the great bulk of the literature in this field, namely, that the positive heritability values of political traits are most likely linked to the heritability of different personality traits (Friesen and Ksiazkiewicz, 2015). This is at least consistent with the data showing that, whereas personality differences develop in early life, political attitudes and preferences reveal no heritability until adulthood after twins have left home.

This last observation is again reminiscent of the data on religiosity. In one study, no heritability of political attitudes was found in adolescents, but measurements of heritability then increased sharply upon leaving home (Hatemi et al., 2009). This longitudinal study of twins involved collecting questionnaire data throughout childhood to reveal the developmental pattern of political attitudes, focusing on 'liberal-conservative orientations on a multi-item index of political attitudes'. During childhood and adolescence, individual differences in political attitudes were accounted for by a variety of environmental influences with the role of shared family environment, including parental socialisation, accumulating markedly between the ages of nine and seventeen. At the point of early adulthood (in the early 20s), for those who had left their parental home, there was evidence of a sizeable positive heritability value on political attitudes, which then remained stable throughout adult life. However, this was not observed in the population comprising those twins who continued to live at home from the age of twenty-one to twenty-five. The authors' conclusions were similar to the interpretation proffered for the comparable findings on religiosity: 'genetic influences are only expressed in early adulthood and only when powerful social pressures such as the parental environment are removed' (Hatemi et al., 2009). Again, it is tempting to speculate that it is the greater social contact between monozygotic as compared to dizygotic twins in adulthood after leaving home that could most simply explain such results. The evidence for genetic influence starting at about age twenty-one to twenty-five arose mainly from a substantial drop in the political attitudes concordant between dizygotic twins while the correlation between monozygotic pairs remained stable. The political attitudes of fraternal twins with less contact after leaving home might well drift further apart when compared to identical twins. Counting against this suggestion, as with the similar suggestion made for religiosity, are measurements of traits such as 'conservatism' and 'right-wing authoritarianism' based on twins 'raised apart', studies that generate positive heritability values (Segal, 2012, pp. 254–5). Furthermore, a method similar to GCTA, which does not depend on twin studies, has been used to assess a range of political attitudes, finding that measurements of immigration, crime and foreign policy attitudes generate positive heritability values (Benjamin et al., 2012); this is in contrast to the case with religiosity, for which, at the time of writing, this approach has not been used. It therefore remains an open question as to the extent to which decreased social contact amongst dizygotic twins in adulthood as compared to monozygotic twins may explain the observed heritabilities of political attitudes that are positive in adults, but zero in children and adolescents.

Given the challenges facing GWAS studies on other behavioural traits as described, it will comes as no surprise that the use of GWAS in an attempt to correlate genetic variants with political ideological traits has failed to identify any SNPs at an acceptable level of significance (Hatemi et al., 2014). Earlier correlations reported between candidate genes and political traits are now generally accepted as being false positives (Charney and English, 2012).

It is noteworthy that in the literature on the genetics of political attitudes and commitments, more than most, the language tends to drift imperceptibly from data and comments on variation in specified populations to inferences concerning the supposed 'strength' of different influences: for example, '[t]o study the relative strength of genetic and environmental influences on individual traits and behaviours, a range of methods have been developed based on the comparison of family members with varying degrees of genetic similarity' (Sturgis et al., 2010). Furthermore, the idea of heritability is often confusingly co-mingled with the idea of inheritance, which is something quite different, as in the title dealing with the heritability of political traits: 'Are Political Orientations Genetically Transmitted?' (Alford et al., 2005). The authors wish to reply in the affirmative, but their data concern heritability and have nothing to say about inheritance. Small wonder, then, that by the time such studies reach the media, the inaccurate inferences become even more distorted, an example already having been cited in the Introduction: 'genes could exert a pull on attitudes concerning topics such as abortion, immigration, the death penalty and pacifism' (Buchen, 2012). It is therefore worth reiterating that heritability studies justify no such inference. Studies on the variation of a trait within a population tell us nothing about the relative contributions of genes and environment within the lives of individuals, and even in a specific individual the term 'relative strength' has little or no meaning.

What Does It All Mean?

It should be noted that the overall attitude of this chapter towards the findings of behavioural genetics has been one of sympathetic caution laced with more stringent critiques where these seem justified. Many in the social and political sciences treat the approaches described here as a waste of time, but there seems no need to adopt such a negative position. The data are in general what might be expected given the biological framework that DICI highlights. Since the development of human traits is, according to this framework, 100 per cent genetic and 100 per cent environmental, any reductionist attempt to use an experimental approach to calve up the relative contributions to the variation found in populations with respect to a particular trait is bound to come up with various forms of 'both-and' answers. It seems perfectly plausible that traits such as measurements of intelligence based on psychometric testing pick up on variations in biological traits such as neuronal processing speed that are in turn dependent on a long developmental process in which multiple genetic, micro- and macroenvironmental effects have been integrated in ways described in Chapter 4 to generate neurosensory systems, of which the brain is a key component, that vary in their cognitive abilities in relation to such tests.

When it comes to more diffuse traits such as levels of religious involvement and political attitudes, there seems no good reason to exclude a role, in principle, for genetic variation during development in helping explain the variation in these traits in populations, albeit with the proviso that such traits are very likely to be linked to underlying 'core' facets of human personhood, most likely personality traits in some cases, that contribute to the measurements of 'upper level' more diffuse traits and attitudes. Traits that display clearly positive heritability values in adulthood but not in childhood should be treated with an extra dose of suspicion because the genetic explanations put forward to explain such findings fail to be supported by any coherent genetic theory having empirical support. The ways in which all the various levels of genetic, epigenetic and environmental development are so thoroughly integrated and entangled, as the DICI framework points out, are well illustrated by the discussion sections found in many papers from the behavioural genetics field. These will often point out that 'of course' adult monozygotic twins share their genes in common so are more likely to seek out similar interests and environments in adulthood, thereby reinforcing their similitude in comparison with dizygotic twins. The influence of the environment is thereby turned into an argument for 'genetic influence'. The point is not that such points lack plausibility, but simply that within a DICI framework they make little sense, given that a complex biological organism like a human being can be defined at any moment by the integration of all of its many components in interaction with many environments. What behavioural genetics seeks to do is to take a unified cake and cut it into proportional slices using the approach of population genetics – as an experimental approach that is a perfectly valid methodology, the kind of reductionist approach that science uses all the time, providing that interpretations are all about variation in populations and not about individual organisms. But it has to be remembered that it is the functioning of the organism as an integrated whole that, in the end, is of most interest to biologists – and indeed to human beings in general – and that questions about genetic determinism relate to the individual and not to the population.

As already emphasised, the question of the 'strength' of the genetic influence on a given trait is not something that can be inferred from population studies. A higher heritability value does not entail a 'greater influence' of genetic variation on a trait than a lower value. A variant gene or a small collection of variant genes, might, in principle, contribute a large amount of variation to a population with respect to a particular trait, and at the same time the genome of each individual might be 100 per cent necessary for the normal development of the 99 per cent of that trait that is invariant between individuals.

The field of behavioural genetics at present is somewhat fragmented into the 'true believers' who will defend traditional methodologies at all costs without really engaging in the critiques of the field and those who adopt a more self-critical stance, recognising that there is a great need for new ideas and new techniques if the field is to make progress. Closer interactions between biologists and philosophers could be very beneficial for all parties, and likewise more inter-disciplinary research between animal and human behavioural geneticists might help move the field forward. Investigating traits using many different experimental approaches in parallel can be informative. In the chapter that follows, an attempt is made to give an inter-disciplinary overview of the causes of a complex human behavioural trait – same-sex attraction – a composite story in which genetic variation represents but one possible contributory factor amongst many.

CHAPTER 9

Gay Genes?

Genetics and Sexual Orientation

> *Is homosexuality hard-wired?*
>
> John Maddox, Editor of *Nature*, 1991

In previous chapters the emphasis has been on the role of genetic variation in potentially explaining differences in traits in human populations. The focus here is broadened to provide a general review of the current academic literature on the aetiology of same-sex attraction (SSA), a review that includes genetic data but which aims to survey a wide range of personal, environmental and biological causes that have been proposed to explain this trait[1].

Many hypotheses about the causes of SSA have been proposed, which can be divided into three broad types of causes: environmental, biological and personal choice. Following some comments on the definition and measurement of SSA and related terms, the personal choice option is briefly discussed, then the environmental explanations, followed by the biological explanations in greater detail. However, in separating in this way the proposed causes, it should not be assumed that a single cause or causal chain is responsible for such a complex phenomenon as SSA. In reality, as will be discussed, it is very likely that many different causes are operating in tandem and that causes are operating in different ways across the cohort of same-sex attracted individuals, in ways that are likely to be gender- and culture-specific. Although considered under separate headings, all influences are in reality completely integrated within the life of a developing individual, meaning that no one should expect to find 'the' cause of an individual's same-sex attraction.

Defining and Measuring Same-Sex Attraction

Before reviewing the causes of SSA, it is important to clarify what is meant by same-sex attraction and acknowledge the complexities of definition and meaning in the field of sexual behaviour studies (Savin-Williams, 2006; Gates, 2011).[2] Sexual attraction refers to erotic desire experienced towards other individuals. Attraction is not a discrete variable and exists along a continuum, from attraction exclusively toward the opposite sex (OSA) to attraction exclusively toward the same sex, with attraction to both sexes equally in the middle. An individual's sexual attraction status is frequently measured with a Kinsey scale, which usually defines seven points ranging from 0 (exclusively OSA) to 6 (exclusively SSA). Although sexual attraction is continuous, in practice almost all the studies reviewed below that use Kinsey scales collapse the continuum into discrete categories to increase statistical power: OSA (0–1), bisexual (2–4) and SSA (5–6), or OSA (0–1) and SSA (2–6), which should be remembered when interpreting empirical data.

Accurately measuring SSA, either demographically or in experimental populations, is not straightforward. Firstly, attraction is often conflated with, or extrapolated from, measures of related but distinct concepts including sexual behaviour, sexual fantasy and self-identity. All these facets together contribute to an individual's sexual orientation. These facets do not always, or even frequently, exactly correlate, but interact in multiple ways in different individuals. For example, a person who experiences SSA may not engage in same-sex behaviour, and someone who self-identifies as heterosexual may well experience some degree of SSA. On average, nearly three times as many people report some degree of SSA than the number of people who self-identify as gay, lesbian or bisexual. Consequently, studies that measure concepts other than attraction should be used to make inferences about SSA with caution. Having said that, in many of the studies reviewed in this chapter, it is a reasonable assumption that same-sex behaviour and/or a homosexual identity accompany SSA (although the converse is not necessarily true).

Secondly, even where attraction is directly assessed, measuring instruments (usually self-report questionnaires or surveys) can differ enormously. Variables include how questions are phrased, the degree or frequency of attraction deemed sufficient to categorise as SSA, the number of possible options offered and the degree of anonymity. Different instruments can thus produce significantly differing results (Gates, 2011). The time period assessed (lifetime SSA vs. current or recent SSA) is critical, as sexual attraction can be a dynamic trait. Long-running longitudinal surveys in the United States and New Zealand have found that although the majority of individuals, between 80 per cent and 90 per cent, have a stable sexual attraction or sexual self-identity across their lifespan, a sizeable minority experience change in their sexual attraction status over time (Ott et al., 2011; Mock and Eibach, 2012). Changes occur in both directions (i.e., SSA

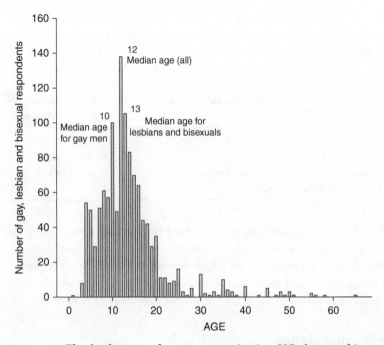

FIGURE 7. **The development of same-sex attraction in a U.S. demographic sample.** The majority of SSA individuals develop stable same-sex attractions in early puberty, with a median age of 10 for men and 12 for women. A minority of individuals first identify their same-sex attraction much later in life. From Pew Research Center, adapted with permission.[3]

to OSA and vice versa) across the lifespan (Figure 7), so it is incorrect to assume that individuals who report current OSA never have or never will experience SSA. In absolute terms, more individuals move from exclusive OSA to degrees of SSA than move in the other direction, although in percentage terms, the reverse is true. There is also a significant sex difference, with female sexual attraction status much more labile than in men. These data are important for interpreting the causes of SSA – the variation in when SSA develops in different people is suggestive that no one cause is sufficient to explain all forms of SSA.

Because of the difficulties of measurement, estimates of the prevalence of SSA, in either national or global populations, vary widely. Table 1 lists some recent prevalence estimates for the national UK population for SSA and behaviour, aggregated from multiple data sources. Across surveys some trends can be identified. Consistently, more women report experiencing some degree of SSA than men. Experiencing some same-sex attraction is more

TABLE 1 *Summary prevalence rates in the United Kingdom for same-sex attraction and behaviour. Data aggregated from multiple data sources. Rates are comparable to other Western societies including the United States and France. Very little data exist for developing societies. Data are derived from Savin-Williams (2009).*

	Men	Women
Any degree of same-sex attraction across the lifespan	6–8%	9–10%
Exclusive same-sex attraction	2%	< 1%
Any same-sex behaviour across the lifespan	5%	2%
Exclusive same-sex behaviour	1%	< 1%

prevalent in the population than undertaking same-sex behaviour, which in turn is generally more prevalent than having a non-heterosexual self-identify (although measures of identity are highly variable across surveys). The prevalence of having exclusive same-sex attraction or behaviour is much lower than the prevalence of having any degree of same-sex attraction or behaviour.

The Question of Choice

Before reviewing in detail putative biological and environmental aetiologies of SSA, it is necessary to briefly consider a third category – that of personal choice. Many would argue that an experience of sexual attraction, by definition, cannot be consciously chosen or willed, as attraction is a fundamental mental state that is not controlled by the conscious mind. Within this model, sexual attraction is a trait to be discovered within oneself, not created by personal choice. This 'standard model' (Wilkerson, 2009), so named because it is the model most commonly held by biologists, sociologists and the general public, underlies many survey instruments of sexual attraction (Savin-Williams, 2009) and popular writing about homosexuality.[4] It is not disputed by these commentators that individuals can make conscious choices about their sexual behaviour and sexual self-identity or that same-sex attraction, once acknowledged, could be strengthened or reinforced by behavioural choices, but attraction itself is thought to be inherent, a state of being that 'happens' to someone rather than being the result of a conscious choice.

However, the standard model is not universally accepted (Wilkerson, 2009), and it has been suggested by a number of groups over the years that SSA is a conscious choice and that individuals can will themselves to be attracted to persons of the same sex. Proponents of a choice-based hypothesis cite several potential reasons why an individual might choose to be attracted to someone of the same sex, including personal politics, restricted opportunity, sociocultural factors or revolt against cultural norms. For example, the lesbian feminist movement of the 1970s and early 1980s argued that women should choose to be attracted only to other women, as a rejection of heteronormativity and patriarchal oppression (Ellis and Peel, 2011). Within the gay community, a group known as Queer by Choice argue that their own experience of SSA is that they made a choice to feel that way (Madwin, 1999)[5]. Campaigners in parts of the United States and Africa portray homosexuality as a 'lifestyle choice' that can be altered simply by choosing to be attracted to the opposite sex. In fact, for the past thirty years between 30 and 40 per cent of respondents to American surveys have cited a choice aetiology when asked about the causes of homosexuality (Lewis, 2009). It is also possible that individuals who support a choice aetiology are eliding attraction and behaviour;[6] as already noted, it is not disputed that sexual behaviour is within a person's conscious control.

There is surprisingly little empirical evidence that directly addresses the choice hypothesis, but much of what is known about sexual attraction suggests that personal choice may be a causal factor for only a small – possibly very small – minority of male SSA individuals, if it can be a matter of choice at all, whereas the scope for choice in females may be somewhat greater. Systematic surveys appear to be lacking, but anecdotal evidence and informal surveys have found that the majority of SSA individuals report that they feel they were 'born gay' or simply became aware of a pre-existing and unconscious attraction (Dahir, 2001). It is noteworthy that the average age at which homosexuals report first thinking they were not straight is twelve (Figure 7); it seems unreasonable to suggest that adolescents of this age are all making conscious decisions to embrace an attraction that they may have seen derided in the popular press and society at large. Overall, if personal choice cannot completely be ruled out as a causal factor (which would involve denying the voices of those SSA individuals who feel that this does reflect their experience), it is unlikely to be the cause of SSA in the large majority of individuals, and there are, in any case, several other plausible aetiologies of SSA, as will be reviewed.

Environmental Causes

Psychoanalysis, Parenting and Phobia

During the middle part of the twentieth century, the most prominent aetiological explanation of SSA was derived from psychoanalysis. Beginning in 1940, neo-Freudian psychoanalyst Sandor Rado and his followers theorised that there was no innate psychological capacity for same-sex attraction (referring exclusively to male SSA) and that SSA resulted from abnormal parent-son relationships. The classic homosexual male was thought to have an emotionally hostile and distant father and a 'close-binding-intimate' mother, known as the 'triangular system'; the smothering relationship with the mother and the lack of a heterosexual male role model caused the son to develop a phobia of having sexual interactions with women (Drescher, 2008). A small flurry of studies during the 1950s and 1960s reported finding this type of triangular relationship in homosexual populations, most famously in Irving Beiber and colleagues' 1962 study of a population of institutionalised homosexuals in New York (Beiber et al., 1962). During this period, the triangular system became psychoanalytic orthodoxy and was incorporated into the first and second editions of the *Diagnostic and Statistical Manual* (DSM) published (in 1952 and 1968 respectively) by the American Psychiatric Association, which listed homosexuality as a 'sociopathic personality disturbance'.

Outside the field of psychoanalysis, however, the triangular system had many critics who highlighted a number of methodological flaws and unproven assumptions inherent to the psychoanalytic model. Sample sizes were always small, generally n < 50, and many studies did not use control groups. Sample cohorts were universally drawn from populations either in prison for homosexual offences or undergoing psychiatric treatment for homosexuality. This introduced a significant source of bias, as probands often presented with multiple co-morbid psychiatric problems. The finding that SSA was associated with mental disorder was thus a reflection of sample bias, not an indication of the underlying cause of the probands' SSA. In 1956, Evelyn Hooker conducted a study with a population of non-criminalised, non-institutionalised gay men in California (Hooker, 1957). She found that these SSA men were otherwise psychologically healthy. When two trained psychiatrists were asked to distinguish between the psychological profiles of these gay men and a group of matched heterosexual controls, they did no better than chance. Beginning in 1966, attempts to replicate the results of Beiber et al. in non-clinical and non-institutionalised populations almost all ended in failure. Only one study supported the triangular system in its entirety, while many others found small or no differences in family relationships between OSA and SSA men (Siegelman, 1974, 1981). In the largest study (n = 979 homosexual, 477 heterosexual), no differences were found between homo- and heterosexual men in terms of having either hostile fathers or overbearing mothers (Bell et al., 1981). In a

1974 survey, Siegelman found that the SSA men were more likely to have more emotionally hostile and rejecting relationships with their fathers only if they also scored highly for neuroticism (Siegelman, 1974), supporting the argument that SSA had been wrongly conflated with other psychiatric problems due to sampling bias.

In conclusion, there is no evidence to suggest that SSA in men is the result of a phobic avoidance of heterosexual relationships caused by abnormal parent-son interactions. Even in studies that seemed to support the theory, sizeable numbers of participants reported perfectly healthy relationships with their parents, and larger, unbiased samples repeatedly found no evidence to support this aetiology. By the 1970s, the case for the triangular model had collapsed, and in 1973 SSA was removed from the DSM-III. It was also removed from the directories of the American Psychological Association in 1975, the World Health Organisation in 1992 and the Chinese Society of Psychiatry in 2001.

Childhood Abuse and Experience of Trauma

A more recent proposal is the theory that SSA develops as a reaction to experiences of childhood sexual abuse. Most of the support for this aetiology comes from the growing body of cross-sectional and clinical studies, mainly from the United States, which report that both male and female homosexuals experience higher levels, as much as two- or threefold higher, of childhood sexual and physical abuse compared to matched heterosexuals (Balsam et al., 2005). A recent U.S. demographic study (n = 22,071) found significantly elevated sexual, physical and emotional abuse during childhood (up to age 18) in adult homosexual and bisexual men and women (Andersen and Blosnich, 2013).

Unfortunately, cross-sectional studies cannot distinguish the temporal order of correlated events, so the direction of causality is unknown. It is possible that experiencing childhood sexual abuse could be a contributing causal factor in the development of a same-sex attraction. If the abuser is the same sex as their victim, abuse could be interpreted as 'confirmation' that the person is homosexual, while an abuser of the opposite sex could provoke a traumatic avoidance of persons of that sex. However, it is equally possible that the effect is causal in the reverse direction. Almost all studies of childhood sexual abuse categorise childhood as occurring up to age sixteen or eighteen, well after the age at which same-sex attraction first appears in the majority of SSA individuals (see Figure 7). Abuse may be a violent reaction to a teenager's 'coming out', or in response to gender-atypical behaviour or appearance. Only two studies have used research designs that can assign a direction of causality, and neither is conclusive. One of these studies (n = 800) investigated historical criminal convictions for child neglect and abuse in children aged eleven or under (i.e., pre-pubertal) in a single U.S. city, then measured adult same-sex behaviour in abused children thirty years later. They found that men sexually abused in childhood reported a higher number of lifetime same-sex partners, but no increase in the likelihood of same-sex cohabitation (Wilson and Widom, 2010).

Much more work is needed to understand the links between childhood sexual abuse and SSA, and the present data are ambiguous. It remains possible that abuse in childhood could be causally responsible for the development of SSA in a small number of individuals, but it is clear that childhood abuse is neither necessary nor sufficient for the development of SSA, as the majority of SSA individuals do not report any form of abuse (Andersen and Blosnich, 2013).

Socialisation

'Socialisation' refers to the hypothesis that pre-pubertal children are 'blank slates' with regard to sexual attraction and that attraction is acquired or learned from sociocultural cues and interactions with parents, siblings, peers, mentors, role models and the wider culture. This theoretical model assumes that SSA develops unconsciously and uncontrollably when an individual is exposed to examples of same-sex attraction and behaviour, presumably via a positive reinforcement mechanism.[7] There is very little empirical evidence to support the socialisation hypothesis and the literature is small. A single study of SSA men with gay brothers found that 83 per cent experienced SSA before becoming aware of their brother's orientation, suggesting that, in males at least, sexual attraction is not acquired from siblings (Dawood et al., 2000). A more recent study found that sexual attraction (though not behaviour) is not 'learned' from adolescent peers and that same-sex attracted adolescents do not predominantly associate in peer groups with other SSA individuals (Brakefield et al., 2014).

A slightly larger number of studies in the United States and the United Kingdom have investigated whether the offspring of homosexual parents are more likely to be same-sex attracted due to parental imitation. Four small quantitative studies looking exclusively at lesbian-headed families have been carried out (Gartrell et al., 2011). Because lesbian parenting is a relatively recent phenomenon, these studies have been confined to adolescents and young adults (15–23 years old). Overall, these studies have concluded that there are no reliable differences in sexual attraction or behaviour between offspring raised by lesbian mothers and offspring raised in mother-father households. However, daughters of lesbian-headed families appear to be significantly more open to possible future same-sex behaviour and more likely to self-identify as other than exclusively heterosexual. It is unknown how far these results are limited by the age of participants – sexuality is known to be particularly fluid in females during adolescence, making it hard to generalise these results to stable sexual attraction in later life. There is thus no evidence currently to suggest that SSA is causatively influenced by parental example, although the small numbers involved in such studies render this conclusion preliminary.

Outside of family and peer networks, no evidence exists about the role that broader cultural factors, like increasing visibility and viability of same-sex relationships in the media or professional sports, might play in the development of SSA. Most studies in this field do not investigate attraction but rather behaviour or identity. In these contexts, it is difficult to distinguish the socialisation hypothesis from the effect of social change. The hypothesis requires that individuals are exposed to mainly positive examples of same-sex relationships to develop SSA, but a society where these examples are available is likely to be a safe environment to publicly identify and 'come out' as homosexual, increasing the number of people who will appear as SSA in surveys and studies. It cannot be argued that a positive correlation between the prevalence of SSA and a liberal society is evidence for a causal connection between the two.

In conclusion, the evidence that socialisation via siblings, peers, parents or culture is a cause of SSA is weak, although there are some dissenting voices to this general conclusion.

Biological Explanations

Genetics – Twin Studies

The theory that SSA is caused by a variant gene or variant genes has become a prevalent hypothesis in the past two decades – 44 per cent of respondents in one U.S. survey mentioned genetics as a cause of SSA (Sheldon et al., 2007). The 'gay gene' hypothesis crops up frequently in media and popular science articles, in song lyrics and at Gay Pride marches. At least two science fiction books take the existence of a 'gay gene' as a central premise.[8] Yet despite its pop-culture prominence, it is clear that there is no single gene for SSA. Children of homosexual parents do not inherit SSA in the strict Mendelian ratios we would expect if this trait were controlled by a single gene, and identical twins are only partially concordant. In fact most studies find that identical twins are significantly more highly correlated than non-identical twins with respect to SSA, albeit with a concordance of less than 50 per cent (Alanko et al., 2010; Langstrom et al., 2010). Systematic twin studies on SSA and sexual orientation date from the early 1990s, and a substantial number have been carried out. An influential early pair of twin studies found heritabilities of 40–70 per cent for both men and women (Bailey and Pillard, 1991; Bailey et al., 1993); these estimates were subsequently reported as 50 per cent in substantial media coverage. However, these studies were potentially heavily affected by bias – they recruited volunteers through gay-friendly publications, making it more likely that concordant pairs of twins would sign up. Follow-up studies have therefore used population-based twin registries from Australia, Sweden, Finland, the United States and the United Kingdom to generate systematic cohorts with typically several thousand participants. These studies have used a range of different measures of attraction, behaviour and identity in male and female cohorts. Overall, twin studies have found significant heritabilities for all the different facets of sexual orientation, including SSA, but with substantial differences in estimated heritability values between studies (Kendler et al., 2000; Alanko et al., 2010; Langstrom et al., 2010; Burri et al., 2011). In both men and women, heritability ranges between around 15 per cent and around 50 per cent. Studies disagree on whether particular facets of sexual orientation are more heritable in men or in women. Heterogeneity of the measured variable is likely to be a key component in the poor reproducibility of these data (i.e., studies are measuring different constructs with different experimental methodologies). However, at face value, these studies point to environmental influences as the largest contributing factor to the variation in specific populations with regard to SSA and related constructs. It is important to remember in this context that twin studies define any non-genetic but nonetheless biological factors, such as hormones, as 'environmental', so this finding does not rule out non-genetic biological aetiologies. As usual in twin studies, various caveats also need to be kept in mind. For example, the usefulness of data from these twin studies is somewhat limited by their very large confidence intervals, which often include 0 per cent heritability as the lowest boundary. Of particular relevance in the present context is the fact that the fetal environment will be experienced more similarly by identical twins sharing the same placenta (about two-thirds of all identical twins) than by twins with individual placentas (one-third of identical twins and all fraternal twins). This has particular significance for hormonal theories of SSA, as will be discussed.

Genetics – Specific Genes

A small number of molecular genetic studies have investigated whether any specific genetic variants are associated with SSA; although no single genetic mutation is causally responsible for SSA by itself, multiple variants, potentially several hundreds or thousands, could be influencing SSA in combination. Furthermore, some preliminary data have been reported on epigenetic differences between male twins discordant for homosexual orientation (Balter, 2015). As with parallel studies on other behavioural traits, as already discussed, research in this area has been contradictory and beset by problems of reliability, replication and validity.

The first molecular genetic study on SSA was published by Dean Hamer and colleagues in 1993 (Hamer et al., 1993); the paper attracted widespread press and public attention and can be credited with bringing the idea of a 'gay gene' into public discourse. Hamer et al. analysed twenty-two genetic markers on the X chromosome in forty pairs of brothers, both of whom were same-sex attracted. Five markers in a region known as Xq28[9] were present in thirty-three of the pairs (83%), suggesting that a gene or genes in this region could be contributing to the brothers' shared SSA. The authors were careful to acknowledge that the study had several limitations, including a small sample size, a sample that was overwhelmingly white and a very stringent definition of SSA. Furthermore, the region of interest identified is several million base-pairs in size and contains more than one hundred genes, any one of which could be the gene of interest. Hamer et al. were also careful not to discount potential environmental factors or to argue that the putative gene was causal in all instances of male SSA (seven of the pairs were discordant for at least one of the five markers). Thus, this first paper did not pinpoint any specific genetic variants contributing to SSA. Nevertheless, Xq28 rapidly came to be known in press reports as 'the gay gene' (Kitzinger, 2005), despite the fact that it is a spatial signifier, not an identified gene, and despite the fact that the study results were preliminary, unreplicated and far from conclusive. Hamer's subsequent book entitled *The Science of Desire: The Search for the Gay Gene and the Biology of Behaviour* (Hamer and Copeland, 1994) only added to the confusion.

A small number of studies have tried to replicate Hamer et al.'s original findings for the X chromosome, but these efforts have mainly ended in failure. Although the original research group managed to replicate their results in an independent cohort (n = 33 pairs), two other studies using the same methodology with larger samples failed to replicate (Rice et al., 1999; Mustanski et al., 2005). One recent linkage study (n = 409 pairs of brothers) has reported finding association with male SSA and a marker in the Xq28 region (Sanders et al., 2015). In addition, pedigree studies, which measure SSA across families, have found inconsistent support for the hypothesis that male homosexuals in one family are related to each other through the maternal line, thereby sharing a genetic variant located on a shared X chromosome, as would be expected if one or more X-linked genes contributed to the trait (Schwartz et al., 2010; VanderLaan et al., 2013). The reason(s) for these various contradictory data are currently unknown.

Four studies have conducted GWAS to locate genetic variants related to sexual orientation; however, none of these have produced any strongly significant or replicated findings. Mustanski et al. (2005) reported a nearly significant correlation between sexual orientation and the 7q36 region in a medium-sized study, but this result was not replicated by another group (Ramagopalan et al., 2010). Sanders et al. (2015) reported a significant association between the 8q12 region and male SSA. In addition, the personal DNA testing company 23andMe conducted a GWAS using its customer database (n = 7887 men; 5,570 women) but found no significant correlation between having a homosexual identity and any genetic marker (Drabant et al., 2012).[10] In addition, three candidate gene studies have been carried out in male populations: two studies on genes involved in sexual differentiation, aromatase cytochrome P450 and the androgen receptor gene, yielded no significant results (Macke et al., 1993; DuPree et al., 2004), but a study conducted in China found significant correlation between a variant of the 'sonic hedgehog' gene, which is involved in morphological patterning and self-identified homosexuality (Wang et al., 2012). However, these results have not as yet been replicated and are limited to a single population, therefore it would be premature to generalise these findings.

In conclusion, no single variant gene, or even genetic region, has been consistently associated with SSA in either men or women.

The Fraternal Birth Order Effect

As far back as the 1930s, it was hypothesised that SSA in males might be related to sibling number or the order in which children are born. The few small studies to investigate this hypothesis during the twentieth century produced inconclusive results, but since the 1990s a Canadian research group has carried out a series of studies into birth order effects on sexual attraction in multiple cohorts. These have consistently found that younger brothers in a family of boys are significantly more likely to be SSA than their older brothers, independent of socioeconomic status, maternal age and overall family size (Bogaert and Skorska, 2011). The total number of siblings is not significant, nor is the effect seen in women; consequently, this phenomenon is termed the fraternal birth order (FBO) effect. These results have been replicated in independent cohorts.

The data are best explained by a theory known as the 'maternal immune hypothesis'. It is known that the likelihood of an individual being SSA increases as the number of older brothers increases (i.e., third sons are more likely to be SSA than second sons). The increasing severity of the phenotype is analogous to an immune response following infection. This led researchers to suggest that male fetal cells carrying Y-chromosome encoded antigens (known as 'H-Y antigens') cross the placental barrier into the mother's bloodstream, provoking an immune response in the mother (because she carries no Y chromosome). In this theory, the antibodies to the Y-chromosome specific antigen(s) then pass back into the fetus and affect either the developing fetal brain and/or the fetal genome and/or fetal epigenome. With each successive male pregnancy, the number and binding efficiency of maternal antibodies increases until it crosses a threshold sufficient to influence the development of SSA. Support for the maternal immune hypothesis comes from the fact that having older sisters (who would not provoke such an immune reaction in the mother in this scenario) does not increase the likelihood of SSA in younger brothers while, at the same time, miscarried male fetuses do increase the likelihood of SSA in subsequent male pregnancies, suggesting a prenatal effect (Ellis and Blanchard, 2001). Furthermore, a study of boys raised with adoptive siblings found that the number of non-biological older brothers had no effect on SSA rates, while the number of biological older brothers did, again suggesting the effect is prenatal (Bogaert, 2006).

However, there are significant problems with this hypothesis. It has never been directly shown that male fetuses can provoke a relevant maternal immune response (although it is known that organs from male donors transplanted into female recipients can cause such a reaction). The mode of action of any putative maternal antibody is also unknown – the uncertainty surrounding other genetic, hormonal and neurological causal pathways to SSA makes elucidation of a presumed mechanism difficult. It is also not clear why the FBO effect is not a universal effect, as the vast majority of younger sons are opposite-sex-attracted. It is possible that birth order acts as a risk factor in combination with another causal mechanism (environmental or biological), but how such interaction might work remains unknown. Furthermore, numerous studies have shown that some first-born males report being SSA. Theoretical attempts to calculate the percentage of the homosexual population who could attribute their SSA to the FBO effect have yielded estimates of between 15 and 30 per cent, which is non-negligible, but also represents a minority of homosexuals. In conclusion, more data are needed, especially data relating to possible molecular mechanisms, before the maternal immune hypothesis can be accepted as valid.

Sex and Gender Atypicality

Humans are sexually dimorphic both physiologically (different sexes) and behaviourally (different genders). Since at least the 1860s it has been hypothesised that SSA arises as a result of deviation from the sex or gender developmental norm – in other words, that SSA males are feminised, and SSA females are masculinised, and thus display a sexual behaviour at variance with their chromosomal sex. This 'intersex'[11] hypothesis is still widely referenced today by the general public; popular stereotypes in both Western and non-Western countries suggest that adult homosexuals are atypical in their physicality, mannerisms and social behaviour, as well as in their sexual behaviour.

Evidence that homosexual men and women are physically sex-atypical has been limited and contradictory. On the one hand, there is no suggestion that any more than a tiny minority of SSA adults are genitally atypical, and no consistent evidence that they differ in height, weight or other physical attributes. On the other hand, a U.S. research team has demonstrated in several culturally diverse groups that an individual's sexual orientation can be accurately assessed at a rate above chance based purely on their physical appearance, speech and movement displayed in short (10s) video clips (Rieger et al., 2010). This assessment of sexual orientation was partially mediated by a judgement of atypicality (i.e., more sex-atypical persons were judged to be homosexual, often correctly), suggesting that some physical differences may exist between SSA and OSA individuals. However, the authors were careful to point out that the prediction rate was not 100 per cent, and sex atypicality is not a necessary trait associated with SSA. Studies by Richard Lippa (2008) and others have also argued that adult homosexuals, both male and female, are more likely to be judged by others as to be gender-atypical for a range of personality traits and to self-identify as more gender-atypical on a variety of measures of masculinity-femininity. The link is even stronger in childhood; numerous studies since the 1960s have demonstrated a robust correlation between gender-atypical behaviour in pre-pubertal childhood and higher likelihood of adult SSA (Zucker et al., 2006). Indeed, this is one of the strongest findings in the field of sexual orientation research, with no contrary findings reported in the past two decades for either male or female cohorts. The majority of studies have been retrospective designs, which are often thought to be subject to recall bias, so it has been suggested that adult homosexuals may falsely remember higher levels of gender-atypicality given their later knowledge of their same-sex sexual attraction. However, both prospective and non-recall studies have produced similar positive findings (Steensma et al., 2013).

Nevertheless, these behavioural data need to be interpreted cautiously. In adults, correlations between sexual attraction status and gender atypicality are generally only weakly significant. The frequent use of small non-typical clinical populations, particularly for children and adolescents, weakens the generalisability of the findings. The link between the most extreme form of gender atypicality, gender dysphoria and SSA is still unclear; it might be expected that gender dysphoria would always be accompanied by same-sex attraction (in terms of the chromosomal sex of the individual), but cases of opposite-sex attracted gender-dysphoric individuals have been documented. Moreover, unlike physical differences, gender presentation is bound up with cultural norms, peer pressures and unconscious social expectations and interactions, making it difficult to draw a direct connection between gender atypicality and a general causative biological difference between SSA and OSA individuals. For example, observational studies in several non-Western cultures have described communities of SSA males who are highly feminised in appearance and behaviour; it has been suggested that in these cultures, atypical sexual behaviour is seen as non-deviant and tolerated only if accompanied by other gender-atypical forms of behaviour, forcing SSA individuals to adopt atypical gender presentations (Lawrence, 2010; VanderLaan et al., 2013).

Overall, although the evidence is somewhat inconclusive and needs further study to tease out potentially confounding factors, it appears that sexual attraction and sex/gender atypicality are correlated in at least a substantial cohort of SSA individuals, and thus it is plausible that atypical masculinisation or feminisation is causal for same-sex attraction in some individuals. In principle, the underlying cause of this correlation could be either psychosocial or biological, or indeed both. The psychosocial explanation is lacking in empirical support and is not further discussed here. Biological causation has been primarily investigated in two interrelated areas: sex hormones and neurology, as will be discussed. As with other aetiologies, causal models based on gender atypicality are neither necessary nor sufficient to explain all SSA, as many SSA adults are gender-typical throughout their lives, and the majority of gender-atypical children go on to develop a heterosexual orientation.

Sex/Gender Atypicality – Hormones

The development of primary and secondary sexual characteristics is controlled by hormones known as 'androgens', such as testosterone, dihydrotestosterone and androstenedione, which are produced by the adrenal gland and the sex organs. Both men and women produce androgens, but men typically have a much higher androgen concentration, which is ultimately caused by the presence of the *Sry* gene located on the Y chromosome. In the vast majority of adults, androgen concentration does not differ between adults who are OSA or SSA. SSA males are not testosterone deficient, nor are SSA females testosterone enhanced. It is therefore not possible to 'treat' SSA with hormones in adulthood, even if it were ethical, and historical uses of hormone treatments were shown to be ineffective.

It is frequently hypothesised that exposure to atypical concentrations of androgens in the womb during fetal development is causal for both atypical sex and gender characteristics and same-sex attraction in childhood and adulthood. However, attempts to demonstrate a correlation between fetal hormone concentration and any form of sex/gender atypicality have been hampered by difficulties of measurement (Berenbaum and Beltz, 2011; Hines, 2011). It is impractical to directly assess fetal blood hormone concentrations routinely (fetal blood-sampling is high-risk and invasive), so a variety of proxy methods have been used to measure fetal androgen exposure, some more valid and reliable than others. These include direct measures of androgen concentration in amniotic fluid and maternal blood serum and the use of the 2D:4D finger ratio (the ratio between the length of the second and fourth finger, a sexually dimorphic trait that appears by the end of the first trimester). These studies have been marked by inconsistency. While a sizeable number report significant correlations between estimated androgen concentration and measures of SSA and other forms of gender atypicality, an equal number report no differences between SSA and OSA individuals, or even findings in the inverse direction.[12]

The most useful data on fetal hormone concentration have been derived from 'natural experiment' studies – rare disease conditions in which androgen concentration is altered by mutations in androgenic hormone or hormone receptor genes. Women with congenital adrenal hyperplasia (CAH), approximately 1 in 5,000–15,000 live births, have deficiencies in cortisol production which causes androgen concentration to be significantly higher than typical for females. Around a dozen studies across multiple age ranges have found that women with CAH are significantly more likely to report experiencing same-sex attraction or fantasy than matched controls (Frisen et al., 2009; Meyer-Bahlburg et al., 2008). The rate of lifetime SSA is much higher in CAH women, between 15 and 40 per cent, compared to the heterosexual control rate of around 10 per cent; this entails that well over half of all CAH-affected women are OSA. In addition, more severe forms of CAH (which raise androgen concentration higher) are associated with higher rates of SSA than less severe forms. This finding is highly reliable, although almost all these studies have been conducted in Caucasian populations and may not be generalisable to other populations. CAH is also associated with atypicality in gendered behaviours in both childhood and adulthood, including toy choice, drawing, interest in rough sports and occupational preferences.

The inverse condition to CAH is androgen insensitivity syndrome (AIS). Complete AIS occurs when the androgen receptor gene is mutated, entailing that although androgen levels are high, they cannot be detected by the body. Various other mutations also lead to reduced ability to detect androgens, causing partial or mild AIS. In complete AIS, male genitalia and secondary sexual characteristics fail to develop, and XY individuals are raised as females, with diagnosis not occurring until puberty. Only two studies of AIS and sexual orientation in adulthood have been performed: both reported that the vast majority of AIS women are attracted to men, meaning that in terms of their chromosomal sex they are same-sex attracted (Hines et al., 2003).

Although they seemingly provide the strongest evidence that hormone concentration is correlated with sexual attraction status, particularly in women, natural experiment studies of this kind are limited by the fact that it is very hard to disentangle hormonal effects from social effects on gender presentation and sexual behaviour in CAH- and AIS-affected populations. CAH women are physiologically distinct from other women, being on average shorter and heavier, and usually have partially formed male genital tissues that are corrected with surgery and hormone treatment at birth. CAH women thus look 'unfeminine', which could prompt feelings of greater masculinity, as could social expectations that CAH women are 'butch'. Furthermore, CAH women tend to find heterosexual sex difficult or unpleasant due to the effects of surgery, which could increase homosexual fantasy, the most commonly measured variable in CAH studies. AIS individuals are even harder to assess. Although their chromosomal sex is male, complete AIS sufferers are phenotypically indistinct from XX women, have female genitalia, and are raised with societal expectations of women. Labelling these women as same-sex attracted when they do not think of themselves as male is problematic.

Overall, the evidence that atypical prenatal hormone exposure underlies either SSA or gender-atypical behaviour is limited by the difficulty of assessing fetal androgen concentration. Results using proxy measures are unreliable and inconsistent, and rare disease conditions, whilst seeming to support the hypothesis, are not clear-cut due to possible social effects. It is important to emphasise as well that all these studies are purely correlational, not mechanistic; the way(s) in which androgenic hormones potentially affects the development of sexual characteristics and behaviours, presumably via a combination of genetic, epigenetic and/or neurological effects, is still unclear. It is thus difficult to draw firm conclusions from these data, although it does seem that any hormonal effect on SSA is more readily observable in women than in men.

Sex/Gender Atypicality – Neurology

The human brain is sexually dimorphic in structure, functionality and disease susceptibility (although, see Joel et al., 2015). It has been hypothesised since the early 1990s that SSA individuals may have brains that are sex-atypical, leading to attraction to the same sex, and a fairly small number of studies have investigated structural and functional differences, primarily in men.

Beginning in 1990, three now infamous studies using the technique of post-mortem brain dissection reported that homosexual men had significantly different volume or area measurement (compared to heterosexual men) in three specific regions of the brain: the anterior commissure, suprachiasmatic nucleus and a hypothalamic area known as 'INAH3' (Swaab and Hofman, 1990; Levay, 1991; Allen and Gorski, 1992). However, these findings have not stood up over time, as they had a significant number of limitations. All three studies used homosexual probands who had died of AIDS. Although attempts were made to control for possible impacts of the disease on brain structure, not enough attention was paid to the possible effects of pharmaceutical treatments or lifestyle factors like drug use. Same-sex attraction was not verified with the probands directly, but taken from hospital records. Studies that have attempted to replicate these results have failed to find any significant differences in these brain regions between SSA and OSA men (Byne et al., 2001; Lasco et al., 2002).

More recently, magnetic resonance imaging (MRI) has been used to map structure in living brain tissue in two studies. One study reported a significant difference in the volume of the isthmus region of the corpus callosum between homo- and heterosexual males (Witelson et al., 2008). The other study measured grey matter (GM) density in heterosexual and homosexual men and women and found only one strongly significant difference: homosexual women had reduced (male-typical) GM density in the left perirhinal cortex (Ponseti et al., 2007). This area is associated with olfactory processing, leading the authors to speculate that olfactory cues may be relevant to the development of SSA in women, but this remains speculative. These results have yet to be replicated and should be treated as preliminary.

In conclusion, the neural mechanisms involved in sexual attraction remain very uncertain, and it is incorrect to say that same-sex-attracted men have female brains, and vice versa. No consistent and reliable differences have been found that distinguish SSA and OSA individuals. Neurological studies are limited by small sample sizes, lack of replication and inconsistency in how sexual attraction is measured. The regions of the brain involved in sexual attraction are not clearly defined (in contrast to traits like speech and visual acuity), so any work involves some degree of groping in the dark. The ultimate causes of sexual differentiation of the brain also require further elucidation; both genes and hormones are known to influence neural development, but (as discussed previously) no specific genes have been associated with SSA that might work via altering neural development, and the relationship between androgenic hormones and SSA is still unclear.

A further crucial problem with all neurological studies is that they cannot distinguish between innate and learned effects. Because the brain is plastic and new neural circuits are continuously forming in response to experience, it is not unreasonable to suggest that the experience of feeling SSA, practising same-sex behaviour and associating with different groups of peers will have a neurological impact. Therefore, it is impossible to conclude that such differences are causal for SSA, as the exact reverse might be the case.

Conclusions

As surveyed here, there are a wide variety of hypothesised causes for same-sex attraction. Some causal models have received more empirical support than others, but no cause has yet gained sufficient support to provide a compelling explanation. In men, genetic influences on same-sex attraction cannot be ruled out, although any genetic variants that do influence SSA will be of small effect, and there is certainly no single 'gay gene'. The fraternal birth order effect is well supported, but there is no firm evidence for an immunological explanation for this observation. In women, almost the only positive evidence comes from hormonal studies, suggesting that exposure to elevated androgen levels can lead to developmental changes causing an attraction to women, but these data need to be treated with caution. Certain causal pathways, such as those resulting from childhood abuse, may be significant for a small percentage of the SSA cohort. There is no positive evidence that same-sex attraction in either men or women is caused by socialisation effects, personal choice or poor parenting or by having a brain of the 'wrong sex'.

It may therefore be concluded that no one causal mechanism is both necessary and sufficient to explain the whole gamut of human sexual attraction. Sexual attraction is a highly complex trait, and it seems likely that across the variety of human sexes and cultures, different influences are more important at different times. Not all homosexual men will be carrying the same variant genes. Not all homosexual women are masculinised. The social and cultural environment in which people live is constantly changing, including friends and partners, together with personal motivations and aspirations, creating a complex system in which biological make-up is integrated with multiple environmental, social and cultural factors. Thus, there is no point in looking for *the* cause of same-sex attraction – it does not exist. This negative conclusion is important, because many assume that the aetiology of SSA is known and straightforward. It is not.

CHAPTER 10

Not My Fault?

The Use of Genetics in the Legal System

> *The safe development of eugenics is indeed assured by [our] insistence that we should not let application outstrip knowledge.*
>
> Robert Yerkes, in an address to the
> American Eugenics Council, c. 1922[1]

> *It is better for all the world, if instead of waiting to execute degenerate offspring for crime, or to let them starve for their imbecility, society can prevent those who are manifestly unfit from continuing their kind.*
>
> Justice Oliver Wendell Holmes,
> *Buck v. Bell* (274 U.S. 200, 1927)

> *Liberty means responsibility. That is why most men dread it.*
> George Bernard Shaw, *Man and Superman*, 1903

Epidemiological research estimates that 10 per cent of the families in a given society are responsible for more than 50 per cent of crime. Small wonder, then, that the topic of genetics continues to be actively discussed within the legal profession.

Some information concerning the genetics of aggression has already been discussed in Chapter 7. The aim of this chapter is rather different: to investigate the degree to which genetic variation is currently being used to assess criminal responsibility, and the allied question as to what this tells us about the legal profession's attitudes not only towards genetics but also towards human free will. Most publications addressing these questions provide examples from the U.S. legal system, which therefore dominate the discussion that follows.

Various types of evidence have been introduced in the courts as behavioural genetic evidence, including single-gene variant testing, neuropsychiatric evidence and family history evidence. Not all of these are strictly what we would call genetic evidence. The use of such evidence has been limited. A review of U.S. cases found only forty-eight examples during the period 1994–2007 during which behavioural genetic evidence had featured in the case (Denno, 2009), and a further thirty-three cases from 2007–11 (Denno, 2011). So, worldwide the total number of cases is unlikely to be more than a few hundred at most. However, it does appear that the rate of use of genetic arguments in the courts, in the United States at least, is steadily increasing. This is in sharp contrast to the attitudes expressed in 1992 when the National Institutes of Health withdrew its funding for a conference on genetics and crime out of concerns that the topic itself stood for eugenics and racism.[2]

Genetic evidence can be used in a number of different ways within the context of a trial. During the penalty phase of a trial (when guilt is determined), genetic evidence can be used to challenge the guilty mind requirement (mens rea) of criminal liability. Genetic evidence has also been used to support an insanity defence and to provide mitigating evidence in cases involving alcoholism, drug abuse and impulsivity. The most common use of behavioural genetic evidence in recent years (within the U.S. court system) has been as mitigating evidence during the sentencing phase of capital cases. The reaction of the courts to these attempts has been mixed. While in some cases such evidence has been considered mitigating, and sentences have been reduced, other courts have ruled that the evidence is not sufficient to overrule the aggravating factors and have handed down a death sentence. These points will now be reviewed in greater detail: first, the types of evidence considered as 'genetic' by the courts will be reviewed; second, the various types of legal defence in which genetic arguments have been invoked will be considered; third, the role of specific genes in defence arguments will be summarised; and fourth, some more general comments will be made concerning the attitudes of the legal profession towards genetics.

Types of Genetic Evidence Relevant to Criminal Trials

In the legal context there seems to be no universally accepted definition of the broad term 'behavioural genetic evidence', and what counts as 'genetic evidence' is not always clear. Different reviewers, and indeed participants within the legal system, have defined this type of evidence in various ways, some of which are outlined in the following sections.

Single-Gene Genotyping

The 'classic' genetic evidence involves genotyping the defendant for a particular genetic variant and arguing that it has been associated with a certain behaviour in a GWAS, pedigree study or G x E investigation. The number of specific genes that have been introduced into court cases appears to be very small, partly because the evidence is not always admissible, or has been mostly rejected when it has been used, partly also because it is relatively expensive to test a defendant (Appelbaum, 2014). The Monoamine Oxidase-A gene variant discussed in Chapter 7 is by far the most utilised single gene in a legal context and will be discussed further.

Family History/Pedigree Evidence

The most common form of genetic evidence is the family history form of argument (Denno, 2011). This type of evidence involves showing that the trait of interest in the defendant (usually alcoholism or drug abuse, but also violence or criminal tendencies) is present in multiple generations of the family. The fact that the trait appears to be inherited is considered as evidence of a genetic tendency to the same behaviour in the defendant. A typical case study of this is *United States v. Loughner* (2011). Jared Lee Loughner was an Arizona shooter who killed a federal judge and injured a congresswoman, and his defence team subpoenaed the medical records of four generations of his relatives (Denno, 2011). In this and other cases, medical records showing that the defendant had displayed the trait throughout their life have been cited as 'genetic evidence'. Although frequently referred to as 'genetic evidence', no genetic tests are carried out on the defendant and no specific genes are implicated in this category of cases. The legal interpretation of the meaning of 'behavioural genetics' tends to be much broader than found within the academic discipline of that name: 'Behavioral genetics evidence includes, but is not limited to, the study of the defendant's family history as well as direct testing of the defendant's physiological makeup by way of brain scans, for example. Both approaches are informative for explaining why a defendant engaged in criminally violent behavior' (Denno, 2011). Arguments made in court sometimes mingle genetic and neuropsychiatric data in a rather confusing way. For example, scans showing that defendants had frontal lobe damage or otherwise abnormal neural structure in the prefrontal cortex have been introduced a number of times and linked to a generalised 'genetic defect' (Farahany and Coleman, 2006). Other defendants have argued that the plaintiff had low serotonin neurotransmitter levels at the time of the crime, which rendered them less capable of controlling their actions, and this has become mingled with strictly genetic arguments (Farahany and Bernet, 2006).

Karyotyping

The karyotype is a way of measuring the number and integrity of the twenty-three pairs of human chromosomes, revealing in the process any detectable departures from normality. In conditions such as Down's syndrome there are three copies of chromosome 21 (trisomy 21), and hundreds of other chromosomal abnormalities have been reported in the literature. However, it is rare that any of these abnormalities are cited in a legal context. An exception is of historical interest only. In 1965 it was reported that aggressive males of low IQ with an abnormal XYY karyotype were found with a higher than expected incidence in a 'survey of mentally subnormal male patients with dangerous, violent or criminal propensities in an institution where they are treated under conditions of special security' (Jacobs et al., 1965); this report was followed up with confirmations from other similar institutions. However, many further studies showed that the supposed 'explanation' for the aggression by the XYY karyotype was quite fallacious and could best be explained by factors such as differences in socioeconomic status (Stochholm et al., 2012). In earlier years, however, this did not prevent the 'XYY story' taking on a life of its own. In the mid-1960s, the Chicago mass murderer Richard Speck was karyotyped under suspicion of being XYY with the aim of using the data to support an insanity plea. Speck's karyotype showed that he was the normal XY, and thus the evidence was never introduced into court. However, it attracted a large amount of media attention, and during the 1970s a number of defendants sought to introduce evidence that they were XYY, so-called super-males, and that the chromosomal imbalance meant that they had a predisposition to impulsive and extreme violence. In general, these attempts were thrown out before they even reached the courtroom, as it was (rightly) pointed out that being XYY is not deterministic for violent and illegal behaviour and that males who were XYY seemed to be capable of understanding the law and appreciating the wrongfulness of their actions. XYY data thus failed to demonstrate that defendants suffered from a mental defect. As it turned out, the court's scepticism was fully justified. Nevertheless, XYY data continued to be sporadically introduced until *State v. Thompson* (2004) finally ruled that an XYY karyotype was not considered sufficient grounds for an insanity defence, the judge arguing that the link between a potential behavioural predisposition and the defendant's actual mental state at the time of the crime was not sufficiently strong to provide solid evidence that the defendant was insane (Farahany and Coleman, 2006). Although of historical interest only, the XYY story provides a potent warning against taking genetic data seriously in the legal system until such time as the science is very well established and, in addition, relevant to the case.

Legal Contexts in which Behavioral Genetic Arguments Are Invoked

Behavioural genetic evidence can be used in a variety of ways during a trial. The principal distinction is between negating and mitigating evidence. Negating evidence is used to try and remove criminal liability altogether from the defendant by proving that the defendant did not possess the necessary mens rea. Mitigating evidence, by contrast, lessens but does not remove criminal liability from the defendant. Behavioural genetics evidence has also been used as support for an insanity plea. The case reviews and analyses that follows are based on Farahany and Coleman (2006), Farahany and Bernet (2006); Bernet et al. (2007) and Denno (2009, 2011).

Attempts to Negate the mens rea or 'Guilty Intent'

To be criminally liable, the accused must have committed the criminal act voluntarily: the behaviour must not be committed under duress, or in self-defence, or otherwise involuntarily or by compulsion. A genetic defence therefore needs to demonstrate that the act was not voluntary due to the genetic make-up of the person even if they committed the act. 'Voluntary' means not that the person necessarily wanted to do it, but that the act was carried out under the person's control. In the legal system, 'people are viewed as autonomous actors not because of a preference for arguments in support of theoretical free will, compatibilism, or determinism, or because it is empirically verifiable. Instead, the criminal law recognizes the autonomy of human choice as fundamental to the operation of a modern system of laws' (Levitt, 2013).

This form of genetic defence has been used a number of times since as far back as the 1970s, with the usual strategy attempting to demonstrate that the criminal behaviour was compulsive or impulsive, and therefore carried out involuntarily. The compulsion is said to arise from a genetic predisposition to some form of addictive or impulsive behaviour, which is out of the control of the actor. Because of their addiction, the free will of the actor is said to be compromised. The compulsion defence is most often used in the context of drug or alcohol addiction. It is not enough to demonstrate that the defendant was intoxicated or otherwise incapacitated at the time of committing the act as the initial act to consume alcohol or drugs is considered to be a free choice. The defence becomes relevant in trying to prove that the initial decision to become intoxicated was outside the defendant's control. In such cases, typically the defendant would not be genotyped for specific markers for alcoholism or addiction, but the defence is based on the present evidence of a family pedigree of alcoholism or addiction.

Overall the mens rea defence seems to have been entirely ineffective within the context of the U.S. court system. Courts have universally rejected this defence, as it would lead to a great number of people claiming that they are not criminally liable due to a predilection for drugs or alcohol. The main concern appears to be the lack of a direct, causal link between genetic predisposition to addictive behaviours and the final criminal act. A predisposition to a behaviour does not necessarily remove volition entirely, as is required for a successful defence: 'courts have recognized that behavioral genetics does not explain the causal relationship between a defendant's genetic profile and his behavior, nor does it provide a biological explanation of the defendant's criminal act' (Farahany and Coleman, 2006).

As Supporting Evidence for a Traditional Insanity Defence

It is well established in legal practice that criminal liability is avoided if the actor is insane at the time of committing the crime. According to the McNaughton (or M'Naghten) rules that govern insanity pleas in both the United States and the United Kingdom, an insane person must be unable to distinguish criminal from non-criminal behaviour. Although insanity is related to the mens rea defence in that it concerns the defendant's mental state at the time of the crime, it is subtly different, as an insane person may well be able to form a state of mind in which they intended to commit the criminal act but did not realise at the time that the act was wrong because of some form of cognitive impairment. So whereas a mens rea defence argues that the murderer did not know it was murder, an insanity defence argues that the person did know it was murder but did not realise or understand that murder was the wrong thing to do. An insanity defence also demands that the defendant suffer from a 'disease of the mind', an empirically demonstrable cognitive impairment that affects their daily actions and interactions.

As described, one of the earliest uses of genetic data within the court system was within the context of an insanity defence based on the (fallacious) claim that the murderer Richard Speck might have an XYY karyotype. But apart from this case, most genetic evidence is presented in support of the traditional methods for establishing insanity, such as clinical diagnosis and psychiatric assessment. '[C]riminal defendants have had some success using evidence of a genetic predisposition to bolster expert diagnosis of a mental condition. In light of current standards for admission of scientific evidence, this use of behavioral genetics will likely be of more immediate benefit to the criminal defendant than a theory of insanity grounded solely in a genetic predisposition to violent, aggressive, or antisocial behavior' (Farahany and Coleman, 2006). Success in this context may be because genetic data are viewed as more reliable, robust and objective than other forms of expert testimony, which could be viewed as subjective. For example, where an expert testifies that a defendant has schizophrenia based on the presentation of their clinical symptoms, that diagnosis would be supported by presentation of evidence that the defendant was carrying genetic variants that predisposed them to schizophrenia. This is the context in which future genetic evidence is likely to be presented, thereby addressing the problem of 'malingering', the fabrication or exaggeration of symptoms to obtain a favourable verdict. Indeed, faked psychiatric symptoms can be hard to detect with a clinical evaluation alone.

> The use of genetic and neuroimaging data cannot change the rationale underlying the determination of criminal liability; this must be based on a causal association between a mental disorder and a crime. However, the use of such data may be crucial in providing objective, biological data which can be used to reduce uncertainty in forensic psychiatric evaluations and validate symptoms as non-faked, thereby minimizing the risk that psychiatric symptoms are the result of malingering. In no way are we claiming that the presence of risk genes or neuronal abnormalities could be the basis of any mental insanity assessment in the absence of clinical manifestations; rather, our point is that this biological information may be useful in forming psychiatric assessment. (Sartori et al., 2011)

As Mitigating Evidence

The vast majority of behavioural genetic evidence that has been presented in U.S. courts has been presented as mitigating rather than as negating evidence and has mostly had to do with demonstrating a loss of control. The usual form of the argument is to suggest that the defendant has a genetic predisposition either to impulsive violence, or to addiction to drugs or alcohol, which substantially affected their volitional control and made them less capable of making rational behavioural choices. This argument strongly resembles arguments made when presenting a mens rea defence, but the crucial difference is that mitigating evidence only requires the defendant to show that their genetic predisposition renders them *less* capable of making different choices, rather than *incapable* of making them. The threshold for establishing non-responsibility is high, so most commentators are sceptical that genetic predisposition evidence can meet the legal standard for non-culpability, whereas the standards used for mitigation are more flexible: 'The usual basis for a claim for mercy in sentencing is that a genetic predisposition makes it harder for the defendant to control his behavior compared with other people, and thus he is not deserving of the most severe punishment for his offense' (Appelbaum, 2014). Thus, while juries and judges have overwhelmingly rejected genetics as a negating factor in criminal cases, they have been much more receptive to genetic data presented as mitigating evidence.

A comparison between the cases reported in Bernet et al. (2007) and the more recent cases in Denno (2011) suggests that courts have been becoming more receptive to the use of genetic evidence over the past decade. Whereas in early years, courts consistently refused to allow genetic evidence, as in *Mobley v. State*(1993), since 2007 U.S. courts are now happy to allow such evidence for a wide range of different conditions: 'substance dependency, alcohol dependency, mental illness, depression, mental retardation, bipolar disorder, schizophrenia, predisposition toward violence, propensity toward criminal behavior, sexual sadism, and family dysfunction' (Denno, 2011). Defendants are also relying more on genetic evidence as a primary source of mitigation, rather than as just one factor among many. Nevertheless it remains clear that behavioural genetics evidence is most effective when it is presented as part of an overall strategy of mitigation that calls on multiple other sources, including socioeconomic background, peer influences and so forth. It is also most effective when used to support mitigating factors that are well known and can be easily conceptualised by the jury, such as arguing that a genetic predisposition makes a defendant more likely to have an extreme reaction to provocation, as provocation is a well-established mitigating factor. '[T]he successful genotyping defense employs an interdisciplinary approach that relies on the genetic test, psychological examination, and prior history of abuse. Combining these two theories provides one potential mitigation theme. The theme would consist of describing psychopathy as a mental illness, disorder, or condition and using the genotyping evidence in conjunction with a myriad of witnesses to build a story of an individual who has always been afflicted with this problem' (Walker, 2013).

Mitigation in Capital Cases

It is one thing for a sentence to be reduced from seven years to five years based on various mitigating factors, including broad appeals to genetic data, but quite a different matter for a death sentence to be mitigated to life imprisonment. Within the context of the U.S. court system, at least, the overwhelming use of behavioural genetic data has been used as mitigating evidence in death penalty cases. Denno found just one non-capital case in a review of thirty-three cases from 2007 to 2011: 'It is striking … that behavioral genetics evidence is of significance nearly exclusively in death penalty cases, and it is applied in no case involving less than a life sentence. Thus, discussions of the effects of such evidence in the guilt-or-innocence phase of a trial, while conceptually important, are not directly applicable to situations where genetics factors are instead used as mitigation evidence in the penalty phase of a capital trial' (Denno, 2011). This bias is likely due to a peculiar feature of capital cases compared to non-capital cases. The list of mitigating and aggravating factors which can be presented to the jury during the guilt phase of a trial is fairly well established and relatively constrained. However, when the death penalty is considered, the number of mitigating factors permitted by the court is greatly increased; essentially, the defence team is allowed to introduce 'any aspect of the defendant's character or record' that they feel may have a mitigating effect. 'Whereas aggravating factors (e.g., multiple victims, murder of a child or police officer) are typically defined by statute, in most states mitigating factors are not limited in any way and can include any aspect of a defendant's character or record and any circumstances of the offense that the defendant proffers as a basis for a sentence less than death' (Greene and Cahill, 2012). Indeed, not only is such evidence allowable but it is also mandatory. Under the Eighth Amendment to the Constitution, the deciding opinion (either judge or jury) must be allowed to see all relevant mitigating evidence before passing sentence. In practice, this law has led to a number of appeal claims based on 'ineffective assistance of counsel' and other forms of post-conviction relief. A sizeable number of defendants who have been convicted and sentenced to death have claimed during the appeals process that, because behavioural genetic evidence was not presented during the original trial, the sentence is invalid, and have applied to have the death sentence vacated or the trial reheard. For example, during *Stewart v. Landrigan*(2001), Landrigan, a convicted murderer, filed a petition for 'ineffective assistance of counsel' on the grounds that his lawyers had not offered any form of genetic mitigating evidence (based on Landrigan's own instructions to counsel). 'Four years after sentencing, however, Landrigan argued that notwithstanding his instructions at trial, he would have cooperated had his attorneys attempted to offer mitigating evidence demonstrating that his "biological background made him what he is"' (Farahany and Coleman, 2006). It is not clear precisely what form that genetic evidence would have taken. The court of appeal denied Landrigan an evidentiary hearing for his behavioural genetic evidence on the grounds that it would not introduce any information that would have altered the verdict. The original Ninth Circuit panel was not persuaded that such evidence would have helped with Landrigan's sentence: '[We find it] highly doubtful that the sentencing court would have been moved by information that Landrigan was a remorseless, violent killer because he was genetically programmed to be violent, as shown by the fact that he comes from a family of violent people, who are killers' (Farahany and Coleman, 2006). This decision was upheld by a higher appeal court in 2007, who ruled that Landrigan's mitigating evidence was 'weak': 'Addressing Landrigan's alleged genetic predisposition to violence, for example, the Court found it "difficult to improve upon the initial Court of Appeals panel's conclusion" that Landrigan "not only failed to show remorse or offer mitigating evidence, but he flaunted his menacing behavior"; therefore, "assuring the court that genetics made him the way he is could not have been very helpful"' (Denno, 2011). Landrigan was executed in 2010.

Such cases opened the doors to a number of other appeals. According to Denno (2011), twenty-six cases between 2007 and 2011 applied to have death sentences struck down due to 'counsel's failure to present behavioral genetics evidence adequately'. This was 79 per cent of all the cases in her analysis that presented genetic evidence in that time frame, indicating that this rationale is a principal driver of the increased use of such evidence. Many of the cases combined this form of defence with an attempt to prove that the genetic evidence was also a genuine mitigating factor, arguing that not only should the death sentence be struck down but the punishment reduced as well. The reaction of courts to such 'ineffective assistance of counsel' claims has been mixed. Most have been dismissed on similar grounds to the *Landrigan* case – that the new evidence presented is not sufficiently compelling to alter the verdict. Some have also failed because the appeal court deemed that the original counsel was not ineffective and that defenders were able to present the genetic information sufficiently without the need for an expert. However, there have been a number of sentences that have been reversed based on new genetic evidence. In *Von Dohlen v. State* (2004), Von Dohlen presented evidence of a genetic predisposition to severe depression, which he said influenced his mental state at the time of the murder. 'In support of his claim for post-conviction relief, a psychologist testified that as a result of "his altered mental state" [the murder] was not a volitional thing but "out of his conscious awareness or control"'. On appeal, the court reversed the earlier court's denial of post-conviction relief, finding instead that the psychological testimony created a 'reasonable probability the outcome of the trial might have been different had the jury heard the available information about [the defendant's] mental condition' (Farahany and Bernet, 2006). In *Hawkins v. Wong* (2010), Hawkins petitioned for post-conviction relief on the grounds that his original counsel had 'fail[ed] to hire a social historian who could have explained how the defendant's background influenced his behavior.... The mitigating evidence would have shown that Hawkins was genetically predisposed to alcoholism and mental illness. A social historian could have also testified about Hawkins's family tree, which included many alcoholics and indicated a family genetic predisposition to alcoholism, as well as included a range of violent, abusive and mentally ill or handicapped persons' (Denno, 2011). Interestingly, it was the same Ninth Circuit Court of Appeal that a few years earlier had denied an evidentiary hearing to Landrigan who ruled in Hawkins' favour. Although this might represent an increased willingness by courts to allow genetic evidence, it might also be due strictly to the facts of the case – *Landrigan* featured a number of especially aggravating factors which meant that a jury was highly unlikely to return a non-death sentence, a situation that was not the case in *Hawkins*.

Behavioural Genetic Evidence as an Aggravating Factor

The examples in the previous two sections all discuss the use of genetic evidence as a mitigating factor. However, it is theoretically possible that genetic evidence could be construed as an aggravating factor. Rather than showing that the defendant is less culpable, the demonstration that the defendant lacks normal volitional control might lead a jury or judge to impose a harsher sentence on the grounds of 'future dangerousness' (an ugly phrase that will nevertheless be used here, as it is the phrase often used in the courts). The high likelihood of recurrence of the behaviour could be considered good grounds for imposing a death sentence or life imprisonment for all criminals so affected. In many U.S. states, future dangerousness is indeed considered as a statutory aggravating factor, although assessments of future dangerousness typically do not depend on any genetic evidence but are assessed through expert psychiatric testimony (Greene and Cahill, 2012). This has been called 'the double-edged sword' of genetic evidence.

The possibility of the double-edged sword has been noted by several courts. For example, in its decision to deny an evidentiary hearing in the *Landrigan* case, the Ninth Circuit Court of Appeal wrote 'although Landrigan's new evidence can be called mitigating in some slight sense, it would also have shown the court that it could anticipate that he would continue to be violent.... As the Arizona Supreme Court so aptly put it when dealing with one of Landrigan's other claims, "[i]n his comments, the defendant not only failed to show remorse or offer mitigating evidence, but he flaunted his menacing behavior." On this record, assuring the court that genetics made him the way he is he could not have been very helpful' (quoted in Farahany and Coleman, 2006). Similarly, in *State v. Creech* (1999), Creech submitted evidence in mitigation that he had a genetic propensity to impulsive violence. This evidence was used, along with evidence of Creech's violent history, to show that he possessed a high 'propensity to kill' (the standard aggravating factor), and a death sentence was duly handed down. The danger that genetic evidence could work against the defendant was also noted in the 2004 TV programme *Genes on Trial* in which expert geneticists, legal scholars and ethicists discussed the implications of behavioural genetics in the legal system (Rothenberg and Wang, 2006). However, despite these concerns, use of genetic evidence as an aggravating factor has not been forthcoming. Denno (2011) notes that genetic evidence has never been introduced by the prosecution in any real-life case she reviewed and, over the last decade, has not been mentioned by judges or juries as an aggravating factor in the manner of the *Creech* judgement. Denno theorises that this might be due to the increasing sophistication of the expert testimony in such cases, which refuses to paint a criminal as 'hardwired' to be always violent or aggressive; as has been pointed out, '[b]oth sides of the double-edged sword rely in part on hard biological determinism, or the idea that once we know something about an individual's genes or brain, we can predict or explain his or her behavior' (Aspinwall et al., 2012). While the defence may be able to get away with vague suggestions about the effect of genetic predispositions on the probability of criminal behaviour, for the same evidence to work for the prosecution the causal link must be much more robust, as the burden of proof is substantially higher for the prosecution, especially in capital cases. To be considered as a genuinely aggravating factor that might lead a jury to sentence someone to death, the scientific evidence must be unassailable, something that is clearly not the case. 'While behavioral genetics evidence is viewed as a double-edged sword, each side of that sword is not the direct flip of the other. The hurdles for the State are substantially different from those for the defense and their evidence and arguments may not be comparably compelling' (Denno, 2011).

Molecular Genetics in the Courtroom

Monoamine oxidase A (MAOA) is the only specific gene to be cited in multiple court cases and therefore provides a good case study of the use of genetic evidence, with the data already presented in Chapter 7 providing some background. For a comprehensive review of the role of MAOA in various legal contexts, see Baum (2013).

The first use of MAOA arguments in the courtroom was in *Mobley v. State* (1994). Stephen Mobley was tried for the murder of a Domino's restaurant manager in Georgia in 1991. The murder was aggravated by the fact that the it was committed during the course of an armed robbery and was thus eligible for the death penalty. During the guilt phase of his trial, Mobley's defence submitted a request for funds to have him tested for the MAOA-null variant, reacting to the recent publication of the Dutch family study by Brunner et al. (1993a) described in Chapter 7 which found that males lacking MAOA activity were prone to a range of criminal behaviours including arson, murder, indecent exposure and antisocial behaviour. The defence also submitted family history evidence that Mobley came from four generations of violent men. The initial request was denied by the trial court, on the grounds that 'the link between the MAOA gene and violence lacked scientific verifiability sufficient for it to be introduced' (Bernet et al., 2007). In other words, the court was wary of introducing such an 'unorthodox' novel defence which was not at a stage of 'verifiable certainty'. Mobley appealed the decision to deny genetic testing, but the Georgia Supreme Court rejected the appeal, ruling that the family history evidence was sufficient to demonstrate the point without the need for additional testing. Mobley was executed in 2005.

Over the following decade, appeals to MAOA evidence in the U.S. courts appear to have been rare. Nevertheless, Bernet et al. (2007) report that from 2004 to 2006 their laboratory at Vanderbilt University, Tennessee, tested fifteen defendants on capital cases for both the MAOA and SLC6A4 common mutations. The reactions of the Tennessee courts to the attempts to introduce such evidence were mixed, but mostly negative. Some courts refused to allow the evidence, whereas others allowed it in a limited fashion, although it had no effect on the outcome of the trial.

The next notable use of MAOA evidence was *Bayout* (2009), an Italian court case. Abdelmalek Bayout, an Algerian citizen living in Italy, was attacked and beaten by a gang of Hispanic youths in Trieste in 2009. Following the attack, he shot and killed a random stranger, Walter Perez, in the mistaken belief that he was one of the attackers. Although there was no death penalty consideration under Italian jurisdiction, the defence team submitted mitigating evidence to try to reduce the sentence. In addition to presenting clinical evidence that Bayout suffered from a mental disorder, Bayout's defence had him tested for the MAOA low activity polymorphism identified as a key risk factor for impulsive violence in the G x E study of Caspi et al. (2002), described in Chapter 7. Somewhat surprisingly, the court allowed the evidence. The defence presented no evidence about an adverse childhood environment, but instead argued that Bayout's move from his native Algeria to Italy had induced a state of culture shock that could count as an adverse environment, although Caspi et al. never investigated environmental factors in adulthood. Ignoring the lack of replication studies to support the G x E effect, the court also argued that the plaintiff's genetic predisposition made him more susceptible to impulsive/ reactive violence, and thus his violent response was provoked and non-volitional due to the earlier attack, although the original Caspi study used an antisocial behaviour outcome variable, not a measure of violence. It is unclear how much influence the genetic evidence had on the court's decision to reduce Bayout's sentence by several years from the recommended maximum. However, it does seem that the MAOA evidence was judged to be genuinely mitigating and played a significant part in the reduction of the sentence.

MAOA evidence was also used in a U.S. case the same year during *Tennessee v. Waldroup* (2009). In 2006, Bradley Waldroup got into a violent argument with his wife and her female friend, as he believed they were having an affair. In a sudden escalation of violence, he shot and killed his wife's friend, shot his wife in the back as she tried to flee, then dragged her inside and cut her with a knife and a machete, before she was saved by the intervention of the police. Waldroup was charged with first-degree murder, attempted first-degree murder and aggravated kidnap. During the penalty phase of the trial, the defence team introduced expert genetic testimony, seemingly without any objection from the court, showing how far acceptance of such evidence had come since the *Mobley* case. Once again the rationale was primarily based on the Caspi et al. study, although unlike the *Bayout* case, Waldroup could present evidence of a severely adverse childhood and long-term childhood abuse. In addition, his violence was much more self-evidently reactive, immediate and impulsive in the face of provocation, so it was a much better fit to the model proposed by Caspi et al. The defence argued that the genetic evidence acted as a partial defence: although the evidence did not completely negate the *actus reus* requirement, it argued that Waldroup was not able to act as a 'reasonable man' might have acted in the same situation due to his genetic predisposition, making the act less volitional. The jury proved receptive to this argument. Waldroup was cleared of first-degree murder and found guilty of voluntary manslaughter of the friend, attempted murder of the wife and especially aggravated kidnap, meaning he was no longer eligible for the death penalty. Comments made by the jurors afterwards showed that the genetic evidence was a crucial part of this decision. When one member of the jury was asked if a decision was swayed by Waldroup's genetics, the juror responded, 'Oh I'm sure.... And his background – nature vs. nurture' (Denno, 2013). 'One juror even remarked that the MAOA evidence suggested – [e]vidently it's just something that doesn't tick right.... Some people without this would react totally different than he would' (Walker, 2013).

Genetics and Human Responsibility in the Legal Context

The comments made by jurors following the *Tennessee v. Waldroup* (2009) case raise the question as to how widely the lay public, many of whom will serve as jurors, are influenced by genetic arguments that tend to reduce the idea of human responsibility for crime. Various experimental studies have been carried out to address this question.

For example, one study conducted an online survey with a representative sample of the U.S. population (n = 191) that presented participants with a vignette describing how a young man impulsively killed another young man in a quarrel over his girlfriend's affection (Appelbaum and Scurich, 2014). The guilt of the perpetrator was not questioned, and the genetic evidence was presented as entirely mitigating rather than negating. Participants were presented with four possible versions: either with no miti- gating evidence, with a history of childhood abuse, with a genetic predispo- sition to impulsive violence or with a combination of abuse and a genetic predisposition. Participants then gave judgments about what conviction the murderer should receive (murder or manslaughter) which was an implicit judgment of the state of mind of the murderer at the time of the crime, and the length of sentence the murderer should receive. The study found no difference between the respondents in their judgments of the perpetrator as either more or less guilty of murder. Similarly, two-thirds of respondents across all conditions recommended a mid-range sentence length, between twelve and twenty-five years, with no differences between conditions. The authors concluded that the study showed that genetic evidence was not impacting strongly on the lay public's decision-making process, arguing that these results suggest that the lay public has a nuanced understanding of causality, risk and uncertainty when it comes to genetic evidence, rather than a strong sense of determinism. However, they also recognised that an online survey presenting theoretical scenarios was necessarily different from the reality of the courtroom.

Another study examined the effect of genetic evidence on practising judges, rather than juries or the lay public (Aspinwall et al., 2012). Online surveys were sent to 181 trial court judges across the United States asking them to examine a fictional case. The vignette in this case featured a violent theft at gunpoint, leading to a conviction of aggravated battery. All partici- pants were provided with expert testimony from a psychiatrist explain- ing that Donahue, the plaintiff, was a diagnosed psychopath. However, whereas one group of respondents received just the expert testimony from a psychiatrist, the other group, in addition, received testimony from a neu- robiologist who explained the factors contributing to the development of psychopathy (low MAOA activity, atypical amygdala function and other neurodevelopmental factors). In the prosecution condition of the vignette, prosecutors argued that the evidence should be considered aggravating because the crime and Donahue's actions afterward all pointed to his being a psychopath and posing a continued threat to society. In the defence con- dition, the defence counsel argued that the evidence should be consid- ered mitigating because the crime and his actions afterward all pointed to Donahue's having a harder time controlling his impulses due to his dis- order. The judges were then asked whether they considered the evidence presented to be aggravating or mitigating and to recommend a sentence for the crime. The study reported only one significant difference between the two groups: if genetic evidence was presented, judges recommended reduced sentences – twelve years as opposed to thirteen years. Although small, this difference was significant as the judges viewed genetic evidence as entirely mitigating. However, it should be pointed out that this study has been criticised by Denno (2013), who has argued that the facts of the fictional case presented to the judges are not reflective of real-life cases, and therefore the results cannot be generalised to the actions of real judges. This reflects the broader concern that there is no substitute for assessing what judges and jurors actually think and do in real courtroom contexts.

As far as academic legal opinion on human responsibility in the courtroom is concerned, there is a striking division of views. Real-life cases, some of which have been cited previously, have acted as a catalyst for broader discussions of the role of free will, responsibility and agency in the current Western legal system in how the law is constructed and how it might be reformed. There appears to be somewhat of a disconnect between academic theory and actual practice. Even those who think that genetic evidence is not useful in a courtroom nor that it is particularly persuasive are nonetheless confident that this will not stop defence teams from introducing it. There is also a consensus that the legal system must reflect the attitudes and opinions of the society it purports to serve, and so the undergirding theory of criminal law must be 'folk-psychological' in nature. Where theorists disagree is exactly what that folk-psychological theory represents.

Positions on genetic evidence and the legal system can be broadly divided into two opposing groups: the reformists and the compatibilists (Kolber, 2014). The reformists, as the label indicates, believe that the current legal system is in need of reform and tend to be *incompatabilists*, regarding a deterministic universe to be incompatible with the idea that humans possess free will. As well as being free will sceptics, most reformists are also *consequentialists* with regard to the legal system. Consequentialists argue that the only proper moral basis for a system of punishment is the utilitarian principle of the greatest good for the greater number. Punishment should therefore be determined by what the outcomes for society will be. Within this understanding, relatively minor crimes could be punished with long sentences if the offender is deemed likely to re-offend. An alternative reformist position places greater emphasis on the treatment of criminals as being sick and in need of therapy, rather than being responsible agents deserving of punishment. Reformist positions are exemplified by Greene and Cohen (2004). Although their review refers mainly to neuroscientific evidence, the reasoning applies equally to genetic evidence. Greene and Cohen argue that as neuroscientific (and genetic) evidence grows in complexity, accuracy and accessibility, it will in the near future begin to alter the folk-psychological consensus on the causes of behaviour. If neuroscience can advance to the point at which neurological functioning can provide a compelling explanation for complex human behaviours, then juries will be much more likely to take the view that no one possesses ultimate responsibility. If a general view of determinism takes hold, this means that the use of genetic evidence will become much more pertinent than currently. Reformists tend to argue that the scientific evidence for causal links between genes or neural structures will be elucidated in the near future and do not question whether such a deterministic framework is scientifically accurate. In short, the reformists argue that genetics and neuroscience will soon demonstrate that all our behaviour is really determined by biological forces, and therefore juries will not be able to hold anyone responsible for their actions, leading to the collapse of the current legal system (Atiq, 2012).

The compatibilist position is the more widely held view, standing in opposition to the reformist view. Compatibilists, as the name suggests, tend to view determinism as being compatible with human free will. Moreover, they argue that the legal system as currently formulated is fundamentally compatibilist, because it does not care about the causes of people's actions. The law, or rather the folk-psychological consensus that makes up the law, cares only whether the person on trial committed the act voluntarily with the right state of mind, not whether this was an inevitable and determined action. Because people have free will, there can be no determined action that negates the responsibility of the person who committed the act. Under this model, the new wave of genetic and neuroscientific evidence is irrelevant to the proceeding of criminal law, except where it serves to support or refute pre-existing conditions that affect judgements of criminality such as insanity or alcoholism.

Compatibilists also tend to be retributive in their theory of punishment, arguing that punishment is predicated on a notion of 'moral dessert'. People should not be punished disproportionately, using as a criterion the way in which their crime would reasonably rank in the moral hierarchy of crime. In short, compatibilists argue that the law will continue much as it ever has and that jurors will continue to assign moral responsibility to defendants even in the face of behavioural genetic evidence that might suggest (to them) that the defendant had no volitional control over their actions. The compatibilist position is exemplified by Stephen Morse (2011) who comments:

To make a useful internal contribution to criminal responsibility, the genetic data must be 'translated' into the law's folk psychological responsibility criteria. It must be shown how, precisely, the genetic data are relevant to whether a defendant acted, whether he or she possessed a particular *mens rea*, and whether the mental states relevant to defenses were present. It is not sufficient to indicate that genetics played a causal role in explaining the criminal behavior, even if that causal role is very powerful. Causation and predictability are not excusing conditions in law and causation is not the equivalent of legal compulsion (most action is not the causal result of dire threats or uncontrollable desires). If they were, no one would be responsible because we inhabit a causal universe, but we nonetheless hold people responsible. A genetic predisposition to criminal conduct does not *per se* mitigate or excuse. Causation is relevant only if it tends to show the presence of a genuine excusing condition, but it is the latter that does the legal work. Believing that causation *per se* mitigates or excuses responsibility is the most pernicious confusion bedeviling the attempt to relate scientific findings to criminal responsibility. I have termed it the 'fundamental psycholegal error'.

So, in Morse's view, 'causation is not compulsion, and predictability is not an excuse', a claim that will lead to a more extended discussion in the following chapter.

An Assessment

Several striking facts emerge from this brief overview of the contemporary role of genetics in the legal system. The first is the very limited role that genetic arguments have actually played in the courts so far, at most the role of a 'supporting cast' for the main arguments used by defence teams. On the whole both judge and jury remain suspicious of arguments that smack of biological determinism, although some exceptions have been noted, and the courts are certainly more open to genetic arguments as admissible evidence than they were two decades ago. Yet despite this very limited current role for genetics in the courtroom, those who publish academic legal articles on the matter seem unanimous in their opinion that the role of genetics in defence arguments seems bound to increase in the coming years.

This latter point appears to be closely linked with the second striking fact that arises from this overview: the sheer naivety of all participants concerning the scope and application of genetics. Judges, and to an even greater extent juries, may be excused for not grasping the finer points of genetics; naturally, they remain heavily dependent on expert testimony during the trial, but one might have hoped for a greater degree of understanding in academic legal texts, an understanding that remains sadly lacking in many cases. For instance, do experts give a highly deterministic reading of their data? Are they allowed space to discuss all the many caveats and limitations that accompany all scientific data? Is there acknowledgement that what may look like a positive finding may be non-replicable or actively disproved? Engagement with such questions is often sadly lacking. In an adversarial system there is a possibility of duelling experts, and there appears to be little discussion as to how the court is supposed to proceed when experts disagree in their presentation of the same data.

This point becomes particularly relevant when considering the legal application of the MAOA results. There seems to be little understanding of the point that around 35–40 per cent of the U.S. population has the so-called low activity MAOA isoform (even though 'low activity' in vivo in humans has never been formally demonstrated). So at least one-third of U.S. defendants in court, on average, will have this variant gene. If the court case was taking place in Beijing, then this proportion would (apparently) rise to more than two-thirds. Furthermore, in several studies, having the low activity version correlates with *lower* levels of antisocial behaviour in the absence of childhood maltreatment, as noted in Chapter 7. So in the Italian court case of *Bayout* (2009), cited earlier, where no data concerning childhood abuse was presented, there was no rational basis for the mitigation of the sentence. In the case of *Tennessee v. Waldroup* (2009), there was at least evidence for severe childhood deprivation, but this in itself correlates with adult aggression in many studies irrespective of the status of MAOA variants. In any event, the vast majority of carriers of the low activity allele of MAOA are not particularly aggressive and never commit a crime, so even if the replication studies of Caspi et al. are accepted, such probabilistic data is of no relevance to a particular individual, with his or her unique developmental history, standing in the dock.

Ironically, the defence lawyers in in *Mobley v. State* (1994) may have been more on the right track in seeking a test for MAOA *deficiency* in the plaintiff, although the court was surely correct to reject the suggestion given the limited extent of the data available at that time. But now let us imagine, for example, that many different families such as the Dutch family reported by Brunner et al. (1993a) have been discovered in different countries. Let us further suppose that in each family study there is a 100 per cent correlation of the total absence of MAOA activity and severe antisocial behaviour in males, but not females, over several generations, in each family. In this imaginary scenario a considerable body of replicated studies has built up in the peer-reviewed literature reporting on this correlation in hundreds of individuals. Now given a plaintiff in the dock arrested for a violent act with a demonstrable absence of MAOA activity, this would surely at least provide grounds for the defence to argue for mitigation, if not more. The problem of course with this example is that it remains imaginary, and the reality is that the genetic data, as already noted, is moving the discussion in quite the opposite direction, especially since the advent of GWAS: hundreds or thousands of genetic variants are contributing in a highly polygenic manner to the population variation in the kind of behavioural traits, such as aggression and addiction, that are generally most relevant in the legal context. So at present there appears to be a yawning gap between the wide expectation expressed by legal commentators, especially those of the 'reformist' school, that genetic arguments will increasingly play a role in the courts and the scientific reality – namely, that whatever gene variant(s) the plaintiff may possess of proven relevance to the generation of variation in a population with regard to a particular trait, the contribution of those genes to that variation will be so tiny that there would be zero hope of persuading a jury of their relevance to the case. If and when the scientific reality filters through to the courts, then there is every expectation that legally relevant genetic arguments will decline rather than increase.

In any case, the reformists are quite mistaken in their proposal that increasing biological knowledge in the neurosciences and genetics necessarily undermines the notion of human responsibility. It is to that particular issue that we now turn.

CHAPTER 11

Causality, Emergence and Freedom?

Tackling Some Tough Philosophical Questions

> *From a given definite cause an effect necessarily follows.*
> Spinoza, 1677, *Ethics* I, axiom III

> *The law of causality, I believe, like much that passes muster among philosophers, is a relic of a bygone age, surviving, like the monarchy, only because it is erroneously supposed to do no harm.*
> Bertrand Russell (1913)

> *Man can do what he wills, but he cannot will what he wills.*
> Arthur Schopenhauer, 1839[1]

> *The very first act of a will endowed with freedom should be to sustain the belief in the freedom itself. I accordingly believe freely in my freedom. I do so with the best of scientific consciences ...*
> William James (1899, p. 187)

> *A symbolizing mind has perhaps the widest possible locus of causal influence of anything on earth.*
> Terence Deacon (2006a, p. 149)

In the Introduction, free will was defined as the 'the ability to intentionally choose between courses of action in ways that make us responsible for what we do', a definition that matches the way in which the concept is generally understood in daily speech (Stillman et al., 2011). It is now time to consider that definition further, in particular asking the question as to whether genetics subverts not merely the experience but also the ontological reality of free will. Some examples were also provided in the Introduction of the negative effects on human flourishing arising from the belief that our actions and choices are in reality determined (Baumeister et al., 2010, 2015).

The aim of this chapter is modest: to postulate a framework that will enable a rational discussion to be carried out concerning the extent to which genetic variation impinges on the human experience of choosing freely. Presenting such a framework will entail incorporating philosophical terms and concepts which often have behind them centuries of reflection and discussion, not to speak of controversy, and through lack of space the only way forward is therefore to provide citations to some helpful references in the field.[2] In places, assertions backed up by citations will therefore have to take the place of rigorous argument. The aim here is to integrate the biological with philosophical perspectives in a harmonious rather than a confrontational manner. Some 'ground-clearing' remarks are necessary before moving to the 'main course'.

There are good empirical data supporting the idea that our brains function on two levels: system 1 is automatic and unconscious (and often plagued by making mistakes), whereas system 2 is deliberate, slow and conscious (Kahneman, 2011). It is system 2 which is most relevant to our definition of free will. Free will is exercised by persons who take into account their beliefs, desires and emotions, and from these, choose which course of action to follow (O'Connor, 2000). If we do not think of a certain course of action, we probably cannot choose it. If we do not have any desire or initial attitude towards a course of action, we probably cannot choose it. And if beliefs and desires are sufficiently strong, these might make it very difficult for us to choose any course of action but one. In some cases, those sorts of situations can undermine freedom (the mugger holding a gun to my head induces very strong desires that undermine freedom). In other cases, they do not undermine freedom: I see my child fall into the river and am overpowered by the desire to save her and so jump in. Differential genetic variation might enter into the process of making different choices by affecting belief, desire and emotion. In so far as situational or somatic states impact those factors, they could in principle constrain or undermine our power of choice. But it is we who allow beliefs, desires and emotions to become effective for action when we choose to follow one course of action over another. I may want to eat the cake and I also have a strong desire to diet, but it is I who chooses which desire will translate into action. Free will is on a spectrum ranging from 'optimal choice mode' to 'minimal choice mode', but wherever the choice might be on the spectrum, it is the human mind that provides the primary relevant data.

The rational cogitating system 2 understanding of free will is illustrated by the following example. Let us imagine a situation in which Parliament needs to decide whether to lead the country into a bombing campaign of another country. Each member of Parliament has a vote which has been left up to the conscience of the individual member without recourse to party politics. Each member ponders – one hopes – long and hard on the various reasons for and against undertaking such an action, taking into account the views of their constituents, mulling over the likely consequences of such a bombing campaign and bringing to bear their knowledge of political history in an attempt to estimate its longer-term political impact. Once the vote is over, no Member would be in the slightest doubt that, whatever the lobbying pressures they might have experienced during this process, they could in principle have voted the other way. Some relevant piece of data could have been texted to them a few seconds before the vote, and they could have changed their mind and voted otherwise. This process of uncoerced, deliberative decision-making typifies the phenomenon here described as free will, a capacity closely linked with the acquisition of lan-guage and with the extension into the surrounding culture that language entails. It is also impossible without consciousness, which is not to say that non-conscious processes are not involved in free will – they almost cer-tainly are – but that the ideas and discourse implicit in the understanding of free will as outlined so far would have no meaning but for consciousness. It is conscious thought that enables the weighing of multiple alternatives.

This bald assertion, sometimes dubbed the 'consciousness thesis', has been critically reviewed by others (Levy, 2014) and space does not allow to do it justice. Suffice it to say that one can fully accept the large role played by non-conscious processes in information processing, and the sub-stantial literature reporting non-conscious influences on human actions, and yet still maintain the position that conscious deliberation is necessary for the kind of free will that generates moral responsibility (Levy, 2014). Consciousness, in this view, plays an integrative role in which awareness allows information to be made available to a broad array of neuronal sys-tems that in turn play complementary roles in behaviour (Morsella, 2005). As Levy points out 'only actions performed consciously express our evalua-tive agency, and that expression of a moral attitude requires consciousness of that attitude' (Levy, 2014, p. 39–40).

Consciousness is inevitably reflected in the dual, albeit complemen-tary, nature of our language from which there is no escape: we can dub the two sets the 'I' language and the 'it' language. As soon as we open our mouths to say 'I', we are presupposing the language of mind and per-sonhood, the world of intentionality, framed within the assumption of consciousness. Scientific discourse depends entirely on the 'I' language with its sophisticated linguistic symbolism and its reason-based methods for assessing and interpreting the data (Searle, 2001). Notions such as the Equal Environments Assumption or many of the human traits, such as the concepts of extroversion or impulsivity, investigated by behavioural geneticists, belong to the language of mind.

In parallel with the 'I' language we have the 'it' language, the language in which the neuroscientist describes your brain, or the language in which the geneticist describes your genome, which in turn can be understood at many complementary levels (Poldrack and Farah, 2015). We are all in 'I-Thou' relationships in which our relational language is imbued with the language of mind and subjectivity, but those of us who are scientists, in particular, find ourselves also in a myriad 'I-it' relationships in which the methodological reductionism that lies at the heart of the experimental biological sciences entails that someone else's brain, or genome, or cancer cells, have to be viewed as an 'it'. In the science of behavioural genetics, as in many other sciences, category errors can readily creep into the discussion coming from those not alert to the very distinctive meanings of words such as 'causality' or 'explanation' when placed in the context of the 'it' language rather than 'I' language, or vice versa. Furthermore, scientists who spend their days studying other people's brains, or genomes or behaviours may slip into the bad habit of thinking that their particular level of explanation of human personhood is the one that 'really counts', when the reality is that the person(s) under study just decided to do something different, and for very good reasons. How the 'it' relates to the 'I making the personal decision' is a topic for further discussion later in this chapter.

It is the assumption of our framework for discussion that as far as the 'I' language is concerned, there are neural correlates of the whole gamut of 'I' language at every moment. 'One and the same sequence of events has both biological features and logical, conceptual, and conscious features' (Searle, 2010, p. 124). For every instance of subjectivity, experience, personal cogitation or rational discourse, there are neuronal mechanisms that could be described by a super-scientist, in principle in great detail, that represent the 'it' language describing the molecular workings of the neuronal system, including the brain, at that very same moment in time. These two language sets are in a complementary relationship with each other: both are essential and are part of our daily discourse. This places the starting assumption of this chapter firmly within the metaphysical framework often referred to as 'non-reductive physicalism', in contrast to physicalists who believe that mind is identical with the brain (Crane, 2003). The following section will develop this particular version of non-reductive physicalism into a developmental dual-aspect monistic emergentism (DAME), but for the moment we need to pause on what exactly we mean by the 'neural correlates' of consciousness.

The neural correlates that are of interest in our present context are much more than those that simply describe the current state of our brain, such as awake, asleep or hallucinating. The correlates of interest refer to the actual contents of our consciousness: that stream of visual content that moves past our eyes in glorious technicolor, that flow of conscious rational thought that leads to making a key decision or that vivid awareness of acceptance and fulfilment as we lie in our lover's arms. It is not necessary that every neural correlate of consciousness be both necessary and sufficient, but it should at least represent a 'minimal sufficient system' (Chalmers, 2010); in other words, there might well be more than one brain state that correlates with one particular stream of consciousness (such as pain in the big toe of the right foot), but the neural correlate in each case should be the one minimally sufficient to function as a correlate.[3] Neither is it necessary that a neural correlate have a particular sub-anatomical location in the brain – it might refer more to a particular brain state, such as a mode of neuronal firing and synaptic activity. These and other considerations lead Chalmers to propose the following definition:

> A Neural Correlate of Consciousness is a minimal neural representation system N such that representation of a content in N is sufficient, under conditions C, for representation of that content in consciousness (Chalmers, 2010, p. 81).

Extensive further discussion of the relationship between consciousness and its neural correlates can be found in the philosophical and neuroscientific literature (Zeman, 2002; Baumeister et al., 2010; Chalmers, 2010; Tse, 2013; Haggard, 2015). For the moment it is worth noting that considerable progress has been made in the neurosciences over the past few decades in elucidating the neural correlates of conscious experience and free will. The neuroscientist Peter Tse, for example, who defends a strong understanding of free will, has presented a detailed neurological 'criterial causation' model with three stages, in two of which randomness plays a role, but not in the third. In this model physical neuronal causal chains are informational causal chains, and vice versa. The 'criteria' refer to the claim that physically realised mental events can change the physical basis of future mental events by triggering changes in the informational criteria that must be met by future synaptic inputs before future neuronal firing occurs (Tse, 2013, p. 19). Tse makes the point that his model bypasses the suggestion of Jaegwon Kim (1989) that a physicalist account of the brain mechanisms underlying free will relies on the 'impossibility of self causation'. The point for our present context is not the precise validity of this particular model, but rather that some progress is being made in elucidating the neural correlates of free will, although all those involved in this research exercise will be the first to admit that there is a long way to go.

The framework assumed here is that the nomic regularity of the physical workings of the brain is essential for free will, referring to the law-like processes described by the natural sciences. In principle such processes might include quantum events or the kind of stochasticity described in Tse's model, although there are good reasons, philosophical and scientific, to suggest that quantum uncertainty plays little or no role in the decision-making process. Philosophically, if quantum events played a decisive role, then there would be no free will. Scientifically, in any case, the role of such events in normal brain functioning is likely to be minimal.[4] The normal causal mechanisms of the brain are not a threat to free will: quite the reverse, as our free willing depends on them.[5] The fact that all the facets of the neuronal system have causal precursors does not necessarily entail that the output of such a capacity is likewise physically causally determined (O'Connor, 2015), a counter-intuitive conclusion that will be addressed further. After all, we can create robots, or chess-playing computers, with the ability to weigh up alternative possibilities before opting for one choice over another based on rational considerations. Even though the brain works in a very different way from such machines, a thought experiment suggests that there is no reason why neuronal systems should not encode a far more sophisticated yet analogous system providing persons with the ability to weigh up alternative possibilities, considering the consequences of the options before them, and then deciding based on reasons. Such a decision-making process is what we experience as free will.[6]

Ah, yes, says the physicalist, but what you are *really* claiming here is that it is the physical description of neural correlates and the like that is doing the 'real work' in this process. What do concepts like mind, consciousness and free will really add to the description of neurons, synaptic connections and neurotransmitter pathways? And how can an abstract concept like mind or the self exert causally efficacious physical events? The following two sections begin to address such concerns.

A Developmental Dual-Aspect Monistic Emergentism (DAME)

The best way to describe DAME[7] is to start by describing what the scientific and philosophical literature mean by the term 'emergentism'. The term is not without its problems and has been extensively discussed by philosophers, neuroscientists and others, who do not always agree (Clayton and Davies, 2006; Bedau and Humphreys, 2008; Corradini and O'Connor, 2010). Given that the position adopted here has already been stated as a non-reductive physicalism and that there have previously been claims that emergentism is incompatible with non-reductive physicalism (Horgan, 1993), it is perhaps worth emphasising at the outset that this claim appears to be mistaken (Crane, 2001).

There are two main types of emergentism – the first obvious, the second somewhat more controversial.[8] In weak emergence, also dubbed 'epistemological emergence', the property of a system or object is determined by the intrinsic properties of the ultimate constituents of the system or object, whilst at the same time it is very difficult to explain, predict or derive the property based only on the observation of those ultimate constituents. 'Epistemologically emergent properties are novel only at a level of description' (Silberstein and McGeever, 1999, p. 186). Umpteen examples can be provided of such a phenomenon. John Stuart Mill was impressed by the fact that both sodium and pure chlorine are poisonous on their own but table salt has an emergent innocuous status. The wetness of H_2O is not predictable by an inspection of its chemical structure. Laminar flow, surface tension, viscosity, transparency and other properties of liquids are all emergent properties, examples of what Terence Deacon has dubbed 'first-order emergence'. Deacon also draws attention to the chemical haemoglobin, the molecule that carries oxygen round the body, and the way in which its emergent properties in biology during the course of evolution could never have been predicted by the mere inspection of its structural components (haemoglobin consists of four polypeptide sub-units, each one consisting of a long string of amino acids) (Deacon, 2006b). The functionality of haemoglobin is an emergent property that only occurs in a particular context, a living body which requires oxygen. Inspection of its components per se provides no help in understanding why that particular structure emerged during the course of evolution and not another one, unless of course one incorporates post hoc functional rationalisations into one's investigation. The structure-function inference is made by biologists so often and so 'naturally' on a daily basis that it is barely noticed that the emergent functional property of the molecule in question only 'makes sense' in the context of its role at a 'higher level' in a living system. This applies to all molecules in living systems. In each example we note that that one cannot even describe the higher-level functions of molecules in terms of the language of atoms and molecular structures. The genetic information encoded in DNA is an emergent property which only makes sense once DNA is conceptually placed within the context of a living cell. Pure DNA chemical sitting on the lab bench does not conceptually possess that property.

So far, so obvious. In our present context, however, it is the claims of 'strong emergence' which are more relevant. This is the claim that there are ontologically emergent features of complex systems that cannot be reduced to any of the intrinsic causal capacities of their component parts, nor to any of the reducible relations between the parts, nor can the 'higher level' emergent properties be deducible, even in principle, from a mere inspection of their parts (Silberstein and McGeever, 1999). This is a bold claim, but it so happens that the relationship between mind and brain, a central theme of this chapter, provides, by common consent, the best exemplar of such 'strong emergence'. As David Chalmers remarks: 'I think there is exactly one clear case of a strongly emergent phenomenon, and that is the phenomenon of consciousness' (Chalmers, 2006). Here, the non-deducibility element of strong emergence, even in principle, is most apparent. When Mary decides to go to the shops to buy some bread, there simply is no way in which the concept of buying bread can be deduced by an inspection of the neuronal networks in Mary's brain, however complete that inspection might be. We could of course, in principle, obtain a complete read-out of the neuronal activity of Mary's brain during this process of deciding to go shopping, but the language and concepts of shopping belong to Mary as a person, not to Mary's brain. It is the emergent phenomenon which 'sets the parameters' for all that follows. This is why we can fully accept the intuition of Descartes that mind is the primary datum of our experience, providing the meaning and framework for all else, yet without any need to follow Descartes into what has become known as Cartesian substance dualism.

What does strong emergence entail? It is generally thought to entail, at the least, that the higher-order emergent properties depend on, or 'supervene upon', the lower-order properties of the system, and yet at the same time the emergent properties are distinct from the properties on which they supervene. The term 'depend on' refers to the particular system being described: in principle, a higher-order property M might supervene on any one of the lower-order properties, P_1, P_2 or P_3, in which case none of these Ps is necessary for M, the one described in fact being P_4. The higher-order properties in this case include the mind, intentionality and decision-making. The lower-order properties focus on, but are not restricted to, the 10^{11} neurons, with their 5×10^{14} synaptic connections, which comprise the human nervous system, the most complex known entity in the universe. It is supervenience which justifies the inclusion of strong emergentism within the framework of ideas known as non-reductive physicalism. Mind supervenes on the neuronal system, including the brain, and is in no way independent of the physical system. Therefore in this version of non-reductive physicalism, mental properties may be distinguished from physical properties, but at the same time mental properties are properties of physical objects (Crane, 2001). This distinguishes this view from property or 'type' identity theory on the one hand and from Cartesian dualism on the other. Indeed 'emergentism can be regarded as the first systematic formulation of non-reductive physicalism' (Kim, 1994, pp. 576–77). And, as Crane points out, 'belief in a supervenience thesis does not require that one's ontological commitment to the supervenience base exhausts one's ontological commitments' (Crane, 2001, p. 31). An emergentist can hold that mental properties supervene on physical properties and yet they are simultaneously something more than those physical properties. As O'Connor comments:

> There is no reason to make an a priori assumption that whatever kind of systemic capacity strongly emerges from the collective activity of an organized system of basic physical entities must be similar in kind, in specific respects, to the capacities that sustain it. Basic physical systems are ateleological, but it is an empirical question whether this is true of the capacities that emerge from them in certain organized contexts (O'Connor, 2015, p.156).

Such comments begin to address the concerns of those who challenge the reality of free will by invoking a Laplacean universe in which, in principle, every event follows another in a deterministic way. Modern physics has shown this idea of the universe to be false even though, as already noted, quantum events are unlikely to be significant players in the neural correlates of the mind. In any case, the main point here is that complex systems are not linear physical systems in which one event leads to another, but ensembles in which every component simultaneously influences every other component, proximally or distally, in a myriad different ways. Out of such systems unexpected properties can arise.

Neuroscientists Brown and Paul describe the kind of neural system that is necessary for the emergence of mind as including '*complexity* ...; a high degree of *interconnectivity* (e.g. axons, dendrites and synapses); *two-way interactions* (recurrent connections; feedback loops); and *non-linear interactions* which amplify small perturbations and small differences in initial conditions', commenting that '[t]he complex and massively interconnected neuronal network that is the cerebral cortex, with hundreds of millions of experience-modulated synapses, is well suited for the emergence of high-level causal properties-of-the-whole through dynamic self-organization' (Brown and Paul, 2015, p. 106).

Monism denies that minds and their bodies are distinct substances, claiming that the contrast between minds and bodies is drawn in terms of attributes or properties (Crane, 2000). Strong emergentism is therefore consistent with monism, at least of a dual-aspect kind, in emphasising that the contrast between mind and brain is with respect to properties and not to substance. The term 'dual-aspect' in DAME aims to draw attention to the fact that it is only by adopting a perspectival attitude towards the data (the I/it divide) that we can come to the conclusion that the single-substance system displays two distinct properties, the mental and the physical. By analogy, scientists are familiar with the very same substance having different properties: H_2O can have the properties of ice, liquid or steam, depending on context. These contrasting properties all supervene on the chemical properties of H_2O. Graphene is a two-dimensional hexagonal lattice of carbon atoms which simultaneously has a range of properties such as the efficient conductance of heat and electricity; transparency; and immense strength, about 207 times stronger than steel by weight – same substance, different properties. These are rather limited analogies (which in any case belong more to the world of weak emergence) when we consider the immensely more complex structure of the brain in relation to mind, but they highlight the way in which scientific thought is permeated by the concept, at least, of the same substance having different properties.

The claim of the strong emergentist is not only that the emergent property (in this case the mind) supervenes on the lower properties but also that it has causal powers which are distinct from the causal powers on which it supervenes, sometimes dubbed the 'litmus test' which distinguishes strong emergence from its 'weaker' versions (Di Francesco, 2010). Indeed having causal powers is the best criterion of having a distinct property. 'If you give a list of an object's causal powers, listing only the causal powers of the lower-level properties of the objects, then you will not have given a complete list of the object's powers' (Crane, 2001, p. 216). This is consistent with the claim that emergent properties are responsible for 'downward causation' from the mental category to the lower categories of neuronal networks.[9] The language of 'downward causation' is commonplace in the neurosciences when investigating how cortical areas of the brain, for example, exert executive control over other aspects of brain functioning, such as memory retrieval (Rajasethupathy et al., 2015) or the control of anxiety (Adhikari et al., 2015). In the present context, however, we are referring to the downward causation exerted by the 'I' using the language of mental discourse. In practice, downward causation is how we live our lives. I decide to write Chapter 11 of this book, I move to my laptop, I start flipping through dozens of books and articles, and then I start tapping away (with one finger as it happens). My mindful decisions in this process cause the coordination of millions of neurons and neuronal networks, hundreds of muscles and major changes in my body's metabolism. Philosophers sometimes worry about the notion of the mental, the world of ideas, causing changes in the physical world.[10] But biologists have no such qualms, because it is trivial to demonstrate that this happens all the time, including the mental states of animals which affect their behaviours. Think about the holding of beliefs about determinism, summarised in the Introduction, that impact negatively on human sociality. Consider our members of Parliament faced with the choice as to whether their country should bomb another country or not. Their decision might be influenced by the ideologies of ISIS in which a self-proclaimed Caliph threatens to conquer the world based on particular theological interpretations of the Qur'an. Ideas written in a book centuries ago impinge on the minds of members of Parliament who weigh those ideas with other mental factors and, on the vote day, choose to move themselves into the Chamber of the House and, at the requisite moment, pass through into the Ayes lobby rather than the Noes lobby, where they are counted. If the Ayes have it, then within hours, planes are caused to take off from some faraway base to drop bombs on the country in question. This is an example of mental factors having major physical consequences. The mental has physical effects every moment of our lives.

But it is worth emphasising, too, that causal influences are operating in all directions in complex systems, bottom-up, top-down and everything in between. Mind and body are thoroughly integrated as one system (Claxton, 2015). We shall consider ideas of causality in greater detail in the next section. Here we simply note that these various causal pathways and directions are complementary to each other: if that were not so, the complex system would cease to function properly (Murphy, 2006). Furthermore, 'the determining effects at the higher levels on the lower ones are real but different in kind from the effects the parts have on each other operating separately at the lower level' and 'the higher and lower levels could be said to be *jointly* sufficient determinants of the lower-level events'(Gillett, 2003; Peacocke, 2006). Indeed, there are strong grounds for thinking that the very idea of 'upper' and 'lower' levels in a complex system exerting particular dominant effects is redundant (Ladyman et al., 2007). Furthermore, it is often not sufficiently recognised that the properties of the components of complex systems are to a large degree defined by the very fact of their being a component, and their actions are severely constrained, and indeed defined, by such a status. This is particularly the case in biological systems.

Consider a neuron which is, of course, like any other cell, already a highly complex system in its own right. But there are good reasons to believe that the most relevant neural correlates of consciousness are represented by that vast network of synaptic connections whereby neural networks, organised into distinct sub-modules, encrypt the higher functions of the brain. It is here that the top-down as well as bottom-up and lateral causal pathways become most apparent. Each individual neuron is a 'prisoner' of its context. Isolate a single neuron on the lab bench and it will display some interesting neurophysiological properties, as long as you can keep it alive, but those properties are completely different from its 'natural ecological niche' in the brain in which it is embedded in a neural network with 5,000 synaptic connections or more with other neurons. And in this context, top-down causation can readily be shown to change its biochemical and firing properties. As the person in question chooses to learn the piano, swat for their exams at university or improve their skills in basketball, so neural networks change, facilitation of signals across certain synaptic junctions is increased, and biochemical changes take place in the neuronal cell membrane. It is not the same neuron as it was before. And in turn this changed neuron, in coordination with millions of its fellow neurons, generate the relevant bottom-up signals that at the higher emergent level enables the person to be a better piano player, or to pass their exams, or to score more consistently in basketball. The brain is a dynamic system characterised by multi-causality in which its myriad components have the properties they do have by the very fact of being part of the complex system. The bottom-up causal components are not exactly the same today as they were yesterday. This is how dynamic complexity works.

So perhaps top-down causation is not so mysterious as it might initially appear to be. The top-level executive neuronal networks which we experience as consciousness, mind, free will and the like exert organisational constraints on the lower-level neuronal networks in such a way that they coordinate the very actions that we desire. It is persons that make decisions, not brains (Baker, 1995). As already emphasised, in the normally functioning non-pathological brain, it is impossible to believe that there could be some kind of 'conflict' between what we experience as mental events and the so-called purely physical events of the lower neuronal-network levels. If that were the case, then psychiatric illness would surely ensue.

A simple illustration might help at this point. Consider your computer as it works on some data involving a particular statistical package with the aim of calculating significant differences between two data-sets. Now consider the passage of electrons as they zip around the computer's myriad microprocessor chips. Are they zipping along randomly or in a constrained, purposive manner? Clearly the latter: it is the software plus the input data plus the hardware that set the parameters for the unified output, which in this case is a measure of 'significance' or 'non-significance'. The upper-level goal of 'calculating mathematical significance' supervenes on the trillions of electron movements by a process of top-down causation in which the electrons are physically channelled in very specific ways. It is the teleology which controls the physics in this case, not the other way round.

Returning to persons for a moment, Professor Ismael from the University of Arizona helpfully reflects on what we actually mean by the 'I' that emerges from the complex human neuronal system (Ismael, 2015), drawing a comparison between the self-organisation that characterises many of the complex systems that characterise the human body, such as the immune system or locomotion, with the self-governance that is such a striking feature of personhood. Ismael likens the 'I' of personhood to a 'collectivization of epistemic and practical effort among components', using the analogy of a jury which may contain different opinions, but which in the end offers a single voice to the court. The 'I' is then the activity that collectivises deliberations to speak with a single voice, not as a homunculus in the head, but as a language-dependent organising system that allows the integration of the information from many sub-systems to be built into a single representation. The human mind, writes Ismael, 'has a collective voice in this sense. It not only integrates sensory information, it explicitly self-ascribes intentional states.' (p. 283). Language is special in giving us a way to represent our representational states. The 'I' is not something given, but achieved, and it is achieved by forging a collective voice. So it is not surprising that when the neuroscientist peers into the brain she sees no 'I' any more than she might 'find' language. The 'I' is an integrating property of the whole, even though we may not understand, at present, the precise neural systems that encrypt such an integrative process. Peering into the brain and expecting to find an 'I' is rather like taking a computer to bits in a search for the mathematical equation that, one knew, was in the process of being solved by the computer as we carried out its dissection. It is systems analysis that enlightens such 'higher order' processes, not a dissection of the physical components of the system.

Before leaving the topic of emergence, some comment on the word 'developmental' in front of DAME might be appropriate (formally the acronym should be D-DAME, but that seems clumsy). Two points are important here. The first is the obvious fact that babies display little sign of free will and in general are rather helpless. Free will emerges along with language in community along with all the other important developmental traits that were described in Chapter 4. The infant presumably learns a form of voluntary control during early development, involving a transition from a bottom-up to a hierarchical process of control (Haggard, 2015). As Haggard comments: 'We are not aware of this learning, because it occurs early in childhood and is not remembered, and because it is so success-ful: voluntary control becomes, in a sense, automatized' (p. 160). This underlines the importance of the supervenience of the system: mental states only emerge as the necessary neuronal networks are established, but equally we remember the ways discussed in Chapter 4 in which the infant brain only develops normally when there are the appropriate environmen-tal cues. Multiple causality, bottom-up as well as top-down, are essential features of the emergence of the 'self'.

A similar point may be made with regard to the emergence of mind and consciousness during animal evolution. No one doubts that animals with a reasonably complex neuronal system are conscious in some way, not in a human way, but in a way that fits their own neuronal endowment and ways of living. In Chapter 5 we have already made the claim that we are unable to assess properly the question of the existence of free will in animals because neither we nor animals have the wherewithal to know whether they have a comparable type of 'free will experience' or not. Be that as it may, complexity and size of brain, tool use, culture and social practices, together seem to render it highly likely that the extinct homi-nins who preceded us in evolutionary history must have had, if not exactly human free will, then something like it, and it is very reasonable to assume that various levels of the free will experience have been emerging in other primates and in their precursors. For those who think that free will is an 'all-or-nothing' phenomenon, such a position would be difficult to maintain. From the perspective of DAME, it seems obvious, remembering that even free will in humans is no all-or-nothing phenomenon, but rather shows every sign of being on a continuum between a maximum and minimum point on the 'free will scale'.

There is much more that could be said about emergentism but space does not allow. In any event, this introduction has hopefully helped pro-vide a useful framework within which the influence of genetic variation on human behaviour may sensibly be discussed. For those interested in reading more about emergentism and its detractors, many helpful publications are available (Clayton and Davies, 2006; Bedau and Humphreys, 2008; Corradini and O'Connor, 2010).

Complexity and Causality

The analysis of complexity in biological systems has received extensive attention over many decades (Bechtel and Richardson, 1993; Schaffner, 1998). The question of the influence of genetic variation on human behavior raises the question of causality in complex systems. It often comes as something of a surprise to scientists to find out that there is no generally agreed theory of causation within philosophy. David Hume (in his 1739 book A *Treatise of Human Nature*)[11] famously denied the reality of causality altogether, arguing that we only think that one thing regularly causes another because one thing regularly follows another; in other words, it is the temporal juxtaposition of two events that leads us to suppose that they are causally connected. Scientists have tended to ignore Hume on this point, though his views, or rather modifications of them, remain popular amongst some philosophers. And it is certainly the case that providing causality with a firm physicalist account is not always easy. Scientists have tended to go more with Spinoza who maintained that '[f]rom a given definite cause an effect necessarily follows'.[12] But what then of quantum uncertainty? Or the question of what causes the Big Bang? It does not take long in some scientific disciplines, at least, to encounter problems with the notion of causality. These can prove particularly tricky, also, in the context of complex systems (Beebee et al., 2009).

Returning to our illustration of a computer working out the significance between two data-sets, we can ask the apparently simple question: Is it the software package plus data inputs that are, taken together, causing the mathematical problem to be solved? Or is it the trillions of electrons flowing through the microprocessors that is causing the problem to be solved? The answer, clearly, is both, depending on the particular level at which the question is being asked.

Multi-causality is a central feature of biological causal explanations, although there is still no generally accepted theory of causation receiving unqualified assent. Multi-causality in biology has been dubbed 'causal democracy' (Oyama, 2000), highlighting a major theme of this book (framed by the DICI acronym) that far from being a recipe, the genome provides but one of an important repertoire of causes woven together to establish the developmental product. Where the language of 'causal democracy' might become misleading is if the inference is then made that all the various influences in development are causally equivalent, which is not at all the case (Okasha, 2009). Here the insights of the philosopher John Mackie (1917–81) remain of particular relevance. Mackie pointed out that for almost every effect there could be multiple causes, maintaining that a cause 'is an *insufficient* but *non-redundant* part of an *unnecessary* but *sufficient* condition [for the effect]' (Mackie, 1980, p. 62). Many different levels of causation are constantly operating in all biological organisms. Consider, again, the account of Mary going out to the shops to buy some bread. What caused such behaviour? A range of distal and proximal causes might be offered. For a far distal cause, an evolutionary biologist could respond that the tripling of brain size that has occurred over the past few million years of primate evolution is one cause of the behaviour – after all, there are good reasons to think that the kind of communal living implicit in shops and in buying bread (albeit in earlier social groups of limited size) was causally critical in a co-evolutionary kind of way in generating such behaviour. A social anthropologist might proffer as a further distal cause the way in which hunter-gatherers gradually settled down into organised communities at the start of the Neolithic era, correlating with the beginning of organised wheat farming and the making of bread as a regular feature of domestic life. Coming to slightly less distal causal explanations, a sociologist might wish to point to the character of Mary's grandmother who was very partial to freshly baked brown bread, and although Mary's mother tended to buy bread from the shops, Mary's family background was clearly a causal influence in her tendency to maintain bread as an important feature of her diet. A geneticist, of course, would wish to point out that Mary's genome was an essential cause of her bread-buying behaviour, a genome which contains genes encoding taste receptors that make bread taste nice and enzymes to make sure that it is digested properly and provides nutrition for the body. A sophisticated geneticist, alert to the DICI framework, could equally well point out that the many factors involved in Mary's development since fertilisation were all involved as a cause of her ultimate status as a bread-loving, bread-buying human. Meanwhile, to come to the most proximal causes, the physiologist would say that the obvious cause was Mary's hunger and the need to find something to eat for supper. But talk to Mary herself and ask her what caused her to go out and buy bread, and she would tell you simply that she wanted to. Indeed, the day before she had felt equally hungry at the same time but had chosen on that occasion to buy fish and chips. Mary also insists that on the day she chose to buy bread, right up to the moment when she enacted that choice, she could still have chosen fish and chips instead. Mary is also implicitly aware that the best time to make a decision is just moments before the decision needs to be made, for, by definition, that is when the optimal amount of information will be available to make the best decision. Indeed, had her mother phoned on the day that Mary bought bread seconds before her decision, saying that she had just baked a new loaf and was bringing it round, then of course Mary would not have had any reason to buy bread.

We note here not only the multi-causality but also that each causal chain only makes sense within its own timescale and level of explanation. Furthermore, each causal chain is complementary to all the other chains; there is no need to suggest any rivalry amongst the causes as if they were in competition. It would also make no sense (within the framework provided by DAME) to make the dualistic claim that 'Mary's neurons caused her to buy bread'. Mary is her neurons, just as much as she is her genome, and every other aspect of her physical being. Monism insists that personhood is unitary and it is persons who make decisions, not disembodied brains, whilst recognising that persons display many different properties.

We also note that there is nothing a-causal about free will. Free will sceptics sometimes assume that people who believe in free will maintain such a position because they do not believe that there are causes for the choices made. The position adopted here is the opposite of that assumption. There are multiple causes feeding into the decision-making process – environmental, historical, internal and so forth – although it remains the case that it is the person who is the ultimate cause of her decision. There is no reason to think that Mary was incorrect in her intuition that it was she herself who decided to go out to the shops and buy some bread. We remember the concept of the 'I', introduced earlier, that sees the 'I' as a language-dependent organising system that allows the integration of the information from many sub-systems to be built into a single representation. Furthermore, within the framework of DAME, each reason for making a decision to do A rather than B is encrypted in neural correlates that, integrated together into the representational 'I', exert downward causation in such a way that the decision is made and the rest of the body just has to follow along and bear the consequences of that decision (trekking alone across the Antarctic, for example).

Coming to biological causes more generally, it is the interventionist account (sometimes described by the ugly word 'manipulability') that has gained traction over the past few decades, as expounded by James Woodward in his book *Making Things Happen* (Woodward, 2003), although the basic idea goes back much further.[13] For example, the person who did so much to establish the neo-Darwinian synthesis, Sewall Wright, wrote in 1921 that 'the problem as to the relative importance of heredity and environment in determining the characteristics of a single given individual has no meaning. Both are absolutely essential.... On the other hand, the problem as to their relative importance in determining variation or *differences*, within a given stock, has a perfectly definite meaning and can often be solved with great ease' (Wright, 1921, p. 250). Cook and Campbell write that '[t]he paradigmatic assertion in causal relationships is that manipulation of a cause will result in the manipulation of an effect.... Causation implies that by varying one factor I can make another vary' (Cook and Campbell, 1979, p. 36). This account of causation focuses on the idea, in Woodward's words, that 'causal relationships are relationships that are potentially exploitable for purposes of manipulation and control' (Woodward, 2010). In other words, in a claim that has always been intrinsic to the experimental biological sciences, a variable A is a cause of variable B, and not merely a correlation, when a suitably isolated manipulation of A would change B, raising the question as to, what if things had been different? For the experimental biologist this seems obvious, since it is the way in which much research is rationalised. A variant version of the gene *PCSK9* is associated with very low cholesterol levels in human populations (Hall, 2013a). Inject a monoclonal antibody that blocks the action of *PCSK9* into people with high cholesterol and the level is markedly reduced. The title of a *Nature* article reporting these results therefore seems well justified: 'A Gene of Rare Effect' (Hall, 2013a).

'Stability' and 'specificity' are deemed to be of particular relevance in identifying causes that matter as compared to those that do not. In this context, stability refers to whether an intervention continues to hold across a wide range of background conditions, whereas specificity refers to the fine-grained control that an intervention might have, for example, controlling a gradient of change, rather than a simple on/off switch (Woodward, 2010; Griffiths and Stotz, 2013). The psychiatric geneticist Kenneth Kendler addresses the question of stability under the heading of 'non-contingency of association', referring to the situation in which 'the relationship between gene X and disorder Y is not dependent on other factors, particularly exposure to a specific environment or on the presence of other genes' – in other words, the case of a typical single-gene Mendelian disorder (Kendler, 2005). However, in Kendler's view it would be less appropriate to talk about genes 'for' a psychiatric disorder, not just because the genes of influence are in any case unknown in most cases, but because of polygenic-environmental interactionism.

Drawing on these various insights, we are now in a better position to propose an appropriate usage of causal language in considering the relationship between genetic variation and individual human behaviours, language that has already been highlighted in several places in previous chapters. The key idea is 'genes as difference makers'. Where there is good evidence that a genetic variant in person A makes a phenotypic difference as compared to person B who is not a carrier of that variant gene, then it seems perfectly reasonable to claim that the gene in question is a cause (not the cause) of that phenotype. Agreeing with Kendler, monogenic Mendelian disorders provide the best examples. When certain mutations are found in the gene which encodes the enzyme α-iduronidase, then Hurler syndrome is the result, as mentioned in Chapter 7. It therefore seems perfectly reasonable to ascribe causality to the mutant α-iduronidase gene as the cause of Hurler syndrome. If a scientific intervention was made (God forbid) to genetically engineer the human genome in such a way that the mutant α-iduronidase gene was expressed in all members of a family, then all members would develop Hurler syndrome. 'Causal democracy' does not seem the right kind of language to use in such a context.

But what about the Dutch family discussed in Chapter 7 in which a complete deficiency of the MAO-A enzyme highly correlates with increased aggression and antisocial behaviour in males in the family over four generations? Does the mutant MAO-A gene in this case cause the aggression? Far more family studies displaying MAO-A gene mutations which cause a complete deficiency in the enzyme product would be necessary to provide a clear answer. But based on the data available so far (admittedly limited), the answer seems to be yes. Of course there is a very long causal pathway (within the DICI frame) between the mutant gene and the aggressive trait, and equally there might be influential environmental factors at work about which we know little, but it does seem that the mutant gene, as the key 'difference maker', lies 'at the top' of the causal chain.

Now let us consider the case of the BRCA1 mutant genes which increase the lifetime risk of breast cancer to 80 per cent or more and of ovarian cancer to 40–65 per cent (compared to 12% and 1.3%, respectively, in the general population). Despite these striking risk estimates, the mutant versions of the genes involved are far from being deterministic. The risks entailed in carrying the very same BRCA1 mutation vary amongst different populations (Velasquez-Manoff, 2015a). Additional factors, including varying degrees of exposure to oestrogen, affect the disease risk. By age fifty, mutation carriers born in the early twentieth century seem to have a lower risk of cancer than those born later. For Ashkenazi Jewish carriers born after 1940, the likelihood of developing breast cancer by age fifty is nearly triple that of women born before that date. BRCA carriers who lose at least 4.5 kg in weight between ages eighteen and thirty have around half the risk of developing breast cancer by age forty-nine compared with carriers who do not lose weight, clearly implicating hormonal factors (Velasquez-Manoff, 2015a).

Breast cancer, like virtually all cancers, is multi-factorial in its development. So, is the mutant BRCA1 gene *the* cause of breast cancer? Not really. We can say that it exerts a strong contributory influence in those women who are both carriers and who do develop breast cancer, but this clearly cannot be the case in carriers who never develop the cancer. Most likely in those individuals the biochemical effects of the gene's product (a DNA repair protein) is comparable but is compensated for by other proteins, such as other DNA repair proteins (of which there are many).

By this stage it should already be clear that the role of genes in causing phenotypic differences is on a sliding scale, ranging from 'strong causation' to 'having a causal influence' to 'best not use the word causal at all for individual genes'. This latter designation brings us to the great majority of human populations who have no particular pathology but who range rather widely in their behavioural traits, as discussed in Chapters 6–9. Nearly all the data assessed there indicate that, at best, variant genes contribute 0.1 per cent or less to the overall variation in a population in respect to the particular trait under investigation. So there can be no suggestion here that 'gene x causes trait y' nor that 'gene x is the gene for trait y', as previously emphasised. So, what causes behavioural trait y? Again, as previously highlighted, the only answer can be 'multiple factors', including the person's individual genome, developmental history, family environment, set of decisions made and life events. There is no escaping multi-causality in this context with each cause, at least in principle, 'making a difference', even though in practice disentangling the causal factors is virtually impossible at the level of the individual. Here causal democracy does indeed reign supreme.

On the spectrum from 'strong cause' to 'having a causal influence' to 'multiple gene variants, in other words the whole genome, could be assigned a causal influence', do we eventually reach phenotypes where the role of genes can be ignored altogether? The answer is both yes and no. The yes comes from the finding that there is zero heritability in populations assessed with respect to 'traits' such as which language people speak, which football club people attend, which set of religious beliefs people adhere to, which supermarket people use for their weekly food purchase and so forth. In other words, it seems that genetic variants are not really difference makers in this context. But the answer is also no in the DICI sense in which genes and environments are completely integrated during the developmental process that has led to the existence of that particular unique individual. That fact is always present, generating commonality in human phenotypes in which people have the capacity to learn languages, enjoy sport, have religious beliefs and go shopping.

But for those behavioural traits that do have non-zero heritability values in populations, including such traits such as divorce rates, hours a week watching television and being a political enthusiast, it is worth remembering that invoking a genetic aetiology for such traits is meaningless. There are no genes for divorce, such that if you find you carried them would then threaten the stability of your marriage. There are, however, no doubt endophenotypes related to such variables as personality traits which in turn have many 'down-stream' influences on such sub-traits, thereby generating their heritability values in populations. The safest inference from such data is the claim that the genome of person A is different from person B and in person A his genome is associated with a very genial and cooperative disposition which is not the case for Person B. Therefore the genome of person A is, in principle, causally a difference maker, even though we have little idea at present as to how that difference is generated. Such an inference may be 'safe', but it does not really say anything particularly useful. If we then ask the broader question as to what are the causes of person A having a genial and cooperative disposition, then the answer can only be the usual mix of genetic and environmental factors, woven together with the DICI framework, linked to the person's own decisions through life, leading to habits, plus life events, perhaps religious beliefs and much else besides. And if person A decides to be grumpy and non-cooperative on occasion, despite his general demeanour, then there is not much more to say, except of course that he has free will.

Is Free Will Falsifiable?

Scientific data per se cannot demonstrate that anyone has free will. As we have noted, the fact that the brain appears to operate mechanistically in a nomically regular way during the decision-making process has been interpreted as if free will must be an illusion: 'free will is an epiphenomenon of physical neuronal processes'. In the present context we have accepted the opposite interpretation of the very same data, that it is precisely the nomic regularity of the physical neuronal processes that generates not only the experience but also the ontological reality of free will. Different philosophical frameworks and assumptions underpin these differing interpretations. But science is not therefore left out in the cold having nothing else to say on the matter. First, it is conceivable that in some future century our understanding of how the brain works, especially in its higher-level computing and systems analysis levels, might demonstrate that it is not the kind of system that engages in sophisticated reason-based calculations of the kind here envisaged, particularly as reasons are weighed and assessed before the person chooses x rather than y. Personally I think this is unlikely because, from the DAME perspective, there must be neural correlates of what we experience, and so it is only a matter of time before those correlates are elucidated. Nevertheless, one cannot formally exclude the mooted possibility. But in any event, second, if one accepts the DAME framework, within which the ontological reality of free will is maintained, then there must be scientific data that could, in principle, falsify such an understanding. The reason is simply that any form of physicalism, be it non-reductive or not, must be open to falsification using scientific data. So we can ask the question as to whether free will has already been falsified based on neuronal system or genetic data.

As far as the neuronal system data are concerned, some neuroscientists think that the reality of free will has already been falsified as a result of the experiments of the late Benjamin Libet and those who have followed in his steps. However, there are good data and strong philosophical arguments that suggest that the Libet-type experiments do not support such an interpretation (Herrmann et al., 2008; Mele, 2009b; Balaguer, 2010; Clarke, 2010; Trevena and Miller, 2010; Tse, 2013; Alexander et al., 2015; Baumeister et al., 2015; Maoz et al., 2015; Nahmias, 2015; Schlegel et al., 2015). Assessment of these data and the inferences from them lies beyond our present scope, but the Libet experiments are mentioned here to make the point that scientific data can be and have been proposed to count in principle against the ontological reality of free will.

We may therefore ask the parallel question (albeit recognising the very distinctive nature of the data involved) as to whether there might be genetic data that refute free will. The answer is clearly yes, albeit not in the sense that the concept of free will itself is rendered implausible – far from it – but due to the fact that there are many pathologies in which genetic mutations play a decisive role during development, leading to deficits in the neuronal system such that both the experience of free will and its ontological reality are severely curtailed. This is precisely what is expected within the DAME framework. For example, in our paradigmatic example of Hurler syndrome, there is progressive mental retardation during development, with death generally occurring by the age of ten. Free will only manages to develop partially since the accumulating waste products in the neuronal lysosomes that lack the α-iduronidase enzyme severely disrupt brain function. Many other similar examples could be cited arising from other single-gene Mendelian disorders in which normal brain development is severely compromised. There is also the example of the so-called catatonic state in adult schizophrenia in which the patient may be unable to move, speak or respond. In severe clinical depression, likewise, the dark cloud engulfing the patient can render rational decision-making difficult or impossible: both the experiential and ontological aspects of free will are markedly deficient. The genetics of both schizophrenia and clinical depression are complex, but there are sufficient data to indicate that there would be no schizophrenia without certain types of variant genomes, and the same appears to be the case for certain types of familial depression, and indeed for the autism spectrum disorders. So variant genes can play a causal role, albeit via a long chain of developmental effects, in rendering the neuronal 'free will apparatus' dysfunctional. This then also becomes acutely important in the assessment of criminal responsibility, as discussed in the previous chapter, although one should hasten to add that there is no particular link between criminality and the syndromes just mentioned.

We are now in a better position to revisit the definition of hard determinism provided in the Introduction: 'given our particular genomes our lives are not really up to us and are constrained to follow one particular future'.[14] For a wide range of pathologies, albeit rare conditions, hard genetic determinism appears to be the appropriate description, although within this cohort there is a 'severity range'. Absent gene therapy, all those lacking α-iduronidase will develop Hurler syndrome: the gene defect is both necessary and sufficient. But in the case of schizophrenia, concordance in identical twins is only 41–48 per cent, as already noted, so a particular set of variant genes is necessary but not sufficient: on a scale of 1–10 we might wish to score this as 'somewhat less deterministic' from a genetic perspective.

But what about the vast majority of the human population who, happily, are not experiencing pathologies that subvert the normal workings of the human mind? What about that vast array of human traits, discussed in Chapters 6–9, that display positive heritability values? Here we need to distinguish carefully between traits that 'happen to us' with traits that 'we make happen', thereby perhaps becoming habits in the process, but nevertheless we chose them to be so. We may place items such as personality traits and sexual orientation in the former category. As childhood development proceeds, very clear personality traits emerge, not out of the child's choice, but as a consequence of DICI. These traits are not static and can be modified to a greater or lesser extent in adulthood, but they were not chosen in childhood. In the case of sexual orientation, the data presented in Chapter 9 suggest that in the vast majority of males, and a somewhat lesser proportion of females, same-sex attraction happens to people during development, it is not a choice. But in neither the case of personality traits, nor same-sex attraction, can we say that these traits are 'genetically determined', given that variant genes represent but one set of factors amongst many during the developmental process. Do these examples count as evidence against free will? Only for those people who think that we choose our personalities or sexual orientation in the first place, and there are good grounds for thinking that such a view is mistaken.

We can safely ascribe to free will other behavioural traits that vary in healthy human populations, such as taking part in political activism, getting divorced, playing basketball, watching television (or not), committing crimes, going to university (or not) and so forth. In a democratic society, at least, these are things we choose to do – they don't 'happen to us'. But genetic variants can still be 'difference-makers' in that they can lead to the development of people with predispositions to choose in certain ways. If that were not the case, then one would expect their heritability values in populations to be zero, which they are not. Tall individuals are more likely to choose to play basketball, although there are plenty of short people who choose to play as well. The decision to go to university will of course be influenced by one's academic record and, as emphasised in Chapter 8, that certainly has something to do with genetic variation, even though pinning down what that entails at the molecular level has proved extremely difficult to elucidate. But many bright and intelligent people choose not to go to university even when the opportunity is clearly present. So this brings us to the definition of 'soft determinism' provided in the Introduction, which states that 'given our particular genomes, our lives are more likely to follow one particular future' (again, let it be noted, a rather different definition from that used by philosophers). That barely deserves the name of 'determinism', so perhaps it is just safer to say that different genomes tend to correlate with different life outcomes in a probabilistic kind of way, and just leave it at that.

What is quite clear from these reflections is that there is nothing in genetics that falsifies the reality of free will, even though variant genes can, on occasion, subvert the 'I' on whom free will depends. Fortunately, however, in the vast majority of cases, it is precisely because we have human genomes, and not some other genome, that we develop into creatures who do have the massive computing capacities that enable free will to become an ontological reality.

In concluding this discussion on free will, it may, or may not, have been noticed that theology has played no role. Indeed we have cited a wide range of commentators, many of whom I know are at least sympathetic to the views expressed here, some of whom I know to be atheists or agnostics, some theists, while for most I have no idea.[15] There is a very rich theological tradition of engaging with topics of free will and predestination, together with a big literature on the topic stretching back many centuries, which I have no intention of discussing here. Suffice it to say that, in my opinion, those of any faith or none can come to the same or similar conclusions about the topic of free will, simply because experiencing free will is a biological human trait like having two legs – a brute fact. Yes, the further arguments and empirical data supporting the idea that the experience relates to an ontological reality may or may not be adequate to establish the point unequivocally, but they are arguments and data open to all to assess irrespective of their religious world view. The theological reflections on genetics that follow in the final chapter assume rather than discuss the reality of free will.

CHAPTER 12

Made in the Image of God?

A Conversation Between Genetics and Theology

> *Men go abroad to admire the heights of mountains, the mighty*
> *waves of the sea, the broad tides of rivers, the compass of the ocean,*
> *and the circuits of the stars, yet pass over the mystery of themselves*
> *without a thought.*
>
> Augustine, *Confessions*, c. AD 400

> *Nature, or nature's great Author, while giving entire freedom and*
> *vast power to the human will, seldom trusts to its action alone to*
> *secure man's well-being.*
>
> Lord Adam Gifford (1874)[1]

> *Nearly all the wisdom we possess, that is to say, true and sound wis-*
> *dom, consists of two parts: the knowledge of God and of ourselves.*
> *But, while joined by many bonds, which one precedes and brings*
> *forth the other is not easy to discern.*
>
> John Calvin, *Institutes of the Christian Religion*, I.1.1

How do we bring the biologically conceptualised notion of human personhood presented so far into conversation with theology? I would like to
suggest that a useful bridge is provided by the theological idea of humankind made in the image of God. There has of course been a vast amount of
academic reflection on this notion, not least by previous Gifford lecturers,[2]
perhaps unsurprisingly when one considers that it has contributed historically to the shaping of moral values, political systems, the abolition of slav-
ery, medical care, education and more generally the justification of human
rights (Wolterstorff, 2008; Hart, 2009; Witte and Alexander, 2010; Gushee,
2013). Irrespective of one's personal interest in theology, or indeed lack
thereof, no one can deny that at least in western cultures strongly influenced by the Judaeo-Christian world view, the cultural air we breathe still

retains values derived from this revolutionary ideology. Like the universal background radiation that points to the Big Bang origins of our universe, the notion of humankind made in the image of God is constantly there in the background, from a cultural perspective at least, reminding us of the intrinsic value of human personhood.

Interpretations over the past two millennia of what it means for humankind to be made in the image of God have involved substantialist, functionalist, relational, representational, existential and eschatological approaches. I have no intention of reviewing these interpretations here, a task already well done by others (Sullivan, 1963; Clines, 1968; Pannenberg, 1970; Cairns, 1973; Westermann, 1984; Hall, 1986; Hoekema, 1986; Middleton, 2005; Waldron, 2010; Welz, 2011; Thiselton, 2015; Marsh, 2016). But we will draw on some of those insights as we first look carefully at the Genesis texts to see how contemporary scholars now view them. As we do so, it soon becomes apparent that these profound theological essays present a transformed vision of humanity, highly subversive in their time of the status quo of ancient Near East cultures which demoted the mass of human beings to their destined role as servants of the gods and their proxies. It is in this transformed view of humanity that one of the most relevant conversations with genetics begins to emerge.

What Does the 'Image of God' Mean?

A previous era tended to see the 'image of God' language as relating to a list of supposedly unique human capacities, such as rationality, language and morality, that distinguishes humans from other members of the animal kingdom. That tendency has declined for two reasons. The first is the recognition that there is a much greater continuity between us and other animals than had previously been realised. The second is the increased discovery and study of ancient Near Eastern literature that sheds light on how the term 'image of God' was understood by the Biblical authors. In addition to these shifts in thinking has come a new theological trend, informed by anthropology, that understands the idea of humankind made in the image of God as referring to the gradual evolutionary process, stretching back over millions of years, whereby the distinctively human neuronal capacities have emerged that enable moral sensibilities and religious practice. All that we know about evolution suggests that this is indeed the case and genetic variation has of course played a critical role in that story. There is a huge literature on this important topic, including the interactions of theology with emergent human traits such as altruism and cooperation (Van Huyssteen, 2006; Nowak and Coakley, 2013), but that is not our focus here. There is, however, a certain irony in the reflection that if these emergent human properties are relabelled with the 'image of God' language as if distinctive human traits provided the main or only sense of the term, then the clock is being turned back once more to focus on certain human qualities as expressing what the 'image of God' idea is all about (De Smedt and De Cruz, 2014). Here it will be 'taken as read' that the account that evolutionary anthropology and psychology describe for us is necessary but certainly not sufficient to do full justice to the theological notion of humankind being made in the image of God. Furthermore, the reading of the Genesis texts outlined in this chapter opens up the way to a particularly fruitful discussion with genetics, not least concerning some pressing ethical issues. A full discussion would require a book in itself, so what follows should be seen as an hors d'oeuvre.

To do the idea of the 'image of God' full justice, we need to give some attention to the way the phrase was used in its original context, and what emerges from this study is that the phrase carries a primary meaning of status and, thereby, of value and delegated responsibilities (Clines, 1968; Middleton, 2005). The biblical account of the *Imago Dei* only makes sense if there is a God who creates the particular value of humanity and if there is a God who then delegates particular responsibilities to humanity. There are only three uses of the 'image of God' language in the early chapters of Genesis, with the first mention of the phrase in Genesis 1:26–8:

> Then God said, 'Let us make *adam* [humankind] in our image, in our likeness, and let them rule over the fish of the sea and the birds of the air, over the livestock, over all the earth, and over all the creatures that move along the ground.' So God created *adam* in his own image, in the image of God he created him; male and female he created them. God blessed them and said to them, 'Be fruitful and increase in number; fill the earth and subdue it. Rule over the fish of the sea and the birds of the air and over every living creature that moves on the ground.'

Genesis 1 is like a manifesto laying out the 'big picture' which provides a set of keys for understanding the rest of the Bible, and the idea of humankind made in God's image provides one of those keys. The idea is repeated at the beginning of Genesis 5 to emphasise the point that both male and female alike are made in God's image and, further, that this sets the pattern for human family resemblances as the generations continue. Genesis 9:6 then makes a direct link between human value and the status of humans as being made in the image of God: 'Whoever sheds the blood of man, by man shall his blood be shed; for in the image of God has God made man.'

Now how would the first readers of these passages have understood these references to the 'image of God' language? Answering that question requires some comment on the cultural and religious context of ancient Israel. Israel lived in the shadow of Mesopotamian civilisations that stretched from the ancient Sumerians of the third millennium BC through the Babylonian and Assyrian empires of the second and first millennia. The cultural achievements of these civilisations were impressive (Pollock, 1999; Algaze, 2008; Kriwaczek, 2010). Mesopotamian cuneiform was the scribal linguistic choice throughout the region, even during periods of Egyptian or Hittite domination. Urban society in great cities such as Babylon was structured politically, socially and economically round the king's court and the cultic practices of temple worship of the various polytheistic deities of the city. Each temple represented a household or estate of the god whose statue would be set up and cared for within the temple. The statue was treated like the king. Serfs worked on the agricultural lands owned by the temple and a further large cohort was engaged in similar service to the royal court, the king exercising absolute power. In addition there was a landed oligarchy. Society was therefore highly stratified and power was in the hands of the privileged few. Social control was further achieved by the predictions of the astrologers who tracked the baleful destinies determined by the stars (Baigent and Baigent, 2015; Middleton, 2005).

The stratified nature of Mesopotamian society was underpinned by a series of powerful creation myths. In Sumerian texts, as well as widely distributed Akkadian texts such as the *Atrahasis Epic* and the *Enuma Elish*, humans are created to serve the squabbling gods to relieve them of their burdens, this being asserted as providing the very meaning of human existence. As Tablet 1 of the *Atrahasis Epic* declares: 'Let man assume the drudgery of god' (1:197), referring in this case to the god Enlil. In contrast to the drudgery envisaged for the mass of humankind, kings are perceived as being made in the image of a god in much ancient Near Eastern literature, most frequently in Egyptian texts ranging all the way from 1640 to 196 BC. Eighteen different Egyptian pharaohs have been recorded as being referred to as the 'image of a god' (Curtis, 1984). Several times the pharaoh is said to have been begotten by the god of whom he is the image, the pharaoh being 'the shining image of the lord of all and a creation of the god of Heliopolis ... he has begotten him, in order to create a shining seed on earth, for salvation for men, as his living image' (Erman and Ranke, 1923, p. 73).

Abundant inscriptions make clear that the reigning pharaoh was seen as the direct offspring of the gods, authorised by divine birth to rule over not only Egypt but also the entire world on behalf of the gods. Even more relevant to the immediate context of Genesis 1 are the six known Mesopotamian texts which explicitly refer to a king as being in the image or likeness of a god, using in all cases except one, the Akkadian word *salmu*, the cognate of the Hebrew word for image used in Genesis, *selem*. Three of these references come from Adad-shum-ussur (c. 740–665 BC), a court astrologer and cultic official in the seventh century BC royal court of Nineveh whose texts assume that the Assyrian king Esarhaddon is the very image of Bel, otherwise known as Marduk, the top god of that era: 'A (free) man is as the shadow of god, the slave is as the shadow of a (free) man; but the king, he is like unto the (very) image (*muššulu*) of god' (Pfeiffer, 1935, p. 234).

In 670 BC, Adad-shum-ussur assured king Esarhaddon that '[t]he king, the lord of the world, is the very image of Samas (Shamash)', the sun god and god of justice. In another interesting text found on a seventh century BC Assyrian tablet, it is a priest who is said to be the image of a god: 'The spell (recited) is the spell of Marduk, the exorcist-priest is the image [*shalmu*] of Marduk', suggesting that in the act of incantation the priest is acting on behalf of Marduk.

The idea, therefore, that the king, and in some cases the priest, acts as the deputy of the gods on earth stretched back to ancient Egyptian and Sumerian times and is central to the royal ideology of Babylonia and Assyria, demonstrated by texts stretching over a period of thirteen centuries. Would Hebrew thinkers and writers have been familiar with this idea? Almost certainly, yes, since Israel had significant cultural and economic contact with both Egypt and Mesopotamia over prolonged periods, contact clearly reflected in the Genesis creation account, a direct rival to that of its competitors, such as the *Atrahasis Epic*. The Assyrian Empire conquered and deported the northern kingdom, Israel, in the seventh century BC and the Babylonian Empire did likewise to Judah in the sixth century, Old Testament texts such as Jeremiah and Lamentations reflecting the experience of exile. Whenever the final compilation and editing of Genesis took place, the scope for the transmission of Mesopotamian ideas concerning the divine image of the gods was considerable. Indeed, as Schmidt points out (Schmidt, 1983, p. 195), the 'image of god' seems to be a 'fixed, technical term' in both Genesis and a wide selection of Mesopotamian sources where one routinely finds terms such as the 'image of Marduk', 'image of Bel', 'image of Shamash' or 'likeness of a god'.

So when the early readers of Genesis 1 first encountered the language of humankind made in the 'image of God', it is very likely that they would have understood this as the claim that the kingly and priestly roles previously allocated to the privileged few by a pantheon of gods were now being delegated instead by the one creator God to the whole of humanity. Such a role and function are made particularly clear in Psalm 8: 4–8:

> What is humankind that you are mindful of them, human beings that you care for them? You made them a little lower than the heavenly beings and crowned them with glory and honour. You made them rulers over the works of your hands; you put everything under their feet: all flocks and herds, and the animals of the wild, the birds in the sky, and the fish in the sea, all that swim the paths of the seas.

The Hebrew verb for 'rule' in Genesis 1:26 and 28, *rada*, is often linked to kingship in the Old Testament, in Genesis pointing to the purpose for, not simply a consequence of, being made in the image of God. By such language was the *Imago Dei* democratised: from now on it was humankind who were to be the significant players in the arena of earthly life, the mandate to rule underlying their new responsibilities (Middleton, 2005, p. 204). The created order that Genesis 1 describes is that of the cosmic temple in which humanity were to take on delegated responsibilities for the earth, as priests of creation, actively mediating God's care to the living world (Walton, 2009). As Middleton comments: 'Just as no pagan temple in the ancient Near East could be complete without the installation of the cult image of the deity to whom the temple was dedicated, so creation in Genesis 1 is not complete ... until God creates humanity on the sixth day as *imago* Dei, in order to represent and mediate the divine presence on earth' (Middleton, 2005). In a stroke, the entire ruling and priestly structure of Mesopotamian society was delegitimised. One type of creation order was to be subverted by another. Far from being created to serve the gods, here in Genesis it is God who creates a bountiful world, providing for the physical and spiritual needs of humankind, then placing the world in their care. As Clines concludes in a careful study of the Genesis texts: the function of the image is to be '*representative* of one who is really or spiritually present, though physically absent' (Clines, 1968, p. 87).

There are some archaeological and inscriptional data that are relevant to this 'functional/representational status' interpretation of the *Imago Dei* in the Old Testament. For example, the terms 'image' and 'likeness' found in Genesis 1:26 can be used of statues. In the Ancient Near East, rulers set up statues of themselves in various parts of their domain so that their subjects would know who ruled them. In 1979 a farmer at Tell Fekheriyeh (ancient Sikan) in Syria was enlarging his field with a tractor when he unearthed a life-size ninth BC statue representing a governor or king with a lengthy inscription in Akkadian, with translation into the local Aramaic dialect, containing the Aramaic cognate equivalents of both *selem* (image) and *demut* (likeness), together with the Akkadian cognate of *selem* (Millard and Bordreuil, 1982). The inscription in Akkadian (with the Aramaic version in brackets) declares that it is the statue of 'Adad-it'i (Hadad-yis'i) governor (king) of Guzan, (and of) Sikan, (and of) Azran'. It is the statue itself which is the image and likeness of the local king or governor, who is described in the inscription as the 'brother of the gods'. The name of the king in Aramaic, Hadad-yis'i, means 'Hadad is my salvation'. This discovery is consistent with the idea that being made in the image and likeness of God in Genesis 1:26 refers to the delegated ruling status that humankind was expected to exert over the earth. The *adam* (humankind) was being called to take responsibility for the *adamah* (earth).

Many Old Testament themes pick up on the functional understanding of the *Imago Dei* even when the 'image of God' language itself makes no appearance. For example, the account of the giving of the Law in Exodus 20 is generally understood as spelling out the logic of humanity being created to 'image' and thus 'correspond' to God. The Ten Commandments start with God articulating the extent of his faithfulness to his people (verse 2), then spelling out how God's people should image this or correspond to this. The first four commandments articulate what it is to image or reflect God's faithfulness back to God and the rest articulate what it is to image this by being faithful to others – to all those others to whom God is likewise faithful (do not kill, commit adultery, steal, lie, etc., verses 12–17). In short, to be constituted as the 'image of God' involves imaging or reflecting God in one's defining relationships – both in one's orientation to God and to others. Precisely, this theme is succinctly summarised by Christ when asked which is the greatest commandment: 'Love the Lord your God ...', replied Christ, 'and love your neighbour as yourself' (Matthew 22:36–9). The logic is clear: as God loves us so we are to image that love both 'vertically' in our relation to God and 'horizontally' in our relation to others.

In the New Testament it is supremely Christ who is seen as being the perfect embodiment of the image of God. In Hebrews, Christ is presented as 'the reflection of God's glory and the exact imprint of God's very being' (Hebrews 1:3). As Paul writes to the church in Colossae: Christ 'is the image of the invisible God, the firstborn over all creation' (Colossians 1:15; in the cultural context the term 'firstborn' referred to prime status, see also

2 Corinthians 4:4). It is Christ who, through his life, death and resurrection has come to restore people to the kind of humans that God had always intentioned, to restore the practical outworkings of the delegated status and authority of humankind to that which God had envisioned. So Paul tells the Christians in the first century church in Colossae that 'you have taken off your old self with its practices and have put on the new self, which is being renewed in knowledge in the image of its Creator. Here there is no Greek or Jew, circumcised or uncircumcised, barbarian, Scythian, slave or free, but Christ is all, and is in all' (Colossians 3:9–11). It is through Christ, says Paul, that the possibility of the renewal and transformation of the self 'in the image of its Creator' becomes a real possibility, a process accessible to all (see also, 2 Corinthians 3:18 and Romans 8:29). Once again we note that Paul here is referring to the whole person – the 'self' – not to some disembodied component of personhood. All humankind are made in the image of God, and thereby every single person has intrinsic value: the New Testament emphasis is on how that image can be renewed and transformed. For those who find their new identity in Christ, Paul explains that '[j]ust as we have borne the image of the man of dust', at the resurrection 'we will also bear the image of the man from heaven' (1 Corinthians 15:49), where Christ is seen as the 'man from heaven'.

The Conversation with Genetics

Thomas Metzinger has commented in an edited volume entitled *Neural Correlates of Consciousness: Empirical and Conceptual Questions* (2000, p. 6) that '[i]mplicit in all these new data on the genetic, evolutionary, or neurocomputational roots of conscious human existence is a radically new understanding of what it *means* to be human' – going on to claim that this emerging account of the human person is 'strictly incompatible with the Christian image of man'. Here we take precisely the opposite view: never before have the findings of science regarding the human person seemed so compatible with a Christian understanding of human personhood, and never before so relevant to the bioethical challenges arising from advances in genetics. With the brief theological summary now in place, there are five particular strands in this idea of humankind made in the image of God that we will here bring into conversation with genetics.

The Image of God Is the Whole Person

A striking aspect of the Genesis texts that introduce us to humankind being made in the image of God is the sheer physicality and 'earthiness' of the account. Here is humankind, male and female (Genesis 1:27), as the embodied self – a theme picked up and expanded through the rest of the Bible. Humankind are not spirits, nor angels, but made from dust. As the theologian Gerhard von Rad once commented, 'One does well to separate as little as possible the bodily and spiritual; the whole man is created in the *imago* of God' (von Rad, 1972, p. 56).

Despite this insight, it is probably fair to say that the great majority of believers within the Abrahamic faith communities continue to maintain a somewhat fragmented view of human personhood, with a tripartite body-plus-soul-plus-spirit (or at least a bipartite body-plus-soul) understanding of the person being the most popular. Indeed, a sociological study of Americans shows that most Christians hold to such a view (Evans, 2016). For some the soul 'enters' or 'infuses' the body at some particular time, perhaps at conception, or at a somewhat later stage. In some Jewish traditions the soul enters the body at birth. All such understandings of the soul perceive it in neo-Platonic terms as a non-material entity that is separately created by God for each individual and then 'added' to the body to bring about true personhood. Within this framework, beliefs concerning immortality are generally linked to the possession of such a soul.

In contrast to this majority view, there has been an increasing theological trend over the past few decades to view human personhood as a psychosomatic unity with different aspects, drawing on the strong Hebrew tradition which emphasises more the unity of the person rather than its separation into distinctive 'parts' (Jewett, 1971; Brown et al., 1998; Green, 2008; Jeeves, 2011, 2015). Brown and Strawn (2012) write of 'embodied soulishness'. Indeed, the Old Testament gives little support to the idea that the soul is a distinct entity: the words 'soul' and 'spirit' are often used interchangeably (e.g., Isaiah 26:9: 'My soul yearns for you in the night; in the morning my spirit longs for you'). In Genesis 2:7, having 'formed man from the dust of the ground', God 'breathed into his nostrils *nishmath chayyîm* (the breath of life) and the man became *nephesh chayyāh*, a 'living being' or, as some translations have it, 'living soul'. The *adam* became an animated enervated body. The Hebrew word *nephesh* can mean, according to context, life, life force, soul, breath, the seat of emotion and desire, a creature or person as a whole, self, body and, even in some cases, a corpse. In Genesis 1:20, 21, 24 and 2:19 exactly the same phrase in Hebrew – *nephesh chayyāh*, translated as 'living creatures' – is used there for animals as is used in Genesis 2 for 'the *adam* (man)'. And we note also that *adam* became a *nephesh*, he was not given one as an extra, so the text is simply pointing out that the life and breath of *adam* was completely dependent upon God's creative work, just as it was for the 'living creatures' in Genesis 1 (remembering that 'creation' refers to the ultimate dependence of all that exists upon God as creator). In the Hebrew Scriptures, it is not so much that a living body '*has* a soul', as in Platonic thought, but more that 'it *is* a soul'.

Yet Platonic rather than Hebrew concepts of the soul remained generally dominant in Christianised Europe during the era of the so-called scientific revolution. Contrarian voices are therefore of particular interest, one such being Thomas Willis (1621–75), a founding member of the Royal Society and one of the key founders of what we now call 'neuroscience'. Willis was a doctor who taught medicine at Oxford in the earlier part of his career, specialising in brain anatomy and physiology. Willis was also a devout Anglican, 'who allocated his rooms in Christ Church to Anglican services, donated the fees earned on Sunday to the poor and funded evening services at St Martin-in-the-Fields (in London) even beyond his death' (Green, 2003). It was Willis who, in his *Cerebre anatome*(1664), first combined the Greek words νευρον (neuron) and λογος (*logos*, 'word' or 'principle') in an otherwise Latin text to generate the word 'neurology'. In his book on brain anatomy Willis insisted that in both Nature and the Bible 'there is no Page certainly which shews not the Author, and his Power, Goodness, Trust, and Wisdom'. For our present context Willis' understanding of the soul is of particular interest: not content with one, he postulated two, a Corporeal Soul (common to humans and 'brutes') and a superior Rational Soul (found only in humans) whose main role seemed to be the supervision of the Corporeal Soul. In practice, Willis focused mainly on the Corporeal Soul in his writings, identifying it with the neurophysiology of the brain and thereby opening up the way to establishing a more unitary understanding of personhood. It was to be a further three centuries before the burgeoning science of neurology acted as a further stimulus to theologians to re-examine their neo-Platonic inheritance and reconsider the biblical understanding of such concepts.

The unitary concept of human personhood in the Old Testament is underlined by its understanding of *sheol*, that shadowy world of the departed which is the destiny of all following death (Johnston, 2002). It is the person who goes to *sheol* at death, not his soul (Job 30:23: 'I know you will bring me down to *sheol*, to the place appointed for all the living'). 'For *sheol* cannot praise you, death cannot sing your praise; those who go down to the pit cannot hope for your faithfulness' (Isaiah 38:18). The soul is envisaged as the real person. When the psalmist proclaims 'my soul will rejoice in the LORD and delight in his salvation' (Psalm 35:9), this could equally well be translated as 'I will rejoice in the LORD and delight in his salvation'. The same could be said of 'Praise the LORD, O my soul; all my inmost being, praise his holy name' (Psalm 103.1) and indeed of many other biblical passages where (in the NIV translation) the phrase 'my soul' appears no less than sixty-three times. There is a strong resonance here with the DAME view of personhood discussed in the previous chapter. In theological terms, there is a single 'substance', the human person, with a variety of different properties (or perspectives) including the physical body; the soul that expresses the state of being alive, in some biblical passages used in a manner similar to the 'I' of personal biography and personal willing; and the spirit, expressing the spiritual capacity that enables knowledge of and relationship with God. In later parts of the Old Testament we begin to see hints of a future resurrection, but it is an embodied resurrection (as in Daniel 12:2).

In the New Testament the resurrection of the body then becomes a major theme (as in 1 Corinthians 15). When immortality is mentioned in the New Testament, it never refers to the 'immortality of the soul'. Instead, in those places where the notion of immortality is unpacked, it is always in the context of the resurrection of the body (as in 1 Corinthians 15:50–4). And the tendency to think that every mention of the word 'soul' in the New Testament refers to some kind of neo-Platonic entity should be resisted. When Jesus exhorted his followers to 'Love the Lord your God with all your heart and with all your soul and with all your mind' (Matthew 22:37), he was simply referring to the expected response of every possible aspect of human personhood, not to some special 'spiritual component' of the human person. More expansive discussions (including discussion of biblical examples that might at first glance suggest a contrary view) may be found in the literature (Green, 2004, 2008).

In our present context the important point is that there is a recent theological emphasis on a holistic, monistic perspective on human personhood that coheres well with the DICI framework. Genomic variation, environmental influences and personal choice are all integrated in human development to generate the whole person, and it is the whole person who is made in the image of God and the whole person who responds to God, with profound consequences for their well-being, including their health (Koenig et al., 2012), their relationships and the choices they make on how to invest their time and money. This is why the science of human behavioural genetics can readily be welcomed by those who embrace such insights because there is nothing 'unspiritual' about the genome, nor indeed about the synaptic architecture of the brain, nor about sex, nor about any other aspect of our physical being. Likewise genomic variation should be welcomed, not only because of the obvious evolutionary perspective that without it we would not exist but also, as will be discussed further, as a guarantor of individual human uniqueness.

The Value and Status of Each Human Individual

The second strand in the conversation with genetics follows immediately from the theological discussion on the meaning of the 'image of God' in its historical context and relates to the value and status of each human individual irrespective of their genetic status. As already mentioned, the substantialist interpretation of the *Imago Dei* popular in a previous era perceived the image as comprising a list of distinctively human qualities, such as rationality and intelligence. Certainly all those qualities are the shared properties of humankind and are hugely valued. One real ethical weakness, however, intrinsic to the substantialist interpretation, arises from the fact that not all humans necessarily share those qualities, sometimes due to severe pathology, including genetic disease, or accident. The ethical strength of the understanding of the image of God that we have been outlining here rests on its rootedness in God's created order. Humankind's value lies not in some list of intrinsic qualities, but in God's grace, in his bestowal of a kingly, priestly status that we certainly do not deserve. And that status is bestowed on the whole of humankind as a community. Where the individual is unable to express or practise that kind of functionality, human solidarity insists that we care for and protect those less fortunate than ourselves. The care receiver is as much reflecting their status as being made in the image of God as the caregiver.

What happens to the applications of genetics when the idea of humankind being made in God's image is lost or distorted? A book-length answer would be required to do justice to that question. But two disparate exemplars of what *can* happen are informative – one historical, one contemporary. The first comes from Hitler's programmatic tract *Mein Kampf* ('My Struggle'), a book that sold ten million copies and influenced ideas about Aryan racial purity like no other (Hitler et al., 1939). In this evil book Hitler utilises image of God language but in a highly mutated sense, referring to the Aryan race as the 'highest image of God among His creatures', and using this as the basis for arguing against interracial marriage. For Hitler, Jews and Gypsies were created separately from the Aryan race, which of course does not exist, and it was the 'final solution' that would ensure the Aryan race's continuing purity, together with the eugenic elimination of hereditary criminals, the mentally ill and the disabled – a task ably assisted by geneticists who were active in establishing racial categories and in adjudicating on individual cases. Only Aryans were made in God's image in Hitler's racist ideology. There is nothing so powerful as the distortion of a great truth, and in *Mein Kampf* we have an echo of the Mesopotamian insistence that the image of the gods was bestowed on only a tiny elite.

A second very different example is what can be proposed when the idea of humankind made in the image of God is rejected in favour of a purely utilitarian moral philosophy. This is exemplified by a paper which appeared in 2012 in the *Journal of Medical Ethics* entitled 'After-Birth Abortion: Why Should the Baby Live?' (Giubilini and Minerva, 2013). In this paper the authors, expanding on the views of other utilitarian philosophers such as Peter Singer, argue that infanticide is morally justifiable for any trait, including especially genetic traits, that might equally justify abortion. As the authors point out, some genetic abnormalities are not hereditary but occur in the gametes of healthy parents, rendering prenatal diagnosis unlikely, so why not carry out the so-called abortion, meaning infanticide, after birth once the condition is apparent? Since the handicapped newborn is not yet a person because it has not yet formulated aims in life, then killing it is justified. In making their chilling proposal, it is not clear whether the authors are aware that it was precisely the then radically new notion of humankind having absolute rather than relative value, because made in the image of God, that led to the Christian-inspired legislation that outlawed the infanticide that was commonly practised in ancient Greek and Roman societies. In fact, infanticide was so common in the Graeco-Roman world that one contemporary historian, Polybius, writing in the second century BC, concluded that it had contributed to the depopulation that had occurred in Greece at that time (Amundsen, 1987). In Plato's description of the ideal state in *The Republic*, infanticide is essential to maintain the quality of the citizens: 'The offspring of the inferior and any of those of the other sort who are born defective, they will properly dispose of in secret, so that no one will know what has become of them' (Plato, 2012, Book 5, p. 460). In his *Politics*, Aristotle also maintained that no deformed child should be allowed to live: 'As to exposing or rearing the children born let there be a law that no deformed child shall be reared' (Amundsen, 1987, p. 10).

Rejection of the image of God framework that guarantees the value and rights of each human individual does entail certain consequences, which many would view as morally troubling. Singer has defended the practice of infanticide in this way: '[W]e should put aside feelings based on the small, helpless, and – sometimes – cute appearance of human infants . . . laboratory rats are "innocent" in exactly the same sense as the human infant . . . killing a disabled infant is not morally equivalent to killing a person. Very often it is not wrong at all' (Singer, 1993, p. 169). Singer is both accurate and explicit concerning the implications of rejecting a theistic framework for human value:

> If we go back to the origins of Western civilisation, to Greek or Roman times, we find that membership of Homo sapiens was not sufficient to guarantee that one's life would be protected.... Greeks and Romans killed deformed or weak infants by exposing them to the elements on a hilltop. Plato and Aristotle thought that the state should enforce the killing of deformed infants.... The change in Western attitudes to infanticide since Roman times is, like the doctrine of the sanctity of human life of which it is a part, a product of Christianity. (Singer, 1993, pp. 88 and 173)

Singer is correct in his suggestion that infanticide declined due to the Christian insistence on the sanctity of life. After the conversion of the Emperor Constantine to Christianity in AD 313, it was made a punishable offence in the Roman Empire for a father to kill his child in 318, and in 331 Constantine decreed that those who raised exposed children could legally adopt them. In 374, infant exposure was made punishable by law. Already by the end of the fourth century AD Christian hospitals were being established in which there was a section called the Brephotropheion set aside for orphaned children.

As Rachels comments: 'The abandonment of lofty conceptions of human nature, and grandiose ideas about the place of humans in the scheme of things, inevitably diminishes our moral status. God and nature are powerful allies; losing them does mean losing something' (Rachels, 1990, p. 205). The conviction that humankind is made in the image of God provides no simplistic answers to complex ethical questions. But it does act as a bulwark against the steady eroding of the understanding of humankind as having intrinsic value, and this fact has been noted by secular authors as much as those writing from within a religious tradition. In one significant U.S. sociological study, an assessment was made of the relationship between beliefs about human identity and moral attitudes concerning human rights (Evans, 2016). The study found that those who agreed with the view that humans are defined merely by a compilation of biological traits, like rationality, were more likely to think that it is morally acceptable to not try to stop genocides, to buy kidneys from poor people, to help people to commit suicide to save money and to endorse torture. Those who agreed with an extreme biological definition of humans reached similar conclusions, but in addition also agreed that it was justified to take blood from prisoners against their will. However, agreeing that humans are defined by their being made in the image of God tended to result in more endorsement of human rights, such as being more likely to agree with stopping genocides, being opposed to suicide to save money and being against the idea of taking blood from prisoners against their will. As Rachels pointed out, the framing of human value within the image of God paradigm really does make a difference.

Subduing the Genome?

The third strand in the conversation between the *Imago Dei* and genetics relates to the command given to humankind to rule over and subdue the earth. As James Luther Mays comments: 'It is in its theocentricity that the Bible is anthropocentric' (Mays, 2006, p. 27). Now I am well aware of the environmentalist and feminist critiques of this mandate, which blame the current ecological crisis and the oppression of women on the presumed anthropocentric and patriarchal nature of such language. Space does not allow a full commentary on those critiques.[3] But whatever may be the historical abuses of the *Imago Dei* vision, as far as the context of the passage is concerned, there is no doubt that its implications are liberating, indeed revolutionary in its context, for the whole of humankind, female and male, for the reasons already outlined. Humankind is being made in the image of a God whose creative work, as the context describes it, is very different from the *Chaoskampf* or creation-by-combat narratives found in the rival creation epics of the time. Instead, what we find is the language of quiet command and of invitation, against which there is no struggle. Creation is described via the Hebrew jussive, which we do not have in English grammar, but which is normally translated as 'let' – 'Let the water teem with living creatures' and 'Let the land produce living creatures', the language of invitation. The creation of humankind is lumped together with the other animals on Day 6. A separate creative day for humankind would have smacked of divinisation, but the actual Day 6 is a reminder of our unity with the rest of creation. And the creation is repeatedly declared to be good or very good. Here there are no bloody battles as you find in the *Enuma Elish*. God is pictured not as warrior but as artisan or craftsman. So this is the kind of God in whose image humankind is made. And in Genesis 2 we see humankind being called to reflect that image in caring for the environment – notice 'care', not limitless dominion.

So if we take the mandate for creation care as being an important entailment of what it means to be made in the image of God, then how do we apply that with reference to human genomic care? Responsibility for our own genomes and those of others through the healing of genetic diseases is obvious, but where are the limits? Are the transhumanists correct in their assertion that it is fatalistic to accept our genomic inheritance as it is, or should we be manipulating the human genome in an attempt to enhance our present capabilities? Once again our understanding of humankind made in the image of God may supply a needed corrective. For whereas such a framework for human value clearly subverts any fatalistic acceptance of the human lot, providing a powerful motivation for creation care and healing, it falls far short of divinisation. Indeed, Genesis 3 provides a vivid account of what happens when humankind strives for autonomy apart from God, setting themselves up as the final arbiters of what it means to be human. As it happens, the sheer complexity of the genome, as surveyed in earlier chapters, makes the transhumanist hopes for genetic enhancement unlikely to be achieved in practice, apart perhaps from the ability to choose some rather trivial traits for the future child, such as a particular eye colour or hair colour. Change one gene and many others might change their function in the system with unpredictable consequences. As far as serious enhancement is concerned, the complexity of the human genome is its own best defence against meddlers. As already noted, hundreds or thousands of genetic variants are involved in contributing to the variation in complex human traits like intelligence or aggression. Before GWAS and the broader programme of human behavioural genetics and genomics brought some greater scientific detail into view, it was easy to make optimistic predictions about human enhancement based on some supposedly simple manipulations of the human genome. Those earlier expectations have now been shown to be naive.

That comment, however, still leaves some pressing ethical questions on the table with regard to medical genetics. What about those severe pathologies caused by genetic mutations, such as Hurler syndrome, which in earlier chapters were taken as exemplars of genetic determinism, in the sense that they entailed individuals being 'constrained to follow one particular future'? In the current absence of any cure, how are we supposed to 'subdue' the unruly genome that is responsible for such suffering followed by early death? The use of preimplantation diagnosis under such circumstances seems a very reasonable strategy. This involves in vitro fertilisation (IVF) followed by growth of the embryo to the stage at which it contains four to eight cells. One or two cells can then be removed without damaging the embryo and defective genes identified. Embryos carrying the defective gene are discarded and only the healthy embryos are implanted in the mother. In practice, pre-implantation diagnosis is normally offered to parents where there is a known family history of debilitating genetic disease. Around 250 tests for different conditions are available in specialised centres and only a few hundred of such screenings are carried out per year in the United Kingdom. Those who believe that deliberately discarding early embryos is wrong under any circumstances will not embark on such a procedure. Others will think that discarding early embryos represents a lesser evil than bringing a child into the world that is destined to die slowly and painfully. Prenatal diagnosis using fetal cells obtained by amniocentesis some twelve to fourteen weeks into the pregnancy, followed by therapeutic abortion if the fetus is a carrier of a genetic mutation that predicts severe disease, raises yet more ethical dilemmas.

For the moment, it is worth noting that these ethical dilemmas are likely to become more rather than less difficult. The reason for this is a technical advance known as non-invasive prenatal genetic testing (NIPT) which involves taking a blood sample from a pregnant mother when the fetus is just ten weeks old (King, 2012, Brady et al., 2015). Fetal DNA circulating in the mother's blood can then be used to screen for genetic or chromosomal abnormalities, as well as determine the sex of the baby. NIPT is already in routine use as an initial screening method for well-established mutations of known effect. But soon it will be possible to obtain an accurate and complete genome sequence from such samples, allowing calculation of the risk factors for developing a wide range of future diseases, some unlikely in any case to emerge until adulthood, and many never to appear at all. The potential for worried parents to terminate pregnancies unnecessarily in these circumstances seems very high. Can the moral framework that the *Imago Dei* provides give some guidance under such circumstances? The communal aspects of the image of God status have already been emphasised. Even in cases of severe genetic pathology entailing early death, the affected child is (or should be) part of a community of care givers who together in their care demonstrate their commitment to the intrinsic value of the worth of each human individual. It has nevertheless been suggested that if an ethically acceptable way of preventing such a suffering child ever coming into the world exists, then that way should be taken. But what about those with a family history of a BRCA1 mutation that, as previously discussed, generates a probabilistic prediction that, if the female fetus is a carrier of that mutation, the child yet to be born will face an 80 per cent risk of developing breast cancer, most probably after decades of cancer-free living? Many factors will come into play at this point, not least the thought that therapies for breast cancer are becoming ever more efficient, so in forty years' time it is highly likely that they will have improved out of all recognition compared to the present. And regular check-ups to catch any sign of cancer early can also be of benefit. In practice, the use of NIPT for whole genome sequencing, if and when it becomes routine, is far more likely to detect gene mutations that comprise part of a polygenic cohort of genes providing probabilistic lifetime risk estimates far lower than the 80 per cent prediction for BRCA1. Here the moral framework that the *Imago Dei* provides points in a different direction. Whilst being robust in our motivation to cure genetic disease in those already born, and taking steps to prevent the birth of children destined for a painful early death, we can at the same time revel in the sheer genetic diversity of the human population, knowing full well that in the light of DICI, and outside of severe clearly predictable pathologies, the future of the individual growing embryo is in any case indeterminate. The allure of total control is a fantasy to be resisted, and certainly no life is without risk.

A further development that raises a somewhat different, though related, set of ethical dilemmas arises from the application of a genetic engineering technique known as CRISPR-Cas (Maxmen, 2015; Ledford, 2016). This enables potentially precise changes to be made anywhere in the human genome. The molecular toolkit contains a so-called single-guide RNA molecule that targets the system to the precise spot in the DNA where a change is required. The Cas part of the system is a protein that cuts the DNA. The system can be used to repair mutations in a given gene that may be causing disease. In 2015 a Chinese research group reported the genetic modification of human blastocysts derived from eggs that had been fertilised by two sperm simultaneously and so were unable to undergo further development (Liang et al., 2015). The blastocysts were obtained from parents who carried gene defects responsible for β-thalassaemia, a genetic disease of the blood. The researchers then used CRISPR technology to target the defective gene in each case responsible for thalassaemia. The procedure was very inefficient and in only four out of the eighty-six embryos injected – that is, 5 per cent of the embryos – was the mutation corrected. At the same time, there were many off-target changes in the genome; in other words, the system they used was profligate in targeting other parts of the DNA in unwanted places. Nevertheless this is the first report of the genetic modification of human embryos in such a way that, had the embryos been viable, they could have been implanted to generate babies with the future potential to pass the 'corrected' gene down to future generations. This initial report will certainly be followed by many other attempts to 'gene-edit' human embryos to reverse disease-causing genetic mutations.

The scientific community is split on the ethics of the applications of CRISPR technology in terms of both human embryo editing and in terms of its potential to genetically modify organisms followed by release into the wild, there to pass their modified gene(s) to succeeding generations (Lanphier et al., 2015, Reardon, 2015). At present there is a moratorium in the United Kingdom and other countries (but not when using non-governmental funds in the United States) on carrying out human embryo editing with the aim of human reproduction. The main reason is that the technique is still too inefficient and too inaccurate to be safe: unless it can be guaranteed that the CRISPR-Cas construct only repairs the mutant gene without changing other parts of the genome, further progress will be limited. Nevertheless, it seems highly likely that such technical problems will be overcome. Will it then be ethically justifiable to genetically cure 'embryos' in the laboratory of their disease-causing mutation? Pragmatically, one response to that question might be, why not use pre-implantation diagnosis instead? That indeed might be the best solution in many cases, but what about those rare cases where both parents are carrying a mutation for a lethal disease that displays a dominant pattern of inheritance? In that case, 100 per cent of their embryos will carry the lethal mutation. And perhaps there will be some rare mutations that cause other diseases where pre-implantation diagnosis is impractical. At this stage, does the concept of humankind being made in the image of God bring any useful insights into the discussion? We remember the one made perfectly in God's image, Jesus of Nazareth, for whom healing was a central aspect of his ministry. If the issue is the genetic healing of an embryo to enable parents to have a healthy child who might otherwise have been born with a lethal disease (entailing early death), and moreover a healing that will be passed down to subsequent generations, then surely this should be a matter for rejoicing? Two objections to such a conclusion are already rife in the literature. The first is that the child-to-be-born never chose to be born with a modified genome. But this appears to be a weak argument. As the genetically normal sib grows up with an older sib who is dying from a lethal genetic disease, one might guess that the younger sib would sense only relief as she sees from what she has been saved. The second objection is more potent: is this not a slippery slope, so that once the technology is safe to change a DNA sequence back to the one that most people have to avoid lethal disease, the pressure will be on to manipulate the genome in some way to generate enhanced people? This possibility indeed has some traction, but its danger is lessened for the reason already highlighted: the multigenic and probabilistic nature of the development of human traits suggests that enhancement is a non-starter. There is, in any case, a good evolutionary reason why one would not wish to embark on such a risky procedure purely for safety reasons. Consider the α-iduronidase mutant gene, already discussed, that is the cause of Hurler syndrome. This is a gene that in its normal functional form has been tested out in the workshop of life by millions of years of evolution. It is known to function well in many different organisms. Reversing a mutant version of the gene to its normal form would therefore represent a safe risk-free transition. This would not be the case if an attempt were made to increase the efficiency of a gene by introducing a mutation not previously found in human populations. Since a mutation would not have been tested out in human populations, given the known interactions between multiple genes, this would represent a genuine experiment, and one should not experiment with children. In practice, mutations would be sought that are already part of natural human variation and that were predicted to provide some benefit. But the problem, then, as many previous chapters testify, is that this is easier said than done.

Given that the idea of humankind being made in the image of God refers to the whole human community, with mutual care and cooperation being central to that framework, the adoption of a child or, indeed, of an embryo, might provide a better way forward than engaging in the risky genetic modification of embryos. Whilst recognising the often powerful desire of parents to have genetically related offspring, that perfectly valid desire needs to be balanced against the risks and costs of alternative approaches. The focus on genetic parenthood tends to drive new reproductive technologies in costly and risky directions. Embryo adoption from an IVF clinic is certainly worth considering as an alternative (Banner, 2014).

It should be obvious, by now, that the notion of the *Imago Dei* provides no 'magic wand' solution to the tricky ethical dilemmas with which contemporary biology continues to face us. Nonetheless, as a general framework it does often help nudge the conclusions in one direction more than another and, when really horrific suggestions are made (as in the Hitler and infanticide examples), provides much more than a nudge – in those kinds of cases an absolute barrier.

The Celebration of Diversity in Community

The fourth resonance that we note between the idea of the image of God and our genetic identity is its insistence that it involves diversity – male and female he created them, what the theologian Karl Barth claimed as 'the definitive explanation given by the text itself of the image of

God' (Barth et al., 1975, p. 182). In Barth's view, God's image in man *is* the reciprocal relationship of human being with human being. The image describes the I-Thou relationship between person and person and between the person and God. There is a consensus, I think, that Barth somewhat overstated the case, but there is an important truth here. The image is not a static status quo, a list of fixed human characteristics, but a dynamic ongoing developmental process carried out in relationship, less like a mirror, more like a prism in which God delegates responsibilities through the human social community in which male and female equally collaborate (Grenz, 2001). John's Gospel (chapters 14–17) spells this out further by expressing the relational character of God's being as Father and Son, leading to a theology of imaging and corresponding: as the Father loves the Son and the Son loves the Father, so the Son loves his disciples, so the disciples should love each other, so are they called to express that love to the world in community. This whole dynamic is summarised in the concluding verse of Jesus's high priestly prayer (John 17:26). As Clive Marsh comments: 'In the same way that God *is* Trinity, and thus embodies relationality within God's own self, so therefore what it means to reflect God must itself be seen relationally.' Being made in the image of God refers to humanity in community, not to humans as atomised individuals.

This theological framework coheres well, I would suggest, with the biological story that DICI provides, a story that focuses on the development of diversity and that guarantees the uniqueness of every human being who has ever lived or whoever will live on this planet. It is not that cloned human beings could not have fruitful relationships with each other, but I-Thou relationships would certainly be challenging and most likely impossible in a large cloned population. In reality, as the previous chapters have emphasised, the genetic diversity that exists between individuals is far greater than we had imagined even a few years ago. Far from being seen as a threat, my suggestion is that within the matrix guaranteeing human value and dignity that the *Imago Dei* provides, we are then free to celebrate this diversity and the immense variation in the expressions of personhood in community to which it contributes.

Moral Responsibility and Free Will

The fifth point that we notice in the idea of the image of God is its entailment of moral responsibilities and duties placed on humankind, implying genuine choice. This might seem obvious to us now, but the very suggestion posed a profound threat to the reigning Mesopotamian ideology in its original context. In such ancient societies the sociopolitical order was seen as a microcosm of the larger world of the gods. Any deviation from the original divine pattern was seen as regression and so was to be resisted, of particular relevance to the mass of the population who had fixed social roles (Pollock, 1999; Algaze, 2008; Kriwaczek, 2010; Baigent and Baigent, 2015). Free choice did not come into it. By contrast, in Genesis the delegated responsibility of subduing and caring for the earth was given to the whole of humankind. Furthermore, a clear choice was presented. The command is given in Genesis 2 not to eat of the tree of the knowledge of good and evil; this powerful figurative imagery is usually taken to represent the option for humanity to choose its own way rather than to pursue God's priorities. The command is disobeyed; human autonomy seems much more alluring. There are terrible consequences: separation from the Tree of Life. The assumption of genuine human responsibility is inescapable.

There is a certain irony in the reflection, therefore, that at least some contemporary scientists seem to be operating more within the matrix of fate and destiny that characterised the societies of ancient Mesopotamia than within the matrix of freedom inherent in the rival idea of the *Imago Dei*. Of course the contemporary matrix derives from a different pantheon of gods from those of the Assyrians and Babylonians. In the account shaped by biological reductionism, it is reified genomes, reified neurophysiology and reified neuronal networks that are purported to provide the 'real story' of what is going on, but the fatalistic outcome looks rather similar. In the Introduction we have provided some empirical data suggesting that belief in determinism is deleterious for health, for human flourishing and for society more generally. And during the course of the book, we have provided many reasons for thinking that the human genome is intimately involved in the development of human beings whose experience of free will is not illusory, threatened only by severe pathology or by severe constraining circumstances. The responsibility to care for the genomes of others has already been discussed, but let us not forget to care for our own genomes, something that we can do every day of our lives. And as discussed in Chapter 4, our own free choices in turn impact epigenetically not only on our own lives in ways that affect our future health and well-being but also very likely in ways that affect our children and our children's children. No person is an island: all our choices have consequences for those around us whether for good or for ill, both in our generation and in generations yet to come. Cain's plaintive cry haunts us: 'Am I my brother's keeper?'

(Genesis 4: 9). The conviction that humankind is made in the image of God provides a clear yes in response.

Concluding Remarks

There are some who may have appreciated the science in this book, the history perhaps also, and maybe even the philosophy, although the theology in this final chapter might finally have triggered some indigestion. Yet we need to be reminded as to how historically recent is the partitioning of academic disciplines into specialised domains of thought that now rarely find the time or the inclination to talk to each other. Natural philosophers were still dominant in the first part of the nineteenth century, when it was very normal for science, history, geography and theology to form part of the same discourse on some particular topic, but by the closing decades of the century, when a distinct profession of 'scientists' began to emerge, disciplinary boundaries became increasingly demarcated, and academics from the different disciplines began to lose the habit of talking to each other. True, specialisation has been, and continues to be, hugely fruitful in terms of generating advances in the sciences. My research group worked for twenty years on the functions of one molecule, known as CD45, in the immune system (we did have plenty of other projects going in parallel I hasten to add). I am an enthusiast for methodological reductionism – in many branches of science it is the only way to make real progress. But at the same time it is often both academically and personally fruitful to engage in interdisciplinary projects, or at least to focus on some interdisciplinary reading and thinking which extends one's horizon beyond the walls of the laboratory (I speak as a biomedical research scientist).

As intimated at the beginning, this is a 'both-and' not an 'either-or' book. Different disciplines provide complementary narratives to help in the understanding of a single reality. There is no need, most of the time, for such narratives to be in any kind of rivalry with each other: we need them all to do justice to this complex reality that we call 'life'. So some chapters in this book highlight the need for human behavioural geneticists to engage more with analytical philosophers to bring greater conceptual clarity to a field where this is often lacking. Geneticists, philosophers, sociologists and theologians need to engage with each other more on the challenging subject of free will, at least in an attempt to understand each other's 'language game', and again to bring conceptual clarity to the discussion. If scientists think that they have 'disproved free will' and convey their scepticism to the public, whereas meanwhile philosophers and theologians are insisting that the scientists' concept of free will has little to do with 'real free will', then clearly we have a problem. And as genetic advances continue to be made at a breath-taking pace, ethicists and theologians need to be brought to the table to contribute to the discussion. The great fear in the scientific community is that ethical concerns will slow the pace of research or even outlaw certain procedures or lines of enquiry. Would that such retardation had been more potent during the Third Reich! In any case, given that modern science was nurtured, shaped and encouraged within a context strongly influenced by a Judaeo-Christian world view, at least as far as European science is concerned, a world view that continues to give many good reasons for the positive pursuit of science, there need be little concern that the advance of science need suffer in the longer term. And if a more cautious humility becomes the defining characteristic of the scientist who communicates their science to the public, then that is no bad thing.

Just two novel acronyms have been used in this book, both admittedly a bit of a mouthful if spelt out, but both challenging 'either-or' attitudes and substituting them with 'both-and' perspectives. DICI (pronounced 'Dicey') – developmental integrated complementary interactionism – aims to subvert all those dichotomies which tend to fragment human personhood: 'nature-nurture', 'genes-environment' and the like. Its close sister DAME (developmental dual-aspect monistic emergentism) again highlights the integrated unity of human personhood, albeit from more of a philosophical perspective. And in the present chapter we have noted that, in much contemporary theological thought, it is not a fragmented humanity having different components that is made in the image of God, but a psychosomatic unity, the 'real I', who becomes answerable to God for our actions by the exercise of free will. Therefore, in the pursuit of history, science, philosophy and theology, we find that we end up in a similar place. T.S. Eliot made a parallel point many decades ago:[4]

> We shall not cease from exploration
> And the end of all our exploring
> Will be to arrive where we started
> And know the place for the first time.

Notes

Introduction

1   Following a highly critical view of this book in the *New York Times* (10 July 10, 2014), more than 130 leading population geneticists published a letter in the same newspaper on 8 August 2014, writing that '[w]e reject Wade's implication that our findings substantiate his guesswork. They do not.'
2   http://www.bbc.co.uk/news/health-20583113. Accessed 12 March 2013.
3   http://www.huffingtonpost.com/2011/01/15/british-study-links-sprea_n_809394.html. Accessed 12 March 2013.
4   http://digitaljournal.com/topic/caring+gene. Accessed 12 March 2013.
5   http://www.newscientist.com/article/mg21028126.300-teen-survey-reveals-gene-for-happiness.html. Accessed 12 March 2013.
6   McMahon, B. *The Times*, 9 March 9, 2013, Body and Soul, p. 4.
7   *The Times*, 15 January 2016, p. 1.
8   *The Times*, 24 July 2015, p. 18.
9   Interview with Will Hodgkinson, *The Times*, 16 March 2013, Saturday Review pp. 8–9.
10 http://www.nature.com/news/2008/080404/full/news.2008.738.html. Accessed 12 March 2013.
11 http://www.nature.com/nature/journal/v468/n7327/full/4681049a.html. Accessed 12 March 2013.
12 Carmichael, M. 'DNA relax, have a drink'. *Newsweek*, 13 May 2002. p. 11.
13 Stein, L. 'Bad genes'. *U.S. News & World Report*, 31 March 2003. p. 45.
14 This narrative has now been replaced by a different text: 'Your genetics may dictate, for example, what foods you like, what diseases you are prone to develop, how smart you are, and likely factor into nearly every aspect of your being' (http://www.mapmygene.com). Accessed 25 February 2016. The word 'dictate' still has powerful deterministic overtones.
15 *The Times*, 20 November 2014.

1 Human Personhood Fragmented?

1   Plato's Dialogues, Book IV. http://www.sacred-texts.com/cla/plato/rep/rep0400.htm. Accessed 22 October 2015.

2   http://en.wikipedia.org/wiki/Nature_versus_nurture. Accessed 26 February 2016.

3   Most notably by Karl Popper in Volume I of his *The Open Society and Its Enemies*, 5th edition revised, 1966.

4   Others would express this rather differently: the essence of the Forms can only be discerned by reference to the essence of 'sensible things'.

5   Similar claims are made of other disciplines also. See Davison, A. *The Love of Wisdom*, pp. 32–3.

6   Aristotle, *Physics* II.3; *Metaphysics* V.2.

7   That which is 'neither independent of matter nor can be defined in terms of matter alone'. Aristotle, *Physics* II.2.

8   There is no evidence that Galton had ever read *Silence*, which in any case was only discovered in the year that he died.

9   Shakespeare, *The Tempest*, Act 4, Scene 1, p. 9. http://nfs.sparknotes.com/tempest/page_162.html. Accessed 23 March 2017.

10 Descartes, *Meditations* 5, AT 7:64.

11 Descartes, 1643 letter, AT 8b:166–7.

12 Also known as the *Vegetable Lamb of Tartary*. *Scythia* was the name the Greeks gave to what we now call Central Asia.

13 http://galton.org/letters/darwin/darwin-galton.html. Accessed 26 March 2014.

2 Reifying the Fragments?

1   Cited in Jaroff, L. 'The Gene Hunt', *Time*, 20 March 1989, 62–7.

2   In a personal letter to Adam Sedgwick, dated 18 April 1905.

3   Conn. Gen. Stat. sec. 1354 (1902).

4   'Though Raymond Pearl criticized eugenics as far back as 1927 … the data demonstrate no sharp drop in the presentation of the topic through the 30s, 40s and 50s, only a gradual decline. This supports Wendy Kline's claim from *Building a Better Race* (2001) that the eugenics movement was not "weak and discredited after 1930," as many scholars contend, but had worked its way deeply into the popular consciousness'. Ronald Ladouceur, 'Eugenics in 20th Century Biology Textbooks,' posted 9 February 2010, at www.textbookhistory.com/?p=3127.

5   For a good review of the contributions of Lancelot Hogben, see Tabery, 2014.

3 The Impact of the New Genetics?

1   The RNA Modification Database lists 109 different kinds of modification. See http://mods.rna.albany.edu/Introduction/overview. Accessed 18 October 2015.

2   It should be noted that the methodology and interpretation of these results have been critiqued (Brown et al., 2014, with a response by the authors Cole and Fredrickson, 2014, who have since extended their findings ([Fredrickson et al., 2015]).

4 Reshaping the Matrix?

1   As a bonus, DICI also happens to be the Latin word pronounced *deeky* by classicists or *deechy* by theologians, the present infinitive passive, meaning 'to be said,' although that is not why the acronym was chosen. I am grateful to Stephen Halliwell, professor of Latin at St. Andrews University, for picking up my faux pas during my second Gifford Lecture in translating DICI as 'it is said'. It is the benefit of such an inter-disciplinary audience that statements outside of one's own area of expertise can be quickly rectified.
2   ASPM stands for 'abnormal spindle-like microcephaly associated'.

5 Is the Worm Determined?

1   *The Guardian*, 27 October 2008.
2   *The Human Genome: Model Organisms: The Fruit Fly.* http://genome .wellcome.ac.uk/doc_WTD020807.html. Accessed 20 April 2015; *Drosophila as a Model Organism.* modencode.scimagdev.org/drosophila/introduction. Accessed 20 April 2015.
3   *Drosophila as a Model Organism.* modencode.scimagdev.org/drosophila/ introduction. Accessed 20 April 2015.
4   Ceolas.Org. 2008. A *Quick and Simple Introduction to Drosophila Melanogaster* http://www.ceolas.org/fly/intro.html. Accessed 20 April 2015; *Drosophila as a Model Organism.* modencode.scimagdev.org/drosophila/introduction. Accessed 20 April 2015.
5   http://www.newscientist.com/article/dn14641-monogamy-gene-found-in-people.html#.VRQtTChTMiA. Accessed 26 March 2015.
6   http://blogs.sciencemag.org/sciencenow/genetics/. Accessed 26 March 2015.
7   https://duluxdreams.wordpress.com/2012/08/28/infidelity-written-in-the-genes/. Accessed 26 March 2015.

6 Prisoners of the Genes?

1   Lewontin to Dobzhansky, 2 May 1973, Dobzhansky Papers (Lewontin File), APS Library.
2   The authors of this paper have established a very useful website at http://match.ctglab.nl/#/home, which enables the exploration of their data regarding any of these traits.
3   This is a statistical statement, in the sense that an arbitrary cut-off was used to determine 37% instead of 36%, 34%, 30% or even 5%. It is a tail of a distribution where the likelihood of having a certain percentage of alleles in common decreases exponentially away from 50% in both directions, assuming Mendelian inheritance.
4   See Plomin (2008) for the derivation of Falconer's formula.
5   The correlation coefficient, r, can be calculated whenever the scores in a sample are paired in some way. Its sign indicates the direction of the

relationship. If r = 0, there is no correlation at all. If r = +1, then there is a perfect positive correlation. If r= -1, then there is a perfect negative correlation.

7 Behavioural Molecules?

1   http://ghr.nlm.nih.gov/handbook/genomicresearch/snp. Accessed 19 February 2016.
2   http://www.ensembl.org/Homo_sapiens/Info/Annotation. Accessed 19 February 2016.
3   Up-to-date information may be found at http://www.broadinstitute.org/collaboration/giant/index.php/GIANT_consortium. Accessed 28 March, 2017.
4   The Schizophrenia Research Forum provides up-to-date research news: http://www.schizophreniaforum.org/. Accessed 28 March, 2017.
5   http://sfari.org. Accessed 6 July 2015.
6   In fact, as of 9 May 2015, only one GWAS study of aggression had been reported in the international GWAS catalogue at http://www.ebi.ac.uk/gwas/search?query=aggression.
7   'Maori Violence Blamed on Gene', Wellington: *The Dominion Post*, 9 August 2006, Section A3.
8   http://www.prisonstudies.org. Accessed 22 February 2016.

8 Mensa, Mediocrity or Meritocracy?

1   http://www.goodreads.com/author/show/43211.Will_Cuppy, Accessed 21 March 2017.
2   White (2006) contains plenty of interesting historical material relating intelligence testing to UK educational policy, but the overall thesis of the author, that the roots of intelligence testing are to be found in puritan theology, I find, in agreement with Ian Deary (2012), 'not convincing'.
3   However, a Supreme Court ruling in 2014 on the case of Freddie Lee Hall (on death row for 35 years in Florida), based on the report that his IQ scores over the years had ranged from 60 to 80, ruled that a sharp cut-off point could not be made since the testing was too imprecise. See http://blogs.nature.com/news/2014/05/flawed-definition-of-intellectual-disability-struck-down-in-death-penalty-case.html. Accessed 21 March 2017.
4   A death row prisoner, Kevin Green, who in 1991 scored 71 on an IQ test last normalised in 1972, might have scored only 65 on a test normalised to the year he took it. After Green was convicted and sentenced to death in 2000, his lawyers appealed, arguing that the court should correct for the Flynn effect. Nevertheless, Green's score exceeded Virginia's cut-off IQ value of 70, and he was put to death in 2008 (Reardon, 2014b).
5   Dr. Duncan Astle, MRC Cognition and Brain Sciences Unit, Cambridge, UK. Personal Communication.
6   http://match.ctglab.nl/#/ home. Accessed 3 December 2015.

7   Also see http://match.ctglab.nl/#/ home. Accessed 3 December 2015.

9 Gay Genes?

1   Pew Research Center. *A Survey of LGBT Americans: Attitudes, Experiences and Values in Changing Times*, Washington, DC: Pew Research Center (2013).
2   For an expanded discussion, see the longer version of this chapter published online at www.scienceandchristianbelief.org.
3   Pew Research Center. *A Survey of LGBT Americans: Attitudes, Experiences and Values in Changing Times*, Washington, DC: Pew Research Center (2013).
4   For an example, see Strudwick, P. 'You Do Not Choose to be Straight or Gay; It Chooses You', *The Independent*, 26 January 2012. http://blogs.independent.co .uk/2012/01/26/you-do-not-choose-to-be-straight-or-gay-it-chooses-you/. Accessed 7 November 2014.
5   Parris, M. 'Who's Totally Gay? There's No Straight Answer', *The Times*, 21 April 2013. http://www.thetimes.co.uk/tto/opinion/columnists/matthewparris/ article3390885.ece. Accessed 7 November 2014.
6   For examples of elision between attraction and behaviour, see some of the responses to a survey conducted in the United States in 2007 (Sheldon et al., 2007).
7   It is worth reiterating that this chapter discusses the aetiology of same-sex attraction only, not behaviour. It is not disputed that sexual behaviour can be strongly influenced by peer and parental attitudes and by legal and moral codes. The socialisation aetiology posits that attraction as well as behaviour is influenced by these factors.
8   Jude, J. *Gay Gene Rising*, PrideInspired.com (2011); Sones Feinberg, L. *The Gay Gene Discovery*, GLB Publishers (2008).
9   Conventions of chromosomal mapping nomenclature can be found at http:// www.nature.com/scitable/topicpage/chromosome-mapping-idiograms-302. Accessed 21 March 2017.
10   It is worth noting that these results have not been published in the peer-reviewed literature and measured only self-identification of homosexuality, rather than SSA.
11   The term 'intersex' is used very broadly in the literature to describe a range of conditions where elements of male and female are present in one individual, including disorders of sex development like hermaphroditism or chromosomal aneuploidies and behavioural conditions such as gender dysphoria.
12   For an expanded discussion and references, see the longer version of this chapter online.

10 Not My Fault?

1   Yerkes, R. Robert M. Yerkes Papers. Manuscripts & Archives, Yale University.
2   Cookson, C. 'Controversial Search for the Criminal Gene: A Conference the Americans Would not Allow', *Financial Times*, 14 February 1995.

11 Causality, Emergence and Freedom?

1   Schopenhauer, A. 'Über die Freiheit des menschlichen Willens (On the Freedom of the Will)', an essay presented to the Royal Norwegian Society of Sciences in 1839.

2   Helpful books in this context, written from various perspectives, include Mele (2003, 2015), Pink (2004), Kane (2005, 2011), Searle (2007), Holton (2009), Balaguer (2010), Tse (2013), Levy (2014) and Baggini (2015).

3   See also http://consc.net/papers/ncc2.html. Accessed 29 December 2015.

4   The neuronal system has ways of coping with stochastic 'thermal noise' to avoid unwanted perturbation. The energy from a putative neuronal quantum event has been estimated as about a thousand million times smaller than the energy of thermal noise, so potentially functional quantum events would be swamped by the thermal noise and, therefore, insignificant. See Clarke (2010, 2014).

5   The philosopher Mark Balaguer maintains that genuine free will does depend on the presence of indeterministic events in the brain but that we do not yet know how large the role of such events might be (Balaguer, 2010). In any event, the term 'indeterministic' from a biological perspective can readily be subsumed under the notion of 'nomic regularity'.

6   Reason-based conscious decision-making is the subject of a large philosophical literature. See, for example, Mele, 2003, 2009b, 2010).

7   I have derived this term from my old tutor from Oxford undergraduate days, Arthur Peacocke, who propounded his own position as 'emergentist monism' (Peacocke, 2006).

8   Some authors opt for a greater degree of emergentist pluralism, emphasizing the context-dependent understanding of emergentism in each case (Bedau, 2010).

9   An issue of the journal *Interface Focus* was devoted to a series of useful articles on 'downward causation': Vol. 2 Issue 1, 2012.

10 A worry expressed, for example, by Jaegwon Kim (1994a). Many responses to this worry may be found in the literature, such as Murphy (1999). Kim's 'exclusion argument' claiming that supervenience forces one to either deny mental causation or to embrace a reductive physicalism is here rejected on the grounds of insufficient nuance in the understanding of these concepts, including the concept of 'non-reductive physicalism'. See also the discussion on 'Anomalous Monism', proposed by Donald Davidson, at http://plato.stanford.edu/entries/anomalous-monism/. Accessed 30 December 2015.

11 Book I, Part III, Section VI.

12 Spinoza 1677, *Ethics* I, axiom III.

13 For a technical discussion and critique, see the *Stanford Encyclopedia of Philosophy* at http://plato.stanford.edu/entries/causation-mani/. Accessed 21 December 2015.

14 As noted in the Introduction, in philosophy, 'determinism' refers to the belief that any world with the same laws and the same past will have the same future. It has been clear since the advent of quantum theory that such a Laplacean

world does not exist. But in any case, the more restricted definition of 'hard determinism' used here is more relevant to the topic in hand.

15   For example, there are many points of agreement between the views expressed here and those of Julian Baggini, an atheist, in his book *Freedom Regained* (Baggini, 2015).

12 Made in the Image of God?

1   In a lecture delivered before the Greenock Philosophical Society, 1874 (cited in S.L. Jaki. 1986. 'Lord Gifford and his Lectures', Edinburgh: Scottish Academic Press, p. 106).

2   For example, Niebuhr (1945); Moltmann (1981); Barr (1993); Van Huyssteen (2006).

3   There have been many critiques of the suggestion by Lynn White, Jr. in 'The Historic Roots of Our Ecological Crisis' (White, 1967) that the 'dominion' language of Genesis can be blamed for the abuse of the environment (Bauckham, 2011; Marlow, 2009; Attfield, 1991). It is true that the Hebrew word for 'rule' in Genesis 1:26 (*radah*) is generally used elsewhere in the Old Testament to refer to military rule or political authority (language perfectly appropriate for kingship), but at the same time the Old Testament vision for kingship entailed caring for the weak, the needy and oppressed (Psalm 72:12–13) and protection from violence (Psalm 72: 14). The king was supposed to display 'truth, humility and righteousness' (Psalm 45:4). The prophet Ezekiel, for example, lambasts the harsh and brutal rule of the Israelite kings, using the same Hebrew word for rule – *rada* – as used for humankind in Genesis 1, reminding them that they were supposed to be shepherds caring for the flock ('You have not strengthened the weak or healed the sick or bound up the injured. You have not brought back the strays or searched for the lost. You have ruled them harshly and brutally.' Ezekiel 34:4). The portrayal of kingship 'rule' in Genesis 1 is therefore consistent with the description of humanity in action together in Genesis 2 as they name the animals and care for the earth. It should also be kept in mind that in its historical context, it was a far greater challenge to 'rule' over the earth in the Ancient Near East than in high-income countries today.

4   'Little Gidding', 1942.

Bibliography

Abbott, A. 2015. Neuroscience: The Brain, Interrupted. *Nature*, 518: 24–6.

Adhikari, A. 2015. Basomedial Amygdala Mediates Top-Down Control of Anxiety and Fear. *Nature*, 527: 179–85.

Alanko, K., Santtila, P., Harlaar, N., et al. 2010. Common Genetic Effects of Gender Atypical Behavior in Childhood and Sexual Orientation in Adulthood: A Study of Finnish Twins. *Arch Sex Behav*, 39: 81–92.

Alati, R., Davey Smith, G., Lewis, S. J., et al. 2013. Effect of Prenatal Alcohol Exposure on Childhood Academic Outcomes: Contrasting Maternal and Paternal Associations in the ALSPAC Study. *PLoS One*, 8: e74844.

Alexander, D. R. 2011. *The Language of Genetics – an Introduction.* Philadelphia: Templeton Foundation Press.

Alexander, P., Schlegel, A., Sinnott-Armstrong, W., et al. 2015. Dissecting the Readiness Potential: An Investigation of the Relationship between Readiness Potentials, Conscious Willing and Action. *In:* Mele, A. R. (ed.) *Surrounding Free Will.* Oxford University Press.

Alford, J. R., Funk, C. L. & Hibbing, J. R. 2005. Are Political Orientations Genetically Transmitted? *Am Polit Sci Rev*, 99: 153–67.

Algaze, G. 2008. *Ancient Mesopotamia at the Dawn of Civilization: The Evolution of an Urban Landscape.* University of Chicago Press.

Allen, L. S. & Gorski, R. A. 1992. Sexual Orientation and the Size of the Anterior Commissure in the Human Brain. *Proc Natl Acad Sci USA*, 89: 7199–202.

Alquist, J. L., Ainsworth, S. E. & Baumeister, R. F. 2013. Determined to Conform: Disbelief in Free Will Increases Conformity. *J Exp Soc Psychol*, 49: 80–6.

Amundsen, D. W. 1987. Medicine and the Birth of Defective Children, Approaches of the Ancient World. *In:* Mcmillan, R. C., Engelhardt, H. T. & Spicker, S. F. (eds.) *Euthanasia and the Newborn.* Dordrecht: D. Reidel.

Andersen, E. C., Bloom, J. S., Gerke, J. P., et al. 2014. A Variant in the Neuropeptide Receptor NPR-1 Is a Major Determinant of Caenorhabditis Elegans Growth and Physiology. *PLoS Genet*, 10: e1004156.

Andersen, J. P. & Blosnich, J. 2013. Disparities in Adverse Childhood Experiences among Sexual Minority and Heterosexual Adults: Results from a Multi-State Probability-Based Sample. *Plos One*, 8: doi:10.1371/journal.pone.0054691.

Ang, S., Van Dyne, L. & Tan, M. 2011. Cultural Intelligence. *In:* Sternberg, R. J. & Kaufman, S. B. (eds.) *The Cambridge Handbook of Intelligence.* Cambridge University Press.

Appelbaum, P. S. 2014. The Double Helix Takes the Witness Stand: Behavioral and Neuropsychiatric Genetics in Court. *Neuron,* 82: 946–9.

Aristotle & Ross, W. D. 1961. *De Anima.* Oxford: Clarendon Press.

Appelbaum, P. S. & Scurich, N. 2014. Impact of Behavioral Genetic Evidence on the Adjudication of Criminal Behavior. *J Am Acad Psychiatry Law,* 42: 91–100.

Arceneaux, K., Johnson, M. & Maes, H. H. 2012. The Genetic Basis of Political Sophistication. *Twin Res Hum Genet,* 15: 34–41.

Aristotle. 1994. South Bend, USA: Dumb Ox Books.

Aristotle & Lennox, J. G. 2001. *On the Parts of Animals.* Oxford: Clarendon Press.

Aslund, C., Nordquist, N., Comasco, E., et al. 2011. Maltreatment, MAOA, and Delinquency: Sex Differences in Gene-Environment Interaction in a Large Population-Based Cohort of Adolescents. *Behav Genet,* 41: 262–72.

Aspinwall, L. G., Brown, T. R. & Tabery, J. 2012. The Double-Edged Sword: Does Biomechanism Increase or Decrease Judges' Sentencing of Psychopaths? *Science,* 337: 846–9.

Atiq, E. H. 2012. How Folk Beliefs about Free Will Influence Sentencing: A New Target for the Neuro-Determinist Critics of Criminal Law. *New Crim Law Rev,* 16: 449–93.

Attfield, R. 1991. *The Ethics of Environmental Concern.* Athens: University of Georgia Press.

Augustine, S. B. O. H. & Dods, M. 2009. *The City of God.* Peabody, MA: Hendrickson Publishers.

Baggini, J. 2015. *Freedom Regained: The Possibility of Free Will.* London: Granta.

Baigent, M. & Baigent, M. 2015. *Astrology in Ancient Mesopotamia: The Science of Omens and the Knowledge of the Heavens.* Rochester: Vermont Bear & Company.

Bailey, J. M. & Pillard, R. C. 1991. A Genetic Study of Male Sexual Orientation. *Arch Gen Psychiatry,* 48: 1089–96.

Bailey, J. M., Pillard, R. C., Neale, M. C., et al. 1993. Heritable Factors Influence Sexual Orientation in Women. *Arch Gen Psychiatry,* 50: 217–23.

Baillie, J. K., Barnett, M. W., Upton, K. R., et al. 2011. Somatic Retrotransposition Alters the Genetic Landscape of the Human Brain. *Nature,* 479: 534–7.

Baker, L. R. 1995. Need a Christian Be a Mind-Body Dualist? *Faith and Philos,* 12: 489–504.

Bakermans-Kranenburg, M. J. & Van Ijzendoorn, M. H. 2011. Differential Susceptibility to Rearing Environment Depending on Dopamine-Related Genes: New Evidence and a Meta-Analysis. *Dev Psychopathol,* 23: 39–52.

Balaguer, M. 2010. *Free Will as an Open Scientific Problem.* Cambridge, MA: MIT Press.

Balsam, K. F., Rothblum, E. D. & Beauchaine, T. P. 2005. Victimization over the Life Span: A Comparison of Lesbian, Gay, Bisexual, and Heterosexual Siblings. *J Consult Clin Psychol,* 73: 477–87.

Balter, M. 2015. Behavioral Genetics. Can Epigenetics Explain Homosexuality Puzzle? *Science,* 350: 148.

Banasr, M. & Duman, R. S. 2012. Adult Neurogenesis: Nature Versus Nurture. *Biol Psychiatry*, 72: 256–7.

Bandura, A. 1989. Human Agency in Social Cognitive Theory. *Am Psychol*, 44: 1175–84.

Banner, M. C. 2014. *The Ethics of Everyday Life: Moral Theology, Social Anthropology, and the Imagination of the Human*: Oxford University Press.

Barash, Y., Calarco, J. A., Gao, W., et al. 2010. Deciphering the Splicing Code. *Nature*, 465: 53–9.

Bargmann, C. I. 2006. Chemosensation in C Elegans. *Wormbook: Doi/10.1895/Wormbook.1.123.1.*

Baron-Cohen, S. 2005. Testing the Extreme Male Brain (EMB) Theory of Autism: Let the Data Speak for Themselves. *Cogn Neuropsychiatry*, 10: 77–81.

2010. Empathizing, Systemizing, and the Extreme Male Brain Theory of Autism. *Prog Brain Res*, 186: 167–75.

Baron-Cohen, S., Wheelwright, S., Burtenshaw, A., et al. 2007. Mathematical Talent Is Linked to Autism. *Human Nature*, 18: 125–131.

Barr, J. 1993. *Biblical Faith and Natural Theology*: Oxford University Press.

Barres, R., Yan, J., Egan, B., et al. 2012. Acute Exercise Remodels Promoter Methylation in Human Skeletal Muscle. *Cell Metab*, 15: 405–11.

Barrett, C. E., Keebaugh, A. C., Ahern, T. H., et al. 2013. Variation in Vasopressin Receptor (Avpr1a) Expression Creates Diversity in Behaviors Related to Monogamy in Prairie Voles. *Horm. Behav.*, 63: 518–526.

Barriere, A. & Felix, M. 2005a. Natural Variation and Population Genetics of Caenorhabditis Elegans. *Wormbook: Doi/10.1895/Wormbook.1.43.1.*

2005b. High Local Genetic Diversity and Low Outcrossing Rate in Caenorhabditis Elegans Natural Populations. *Curr Biol*, 15: 1176–1184.

Barth, K., Bromiley, G. W. & Torrance, T. F. 1975. *Church Dogmatics*, Edinburgh: T. & T. Clark.

Bates, B. R. 2005. Public Culture and Public Understanding of Genetics: A Focus Group Study. *Public Underst. Sci.*, 14: 47–65.

Bateson, P. P. G. & Gluckman, P. D. 2011. *Plasticity, Robustness, Development and Evolution*: Cambridge University Press.

Bateson, W. 1909. *Mendel's Principles of Heredity*: Cambridge University Press.

Bauckham, R. 2011. *Living with Other Creatures: Green Exegesis and Theology*, Waco, Tex.: Baylor University Press.

Baum, M. L. 2013. The Monoamine Oxidase A (MAOA) Genetic Predisposition to Impulsive Violence: Is It Relevant to Criminal Trials? *Neuroethics*, 6: 287–306.

Baumeister, R. F. & Brewer, L. E. 2012. Believing Versus Disbelieving in Free Will: Correlates and Consequences. *Social and Personality Psychology Compass*, 6: 736–745.

Baumeister, R. F., Clark, C. & Luguri, J. 2015. Free Will – Belief and Reality. In: Mele, A. R. (ed.) *Surrounding Free Will*. Oxford: Oxford University Press.

Baumeister, R. F., Masicampo, E. J. & Dewall, C. N. 2009. Prosocial Benefits of Feeling Free: Disbelief in Free Will Increases Aggression and Reduces Helpfulness. *Pers Soc Psychol Bull*, 35: 260–268.

Baumeister, R. F., Mele, A. R. & Vohs, K. D. 2010. *Free Will and Consciousness: How Might They Work?*: Oxford University Press.

Baumrind, D. 1993. The Average Expectable Environment Is Not Good Enough: A Response to Scarr. *Child Dev*, 64: 1299–317.

Bazak, L., Haviv, A., Barak, M., et al. 2014. A-to-I RNA Editing Occurs at over a Hundred Million Genomic Sites, Located in a Majority of Human Genes. *Genome Res*, 24: 365–76.

Beaver, K. M., Wright, J. P., Boutwell, B. B., et al. 2013. Exploring the Association between the 2-Repeat Allele of the MAOA Gene Promoter Polymorphism and Psychopathic Personality Traits, Arrests, Incarceration, and Lifetime Antisocial Behavior. *Pers. Individ. Differ.*, 54: 164–168.

Bechtel, W. & Richardson, R. C. 1993. *Discovering Complexity: Decomposition and Localization as Strategies in Scientific Research*, Princeton, NJ: Princeton University Press.

Beckwith, J. & Morris, C. A. 2008. Twin Studies of Political Behavior: Untenable Assumptions? *Perspect. Polit.*, 6: 785–791.

Bedau, M. & Humphreys, P. 2008. *Emergence: Contemporary Readings in Philosophy and Science*, Cambridge, Mass.; London: MIT.

Bedau, M. A. 2010. Weak Emergence and Context-Sensitive Reduction. In: Corradini, A. & O'Connor, T. (eds.) *Emergence in Science and Philosophy*. London: Routledge.

Bedny, M., Konkle, T., Pelphrey, K., et al. 2010. Sensitive Period for a Multimodal Response in Human Visual Motion Area MT/MST. *Curr Biol*, 20: 1900–6.

Bedny, M., Pascual-Leone, A., Dravida, S., et al. 2011. A Sensitive Period for Language in the Visual Cortex: Distinct Patterns of Plasticity in Congenitally Versus Late Blind Adults. *Brain Lang*, 122: 162–70.

Bedny, M. & Saxe, R. 2012. Insights into the Origins of Knowledge from the Cognitive Neuroscience of Blindness. *Cogn Neuropsychol*, 29: 56–84.

Beebee, H., Hitchcock, C. & Menzies, P. C. 2009. *The Oxford Handbook of Causation*: Oxford University Press.

Beiber, I., Dain, H. J., Dince, P. R., et al. 1962. *Homosexuality: A Psychoanalytic Study*, New York: Basic Books.

Bejerot, S., Eriksson, J. M., Bonde, S., et al. 2012. The Extreme Male Brain Revisited: Gender Coherence in Adults with Autism Spectrum Disorder. *Br J Psychiatry*, 201: 116–23.

Belay, A. T., Scheiner, R., So, A. K. C., et al. 2007. The Foraging Gene of Drosophila Melanogaster: Spatial-Expression Analysis and Sucrose Responsiveness. *J. Comp. Neurol.*, 504: 570–582.

Bell, A. P., Weinberg, M. S. & Hammersmith, S. K. 1981. *Sexual Preference, Its Development in Men and Women*: Indiana University Press.

Bell, J. T. & Spector, T. D. 2012. DNA Methylation Studies Using Twins: What Are They Telling Us? *Genome Biol*, 13: 172.

Bell, J. T., Tsai, P. C., Yang, T. P., et al. 2012. Epigenome-Wide Scans Identify Differentially Methylated Regions for Age and Age-Related Phenotypes in a Healthy Ageing Population. *PLoS Genet*, 8: e1002629.

Belsky, J. 1997. Variation in Susceptibility to Environmental Influence: An Evolutionary Argument. *Psychological Inquiry*, 8: 182–186.

Benjamin, D. J., Cesarini, D., Van Der Loos, M. J., et al. 2012. The Genetic Architecture of Economic and Political Preferences. *Proc Natl Acad Sci U S A*, 109: 8026–31.

Benyamin, B., Pourcain, B., Davis, O. S., et al. 2014. Childhood Intelligence Is Heritable, Highly Polygenic and Associated with FNBP1L. *Mol Psychiatry*, 19: 253–8.

Berenbaum, S. A. & Beltz, A. M. 2011. Sexual Differentiation of Human Behavior: Effects of Prenatal and Pubertal Organizational Hormones. *Front. Neuroendocrinol.*, 32: 183–200.

Bernet, W., Vnencak-Jones, C. L., Farahany, N., et al. 2007. Bad Nature, Bad Nurture, and Testimony Regarding MAOA and SLC6A4 Genotyping at Murder Trials. *Journal of Forensic Sciences*, 52: 1362–1371.

Bernstein, B. E., Birney, E., Dunham, I., et al. 2012. An Integrated Encyclopedia of DNA Elements in the Human Genome. *Nature*, 489: 57–74.

Berntson, G. G., Sarter, M. & Cacioppo, J. T. 2003. Ascending Visceral Regulation of Cortical Affective Information Processing. *Eur J Neurosci*, 18: 2103–9.

Biehn, K. J. 2009. "Monkeys, Babies, Idiots" and "Primitives": Nature-Nurture Debates and Philanthropic Foundation Support for American Anthropology in the 1920s and 1930s. *J. Hist. Behav. Sci.*, 45: 219–235.

Billig, J. P., Hershberger, S. L., Iacono, W. G., et al. 1996. Life Events and Personality in Late Adolescence: Genetic and Environmental Relations. *Behav Genet*, 26: 543–54.

Birch, E. 2012. Vision: Looking to Develop Sight. *Nature*, 487: 441–2.

Birke, L. 1999. *Feminism and the Biological Body*: Edinburgh University Press.

Bizer, G. Y., Krosnick, J. A., Holbrook, A. L., et al. 2004. The Impact of Personality on Cognitive, Behavioral, and Affective Political Processes: The Effects of Need to Evaluate. *J Pers*, 72: 995–1027.

Boakes, R. 1984. *From Darwin to Behaviourism*: Cambridge University Press.

Boas, F. & Stocking, G. W. 1974. *A Franz Boas Reader: The Shaping of American Anthropology, 1883–1911*: University of Chicago Press.

Bobadilla, J. L., Macek, M., Jr., Fine, J. P., et al. 2002. Cystic Fibrosis: A Worldwide Analysis of CFTR Mutations–Correlation with Incidence Data and Application to Screening. *Hum Mutat*, 19: 575–606.

Boender, A. J., Roubos, E. W. & Van Der Velde, G. 2011. Together or Alone? Foraging Strategies in Caenorhabditis Elegans. *Biol. Rev.*, 86: 853–862.

Bogaert, A. F. 2006. Biological Versus Nonbiological Older Brothers and Men's Sexual Orientation. *Proc. Natl. Acad. Sci. U. S. A.*, 103: 10771–10774.

Bogaert, A. F. & Skorska, M. 2011. Sexual Orientation, Fraternal Birth Order, and the Maternal Immune Hypothesis: A Review. *Front. Neuroendocrinol.*, 32: 247–254.

Bond, J., Roberts, E., Mochida, G. H., et al. 2002. ASPM Is a Major Determinant of Cerebral Cortical Size. *Nat Genet*, 32: 316–20.

Boomsma, D. I., De Geus, E. J., Van Baal, G. C., et al. 1999. A Religious Upbringing Reduces the Influence of Genetic Factors on Disinhibition: Evidence for Interaction between Genotype and Environment on Personality. *Twin research*, 2: 115–25.

Bouchard, T. J. 2004. Genetic Influence on Human Psychological Traits – a Survey. *Curr. Dir. Psychol.*, 13: 148–151.

Bouchard, T. J., Jr. 2014. Genes, Evolution and Intelligence. *Behav Genet,* 44: 549–77.

Bouchard, T. J., Jr., Lykken, D. T., McGue, M., et al. 1990. Sources of Human Psychological Differences: The Minnesota Study of Twins Reared Apart. *Science,* 250: 223–8.

Bouchard, T. J., Jr., McGue, M., Lykken, D., et al. 1999. Intrinsic and Extrinsic Religiousness: Genetic and Environmental Influences and Personality Correlates. *Twin Research,* 2: 88–98.

Bradshaw, M. & Ellison, C. G. 2008. Do Genetic Factors Influence Religious Life? Findings from a Behavior Genetic Analysis of Twin Siblings. *Journal for the Scientific Study of Religion,* 47: 529–544.

Brady, P., Brison, N., Van Den Bogaert, K., et al. 2015. Clinical Implementation of NIPT – Technical and Biological Challenges. *Clin Genet,* 10.1111/cge.12598.

Brakefield, T., Mednick, S., Wilson, H., et al. 2014. Same-Sex Sexual Attraction Does Not Spread in Adolescent Social Networks. *Arch. Sex. Behav.,* 43: 335–344.

Brewer, K. 2006. *The Nature-Nurture Debate on Human Sexual Orientation,* Grays: Orsett Psychological Services.

Brooks-Crozier, J. 2011. The Nature and Nurture of Violence: Early Intervention Services for the Families of Maoa-Low Children as a Means to Reduce Violent Crime and the Costs of Violent Crime. *Connecticut Law Review,* 44: 531–573.

Brown, A. S. 2006. Prenatal Infection as a Risk Factor for Schizophrenia. *Schizophr Bull,* 32: 200–2.

Brown, N. J., Macdonald, D. A., Samanta, M. P., et al. 2014. A Critical Reanalysis of the Relationship between Genomics and Well-Being. *Proc Natl Acad Sci U S A,* 111: 12705–9.

Brown, W. S., Murphy, N. C. & Malony, H. N. 1998. *Whatever Happened to the Soul?: Scientific and Theological Portraits of Human Nature,* Minneapolis, MN: Fortress Press.

Brown, W. S. & Paul, L. K. 2015. Brain Connectivity and the Emergence of Capacities of Personhood: Reflections from Callosal Agenesis and Autism. In: Jeeves, M. (ed.) *The Emergence of Personhood – a Quantum Leap?* Grand Rapids: Eerdmans.

Brown, W. S. & Strawn, B. D. 2012. *The Physical Nature of Christian Life: Neuroscience, Psychology, and the Church*: Cambridge University Press.

Brunner, H., Nelen, M., Breakefield, X., et al. 1993a. Abnormal Behavior Associated with a Point Mutation in the Structural Gene for Monoamine Oxidase A. *Science,* 262: 578–580.

Brunner, H. G., Nelen, M. R., Vanzandvoort, P., et al. 1993b. X-Linked Borderline Mental Retardation with Prominent Behavioral Disturbance – Phenotype, Genetic Localization, and Evidence for Disturbed Monoamine Metabolism. *Am. J. Hum. Genet.,* 52: 1032–1039.

Bubela, T. M. & Caulfield, T. A. 2004. Do the Print Media "Hype" Genetic Research? A Comparison of Newspaper Stories and Peer-Reviewed Research Papers. *CMAJ,* 170: 1399–407.

Buchen, L. 2012. Biology and Ideology: The Anatomy of Politics. *Nature,* 490: 466–468.

Buckholtz, J. W. & Meyer-Lindenberg, A. 2008. MAOA and the Neurogenetic Architecture of Human Aggression. *Trends Neurosci.*, 31: 120–129.

Bundo, M., Toyoshima, M., Okada, Y., et al. 2014. Increased L1 Retrotransposition in the Neuronal Genome in Schizophrenia. *Neuron*, 81: 306–13.

Burkhardt, R. W. 1999. Ethology, Natural History, the Life Sciences, and the Problem of Place. *J. Hist. Biol.*, 32: 489–508.

2015. That Lamarckian Evolution Relied Largely on Use and Disuse and That Darwin Rejected Lamarckian Mechanisms In: Numbers, R. L. & Kampourakis, K. (eds.) *Newton's Apple and Other Myths About Science.* Cambridge, MASS: Harvard University Press.

Burleigh, M. 2000. *The Third Reich: A New History,* London: Macmillan.

Burne, T., Scott, E., Van Swinderen, B., et al. 2011. Big Ideas for Small Brains: What Can Psychiatry Learn from Worms, Flies, Bees and Fish? *Mol. Psychiatr.*, 16: 7–16.

Burri, A., Cherkas, L., Spector, T., et al. 2011. Genetic and Environmental Influences on Female Sexual Orientation, Childhood Gender Typicality and Adult Gender Identity. *Plos One*, DOI: 10.1371/journal.pone.0021982.

Burt, A. 2011. Some Key Issues in the Study of Gene-Environment Interplay: Activation, Deactivation, and the Role of Development. *Research in Human Development*, 8: 192–210.

Buss, C., Davis, E. P., Shahbaba, B., et al. 2012. Maternal Cortisol over the Course of Pregnancy and Subsequent Child Amygdala and Hippocampus Volumes and Affective Problems. *Proc. Natl. Acad. Sci. U. S. A.*, 109: E1312-E1319.

Button, T. M. M., Stallings, M. C., Rhee, S. H., et al. 2011. The Etiology of Stability and Change in Religious Values and Religious Attendance. *Behav. Genet.*, 41: 201–210.

Bygren, L. O., Tinghog, P., Carstensen, J., et al. 2014. Change in Paternal Grandmothers' Early Food Supply Influenced Cardiovascular Mortality of the Female Grandchildren. *BMC Genet*, 15: 12–18.

Byne, W., Tobet, S., Mattiace, L. A., et al. 2001. The Interstitial Nuclei of the Human Anterior Hypothalamus: An Investigation of Variation with Sex, Sexual Orientation, and Hiv Status. *Horm. Behav.*, 40: 86–92.

Byrd, A. L. & Manuck, S. B. 2014. MAOA, Childhood Maltreatment, and Antisocial Behavior: Meta-Analysis of a Gene-Environment Interaction. *Biol Psychiatry*, 75: 9–17.

Cairns, D. 1973. *The Image of God in Man,* London: Collins.

Callaway, E. 2012. Gene Mutation Defends against Alzheimer's Disease. *Nature*, 487: 153.

2014. Geneticists Tap Human Knockouts. *Nature*, 514: 548.

Callier, P., Calvel, P., Matevossian, A., et al. 2014. Loss of Function Mutation in the Palmitoyl-Transferase HHAT Leads to Syndromic 46,XY Disorder of Sex Development by Impeding Hedgehog Protein Palmitoylation and Signaling. *PLoS Genet*, 10: e1004340.

Canny, S. P., Goel, G., Reese, T. A., et al. 2013. Latent Gammaherpesvirus 68 Infection Induces Distinct Transcriptional Changes in Different Organs. *J Virol*, 88: 730–8.

Carey, J. M. & Paulhus, D. L. 2013. Worldview Implications of Believing in Free Will and/or Determinism: Politics, Morality, and Punitiveness. *J. Pers.*, 81: 130–141.

Carlile, T. M., Rojas-Duran, M. F., Zinshteyn, B., et al. 2014. Pseudouridine Profiling Reveals Regulated mRNA Pseudouridylation in Yeast and Human Cells. *Nature*, 515: 143–6.

Carmichael, L. 1925. Heredity and Environment – Are They Antithetical? *J Abnorm. Soc. Psychol.*, 20: 245–260.

Carroll, J. B. 1993. *Human Cognitive Abilities: A Survey of Factor-Analytic Studies*: Cambridge University Press.

1995. Reflections on Stephen Jay Gould's the Mismeasure of Man (1981): A Retrospective Review. *Intelligence*, 21: 121–134.

1997. Psychometrics, Intelligence, and Public Perception. *Intelligence*, 24: 25–52.

Carson, J. 2007. *The Measure of Merit: Talents, Intelligence, and Inequality in the French and American Republics, 1750–1940*: Princeton University Press.

Cases, O., Seif, I., Grimsby, J., et al. 1995. Aggressive Behavior and Altered Amounts of Brain Serotonin and Norepinephrine in Mice Lacking MAOA. *Science*, 268: 1763–6.

Cashmore, A. R. 2010. The Lucretian Swerve: The Biological Basis of Human Behavior and the Criminal Justice System. *Proc Natl Acad Sci U S A*, 107: 4499–504.

Caspi, A., Mcclay, J., Moffitt, T. E., et al. 2002. Role of Genotype in the Cycle of Violence in Maltreated Children. *Science*, 297: 851–854.

Caspi, A., Moffitt, T. E., Cannon, M., et al. 2005. Moderation of the Effect of Adolescent-Onset Cannabis Use on Adult Psychosis by a Functional Polymorphism in the Catechol-O-Methyltransferase Gene: Longitudinal Evidence of a Gene X Environment Interaction. *Biol Psychiatry*, 57: 1117–27.

Caspi, A., Sugden, K., Moffitt, T. E., et al. 2003. Influence of Life Stress on Depression: Moderation by a Polymorphism in the 5-HTT Gene. *Science*, 301: 386–389.

Castelli, F. R., Kelley, R. A., Keane, B., et al. 2011. Female Prairie Voles Show Social and Sexual Preferences for Males with Longer Avpr1a Microsatellite Alleles. *Animal Behaviour*, 82: 1117–1126.

Castera, J., Bruguiere, C. & Clement, P. 2008. Genetic Diseases and Genetic Determinism Models in French Secondary School Biology Textbooks. *Journal of Biological Education*, 42: 53–59.

Castera, J., Sarapuu, T. & Clement, P. 2013. Comparison of French and Estonian Students' Conceptions in Genetic Determinism of Human Behaviours. *Journal of Biological Education*, 47: 12–20.

Cavalli-Sforza, L. & Feldman, M. W. 1973. Models for Cultural Inheritance. I. Group Mean and within Group Variation. *Theor Popul Biol*, 4: 42–55.

Cerasa, A., Cherubini, A., Quattrone, A., et al. 2010. Morphological Correlates of Mao a Vntr Polymorphism: New Evidence from Cortical Thickness Measurement. *Behavioural Brain Research*, 211: 118–124.

Cesarini, D., Dawes, C. T., Fowler, J. H., et al. 2008. Heritability of Cooperative Behavior in the Trust Game. *Proc Natl Acad Sci U S A*, 105: 3721–6.

Chalmers, D. J. 2006. Strong and Weak Emergence. In: Clayton, P. & Davies, P. (eds.) *The Re-Emergence of Emergence: The Emergentist Hypothesis from Science to Religion*. Oxford: Oxford University Press.

2010. *The Character of Consciousness*: Oxford University Press.

Charney, E. 2012. Behavior Genetics and Postgenomics. *Behavioral and Brain Sciences*, 35: 331–410.

Charney, E. & English, W. 2012. Candidate Genes and Political Behavior. *American Political Science Review*, 106: 1–34.

Charollais, J. & Van Der Goot, F. G. 2009. Palmitoylation of Membrane Proteins (Review). *Mol Membr Biol*, 26: 55–66.

Chen, J., Li, X. Y., Chen, Z. Y., et al. 2010. Optimization of Zygosity Determination by Questionnaire and DNA Genotyping in Chinese Adolescent Twins. *Twin Research and Human Genetics*, 13: 194–200.

Chester, D. S., Dewall, C. N., Derefinko, K. J., et al. 2015. Monoamine Oxidase A (MAOA) Genotype Predicts Greater Aggression through Impulsive Reactivity to Negative Affect. *Behav Brain Res*, 283: 97–101.

Cheung, B. H. H., Cohen, M., Rogers, C., et al. 2005. Experience-Dependent C-Elegans Behavior by Modulation of Ambient Oxygen. *Curr Biol*, 15: 905–917.

Cholfin, J. A. & Rubenstein, J. L. 2007. Patterning of Frontal Cortex Subdivisions by Fgf17. *Proc Natl Acad Sci U S A*, 104: 7652–7.

Christian, K. M., Song, H. & Ming, G. L. 2014. Functions and Dysfunctions of Adult Hippocampal Neurogenesis. *Annu Rev Neurosci*, 37: 243–62.

Churchill, F. B. 2015. *August Weismann: Development, Heredity, and Evolution.* Cambridge, Massachusetts Harvard University Press,.

Churchland, P. S. 2006. The Big Questions: Do We Have Free Will? *New Scientist*, 2578: 42–45.

2012. *Braintrust: What Neuroscience Tells Us About Morality*: Princeton University Press.

Cirulli, E. T. & Goldstein, D. B. 2007. In Vitro Assays Fail to Predict in Vivo Effects of Regulatory Polymorphisms. *Hum Mol Genet*, 16: 1931–9.

Clarke, G., Grenham, S., Scully, P., et al. 2012. The Microbiome-Gut-Brain Axis During Early Life Regulates the Hippocampal Serotonergic System in a Sex-Dependent Manner. *Mol Psychiatry*, 18: 666–73.

Clarke, P. G. 2014. Neuroscience, Quantum Indeterminism and the Cartesian Soul. *Brain Cogn*, 84: 109–17.

Clarke, P. G. H. 2010. Determinism, Brain Function and Free Will. *Science and Christian Belief*, 22: 133–149.

Claxton, G. 2015. *Intelligence in the Flesh: Why Your Mind Needs Your Body Much More Than It Thinks*: Yale University Presss.

Clayton, P. & Davies, P. 2006. *The Re-Emergence of Emergence: The Emergentist Hypothesis from Science to Religion*: Oxford University Press.

Clines, D. J. A. 1968. The Image of God in Man. *Tyndale Bulletin*, 19: 53–103.

Cole, S. W. & Fredrickson, B. L. 2014. Errors in the Brown Et Al. Critical Reanalysis. *Proc Natl Acad Sci U S A*, 111: E3581.

Collins, D. 2002. *Nature Versus Nurture: The Long-Standing Debate over What Makes Us the Way We Are*, Tiverton: Clayhanger Books.

Colvert, E., Tick, B., Mcewen, F., et al. 2015. Heritability of Autism Spectrum Disorder in a UK Population-Based Twin Sample. *JAMA Psychiatry*, 72: 415–23.

Condit, C. M. 1999. How the Public Understands Genetics: Non-Deterministic and Non-Discriminatory Interpretations of the "Blueprint" Metaphor. *Public Underst. Sci.*, 8: 169–180.

2010. Public Attitudes and Beliefs About Genetics. *In:* Chakravarti, A. & Green, E. (eds.) *Annual Review of Genomics and Human Genetics, Vol 11*. Palo Alto: Annual Reviews.

2011. When Do People Deploy Genetic Determinism? A Review Pointing to the Need for Multi-Factorial Theories of Public Utilization of Scientific Discourses. *Sociology Compass*, 5: 618–635.

Condit, C. M., Gronnvoll, M., Landau, J., et al. 2009. Believing in Both Genetic Determinism and Behavioral Action: A Materialist Framework and Implications. *Public Underst. Sci.*, 18: 730–746.

Condit, C. M., Ofulue, N. & Sheedy, K. M. 1998. Determinism and Mass-Media Portrayals of Genetics. *Am. J. Hum. Genet.*, 62: 979–984.

Condit, C. M. & Shen, L. 2011. Public Understanding of Risks from Gene-Environment Interaction in Common Diseases: Implications for Public Communications. *Public Health Genomics*, 14: 115–124.

Conley, J. J. 1984. Not Galton, but Shakespeare – a Note on the Origin of the Term Nature and Nurture. *J. Hist. Behav. Sci.*, 20: 184–185.

Consortium. 2004. Finishing the Euchromatic Sequence of the Human Genome. *Nature*, 431: 931–945.

2014. Biological Insights from 108 Schizophrenia-Associated Genetic Loci. *Nature*, 511: 421–427.

Cook, T. D. & Campbell, D. T. 1979. *Quasi-Experimentation: Design and Analysis Issues for Field Settings*, Chicago: Rand McNally College.

Cooke, K. J. 1998. The Limits of Heredity: Nature and Nurture in American Eugenics before 1915. *J. Hist. Biol.*, 31: 263–278.

Cordaux, R. & Batzer, M. A. 2009. The Impact of Retrotransposons on Human Genome Evolution. *Nat Rev Genet*, 10: 691–703.

Cordaux, R., Hedges, D. J., Herke, S. W., et al. 2006. Estimating the Retrotransposition Rate of Human Alu Elements. *Gene*, 373: 134–7.

Corradini, A. & O'Connor, T. 2010. *Emergence in Science and Philosophy*, London: Routledge.

Crane, T. 2000. Dualism, Monism, Physicalism. *Mind and Society*, 2: 73–85.

2001. The Significance of Emergence. *In:* Loewer, B. & G., G. (eds.) *Physicalism and Its Discontents*. Cambridge University Press.

2003. *The Mechanical Mind: A Philosophical Introduction to Minds, Machines and Mental Representation*, London: Routledge.

Cranmer, S. J. & Dawes, C. T. 2012. The Heritability of Foreign Policy Preferences. *Twin Res Hum Genet*, 15: 52–9.

Crews, D., Gillette, R., Miller-Crews, I., et al. 2014. Nature, Nurture and Epigenetics. *Mol Cell Endocrinol*, 398: 42–52.

Cruickshank, M. N., Oshlack, A., Theda, C., et al. 2013. Analysis of Epigenetic Changes in Survivors of Preterm Birth Reveals the Effect of Gestational Age and Evidence for a Long Term Legacy. *Genome Med*, 5: 96.

Crusio, W. E. 2012. Heritability Estimates in Behavior Genetics: Wasn't That Station Passed Long Ago? *Behav Brain Sci*, 35: 361–2.

Cubas, P., Vincent, C. & Coen, E. 1999. An Epigenetic Mutation Responsible for Natural Variation in Floral Symmetry. *Nature*, 401: 157–61.

Curtis, E. M. 1984. *Man as the Image of God in Genesis in the Light of Ancient near Eastern Parallels*: http://repository.upenn.edu/dissertations/ AAI8422896.

Czyz, W., Morahan, J. M., Ebers, G. C., et al. 2012. Genetic, Environmental and Stochastic Factors in Monozygotic Twin Discordance with a Focus on Epigenetic Differences. *BMC Med*, 10: 93.

D'Onofrio, B. M., Eaves, L. J., Murrelle, L., et al. 1999. Understanding Biological and Social Influences on Religious Affiliation, Attitudes, and Behaviors: A Behavior Genetic Perspective. *J. Pers.*, 67: 953–984.

Dahir, M. 2001. Why Are We Gay? *The Advocate, July 17*.

Dambrun, M., Kamiejski, R., Haddadi, N., et al. 2009. Why Does Social Dominance Orientation Decrease with University Exposure to the Social Sciences? The Impact of Institutional Socialization and the Mediating Role of "Geneticism". *European Journal of Social Psychology*, 39: 88–100.

Darwin, C. 1868. *Variation of Plants and Animals under Domestication, Vol. 2*, London: John Murray.

1871a. Pangenesis. *Nature*, 3: 502–503.

1871b. *The Descent of Man*. London: John Murray (2nd edn 1874).

Daston, L. & Park, K. 1998. *Wonders and the Order of Nature, 1150–1750*, New York: Zone Books.

Davidson, J. E. & Kemp, I. A. 2011. Contemporary Models of Intelligence. *In*: Sternberg, R. & Kaufman, S. (eds.) *The Cambridge Handbook of Intelligence*. Cambridge University Press.

Davies, G., Tenesa, A., Payton, A., et al. 2011. Genome-Wide Association Studies Establish That Human Intelligence Is Highly Heritable and Polygenic. *Mol. Psychiatr.*, 16: 996–1005.

Davison, A. 2013. *The Love of Wisdom*. London: SCM Press

Dawkins, R. 1976. *The Selfish Gene*: Oxford University Press.

Dawood, K., Pillard, R. C., Horvath, C., et al. 2000. Familial Aspects of Male Homosexuality. *Arch. Sex. Behav.*, 29: 155–163.

Dawson-Scully, K., Armstrong, G. a. B., Kent, C., et al. 2007. Natural Variation in the Thermotolerance of Neural Function and Behavior Due to a cGMP-Dependent Protein Kinase. *Plos One*, 2: 10.1371/journal.pone. 0000773.

Daxinger, L. & Whitelaw, E. 2014. Understanding Transgenerational Epigenetic Inheritance Via the Gametes in Mammals. *Nat Rev Genet*, 13: 153–62.

De Bono, M. & Bargmann, C. I. 1998. Natural Variation in a Neuropeptide Y Receptor Homolog Modifies Social Behavior and Food Response in C-Elegans. *Cell*, 94: 679–689.

De Smedt, J. & De Cruz, H. 2014. The Imago Dei as a Work in Progress: A Perspective from Paleoanthropology. *Zygon*, 49: 135–156.

Deacon, T. 2006a. *In*: Clayton, P. & Davies, P. (eds.) *The Re-Emergence of Emergence: The Emergentist Hypothesis from Science to Religion*. Oxford University Press.

2006b. Emergence: The Hole at the Wheel's Hub. *In:* Clayton, P. & Davies, P. (eds.) *The Re-Emergence of Emergence: The Emergentist Hypothesis from Science to Religion.* Oxford University Press.

Deary, I. J. 2012. Intelligence. *Annu Rev Psychol,* 63: 453–82.

Deary, I. J., Johnson, W. & Houlihan, L. M. 2009. Genetic Foundations of Human Intelligence. *Hum. Genet.,* 126: 215–232.

Degler, C. 1991. *In Search of Human Nature: The Decline and Revival of Darwinism in American Social Thought:* Oxford University Press.

Del Rio, C. M. & White, L. J. 2012. Separating Spirituality from Religiosity: A Hylomorphic Attitudinal Perspective. *Psychol. Relig. Spiritual.,* 4: 123–142.

Deng, W., Aimone, J. B. & Gage, F. H. 2010. New Neurons and New Memories: How Does Adult Hippocampal Neurogenesis Affect Learning and Memory? *Nat Rev Neurosci,* 11: 339–50.

Denno, D. W. 2009. Behavioral Genetics Evidence in Criminal Cases: 1994–2007. *In:* Farahany, N. (ed.) *The Impact of Behavioral Sciences on Criminal Law.* Oxford University Press.

2011. Courts' Increasing Consideration of Behavioral Genetics Evidence in Criminal Cases: Results of a Longitudinal Study. *Michigan State Law Review,* 2011: 967–1047.

2013. What Real-World Criminal Cases Tell Us About Genetics Evidence. *Hastings Law Journal,* 64: 1591–1618.

Devlin, B., Daniels, M. & Roeder, K. 1997. The Heritability of IQ. *Nature,* 388: 468–71.

Di Francesco, M. 2010. The Varieties of Causal Emergentism. *In:* Corradini, A. & O'Connor, T. (eds.) *Emergence in Science and Philosophy.* London: Routledge.

Dick, K. J., Nelson, C. P., Tsaprouni, L., et al. 2014. DNA Methylation and Body-Mass Index: A Genome-Wide Analysis. *Lancet,* 383: 1990–8.

Dinman, J. D. 2012. Control of Gene Expression by Translational Recoding. *Adv Protein Chem Struct Biol,* 86: 129–49.

Dodge, K. A. & Rutter, M. 2011. *Gene-Environment Interactions in Developmental Psychopathology,* New York: Guilford Press.

Donaldson, Z. R. & Young, L. J. 2013. The Relative Contribution of Proximal 5' Flanking Sequence and Microsatellite Variation on Brain Vasopressin 1a Receptor Avpr1a Gene Expression and Behavior. *PLoS Genet,* 9: e1003729.

Dowling, J. E. 2004. *The Great Brain Debate: Nature or Nuture?,* Washington, D.C.: Joseph Henry Press.

Drabant, E. M., Kiefer, A. K., Eriksson, N., et al. 2012. Genome-Wide Association Study of Sexual Orientation in a Large, Web-Based Cohort. 23andMe.

Draganski, B., Gaser, C., Busch, V., et al. 2004. Neuroplasticity: Changes in Grey Matter Induced by Training. *Nature,* 427: 311–2.

Drescher, J. 2008. A History of Homosexuality and Organized Psychoanalysis. *J Am Acad Psychoanal Dyn Psychiatry,* 36: 443–60.

Duckworth, A. L. 2011. The Significance of Self-Control. *Proc Natl Acad Sci U S A,* 108: 2639–40.

Dugdale, R. L. 1877. *"The Jukes": A Study in Crime, Pauperism, Disease, and Heredity,* New York: G.P. Putnam's Sons.

Dunham, I., Kundaje, A., Aldred, S. F., et al. 2012. An Integrated Encyclopedia of DNA Elements in the Human Genome. *Nature*, 489: 57–74.

Dunn, L. C. & Dobzhansky, T. 1952. *Heredity, Race and Society*, New York: New American Library.

Dupree, M., Mustanski, B., Bocklandt, S., et al. 2004. A Candidate Gene Study of Cyp19 (Aromatase) and Male Sexual Orientation. *Behav. Genet.*, 34: 243–250.

Duyme, M., Dumaret, A. C. & Tomkiewicz, S. 1999. How Can We Boost IQs of "Dull Children"?: A Late Adoption Study. *Proc Natl Acad Sci U S A*, 96: 8790–4.

Eaves, L. J., Hatemi, P. K., Prom-Wornley, E. C., et al. 2008. Social and Genetic Influences on Adolescent Religious Attitudes and Practices. *Soc. Forces*, 86: 1621–1646.

Ebert, D. H. & Greenberg, M. E. 2013. Activity-Dependent Neuronal Signalling and Autism Spectrum Disorder. *Nature*, 493: 327–37.

Editorial 2015. Listen Up. *Nature*, 517: 121.

Edmonds, T. R. 1828. *Practical Moral and Political Economy; or, the Government, Religion, and Institutions Most Conducive to Individual Happiness and to National Power*: London. Effingham Wilson. Royal Exchange.

Edwards, A. C. & Mackay, T. F. C. 2009. Quantitative Trait Loci for Aggressive Behavior in Drosophila Melanogaster. *Genetics*, 182: 889–897.

Edwards, A. C., Rollmann, S. M., Morgan, T. J., et al. 2006. Quantitative Genomics of Aggressive Behavior in Drosophila Melanogaster. *PLoS Genet*, 2: e154.

Einstein, A. 1916. Ernst Mach. *Physikalische Zeitschrift*, 17: 101–102.

Ellis, L. & Blanchard, R. 2001. Birth Order, Sibling Sex Ratio, and Maternal Miscarriages in Homosexual and Heterosexual Men and Women. *Pers. Individ. Differ.*, 30: 543–552.

Ellis, S. J. & Peel, E. 2011. Lesbian Feminisms: Historical and Present Possibilities. *Feminism & Psychology*, 21: 198–204.

Elman, J. L. 1996. *Rethinking Innateness: A Connectionist Perspective on Development*, Cambridge, Mass.: MIT Press.

Elston, R. C. & Gottesman, Ii 1968. The Analysis of Quantitative Inheritance Simultaneously with Twin and Family Data. *Am J Hum Genet*, 20: 512–21.

Eppig, C., Fincher, C. L. & Thornhill, R. 2010. Parasite Prevalence and the Worldwide Distribution of Cognitive Ability. *Proc Biol Sci*, 277: 3801–8.

Erman, A. & Ranke, H. 1923. *Ägypten Und Ägyptisches Leben Im Altertum*, Tübingen: Mohr.

Espinosa, J. S. & Stryker, M. P. 2012. Development and Plasticity of the Primary Visual Cortex. *Neuron*, 75: 230–49.

Estabrook, A. H. 1916. The Jukes in 1915. *The Station For Experimental Evolution At Cold Spring Harbor*, Paper No. 25.

Evans, J. H. 2016. *What Is a Human? What the Answers Mean for Human Rights*: Oxford University Press.

Evrony, G. D., Cai, X., Lee, E., et al. 2012. Single-Neuron Sequencing Analysis of L1 Retrotransposition and Somatic Mutation in the Human Brain. *Cell*, 151: 483–96.

Eysenck, H. J. & Kamin, L. J. 1981. *The Intelligence Controversy.*, New York: John Wiley.

Fancher, R. E. 1983. Alphonse De Candolle, Francis Galton, and the Early History of the Nature-Nurture Controversy. *J. Hist. Behav. Sci.*, 19: 341–52.

1984. Not Conley, but Burt and Others: A Reply. *J. Hist. Behav. Sci.*, 20: 186–186.

Farahany, N. A. & Bernet, W. 2006. Behavioural Genetics in Criminal Cases: Past, Present and Future. *Genomics, Society and Policy*, 2: 72–79.

Farahany, N. A. & Coleman, J. 2006. Genetics and Responsibility: To Know the Criminal from the Crime. *Law and Contemporary Problem*, 69: 115–166.

Faraone, S. V., Tsuang, M. T. & Tsuang, D. W. 1999. *Genetics of Mental Disorders: A Guide for Students, Clinicians, and Researchers*, New York: Guilford Press.

Fass, D. M., Schroeder, F. A., Perlis, R. H., et al. 2014. Epigenetic Mechanisms in Mood Disorders: Targeting Neuroplasticity. *Neuroscience*, 264: 112–30.

Feschotte, C. & Pritham, E. J. 2007. DNA Transposons and the Evolution of Eukaryotic Genomes. *Annu Rev Genet*, 41: 331–68.

Ficks, C. A. & Waldman, I. D. 2014. Candidate Genes for Aggression and Antisocial Behavior: A Meta-Analysis of Association Studies of the 5HTTLPR and MAOA-uVNTR. *Behav Genet*, 44: 427–44.

Fine, I. 2014. Sensory Systems: Do You Hear What I See? *Nature*, 508: 461–2.

Fink, S., Excoffier, L. & Heckel, G. 2006. Mammalian Monogamy Is Not Controlled by a Single Gene. *Proc. Natl. Acad. Sci. U. S. A.*, 103: 10956–10960.

Finn, E. S., Shen, X., Scheinost, D., et al. 2015. Functional Connectome Fingerprinting: Identifying Individuals Using Patterns of Brain Connectivity. *Nat Neurosci*, 18: 1664–71.

Fitzpatrick, M. J., Feder, E., Rowe, L., et al. 2007. Maintaining a Behaviour Polymorphism by Frequency-Dependent Selection on a Single Gene. *Nature*, 447: 210-U5.

Flynn, J. R. 2012. *Are We Getting Smarter?: Rising IQ in the Twenty-First Century*: Cambridge University Press.

Forsythe, P., Bienenstock, J. & Kunze, W. A. 2014. Vagal Pathways for Microbiome-Brain-Gut Axis Communication. *Adv Exp Med Biol*, 817: 115–33.

Fowler, J. H., Baker, L. A. & Dawes, C. T. 2008. Genetic Variation in Political Participation. *American Political Science Review*, 102: 233–248.

Fowler, J. S., Alia-Klein, N., Kriplani, A., et al. 2007. Evidence That Brain MAOA Activity Does Not Correspond to MAOA Genotype in Healthy Male Subjects. *Biological Psychiatry*, 62: 355–358.

Fox, S. E., Levitt, P. & Nelson, C. A., 3rd 2010. How the Timing and Quality of Early Experiences Influence the Development of Brain Architecture. *Child Dev*, 81: 28–40.

Fraga, M. F., Ballestar, E., Paz, M. F., et al. 2005. Epigenetic Differences Arise During the Lifetime of Monozygotic Twins. *Proc. Natl. Acad. Sci. U. S. A.*, 102: 10604–10609.

Fredrickson, B. L., Grewen, K. M., Algoe, S. B., et al. 2015. Psychological Well-Being and the Human Conserved Transcriptional Response to Adversity. *PLoS One*, 10: e0121839.

Fredrickson, B. L., Grewen, K. M., Coffey, K. A., et al. 2013. A Functional Genomic Perspective on Human Well-Being. *Proc Natl Acad Sci U S A*, 110: 13684–9.

Friesen, A. & Ksiazkiewicz, A. 2015. Do Political Attitudes and Religiosity Share a Genetic Path? *Political Behavior*, 37: 791–818.

Frisen, L., Nordenstrom, A., Falhammar, H., et al. 2009. Gender Role Behavior, Sexuality, and Psychosocial Adaptation in Women with Congenital Adrenal Hyperplasia Due to CYP21A2 Deficiency. *J Clin Endocrinol Metab*, 94: 3432–9.

Fromer, M., Pocklington, A. J., Kavanagh, D. H., et al. 2014. De Novo Mutations in Schizophrenia Implicate Synaptic Networks. *Nature*, 506: 179–84.

Gage, F. H. & Muotri, A. R. 2013. What Makes Each Brain Unique. *Sci Am*, 306: 26–31.

Galton, F. 1865. Hereditary Talent and Character. *Macmillan's Magazine*, 12: 157–166.

1871. Pangenesis. *Nature*, 3: 5–6.

1874. *English Men of Science: Their Nature and Nurture*, London: Macmillan.

1876. A Theory of Heredity. *Journal of the Anthropological Institute*, 5: 329–348.

1890. Criminal Anthropology. *Nature*, 42: 75–76.

1908. *Memories of My Life (2nd Edn)*,London: Methuen.

Garrod, A. E. 1902. The Incidence of Alkaptonuria: A Study in Chemical Individuality. *Lancet*, 160: 1616–1620.

Gartrell, N. K., Bos, H. M. W. & Goldberg, N. G. 2011. Adolescents of the US National Longitudinal Lesbian Family Study: Sexual Orientation, Sexual Behavior, and Sexual Risk Exposure. *Arch. Sex. Behav.*, 40: 1199–1209.

Gates, G. J. 2011. *How Many People Are Lesbian, Gay, Bisexual and Transgender?* Los Angeles: Williams Institute.

Genomes Project. 2015. A Global Reference for Human Genetic Variation. *Nature*, 526: 68–74.

Gervin, K., Hammero, M., Akselsen, H. E., et al. 2011. Extensive Variation and Low Heritability of DNA Methylation Identified in a Twin Study. *Genome Res*, 21: 1813–21.

Geschwind, D. H. & State, M. W. 2015. Gene Hunting in Autism Spectrum Disorder: On the Path to Precision Medicine. *Lancet Neurol*.

Ghiglieri, M. P. 1999. *The Dark Side of Man: Tracing the Origins of Male Violence*, Reading, Mass.: Perseus Books.

Gibbons, A. 2004. American Association of Physical Anthropologists Meeting. Tracking the Evolutionary History of a "Warrior" Gene. *Science*, 304: 818.

Gilbert, W. 1992. A Vision of the Grail. In: Kevles, D. a. H., L (ed.) *The Code of Codes: Scientific and Social Issues in the Human Genome Project*. Harvard University Press.

Gillett, C. 2003. Strong Emergence as a Defense of Non-Reductive Physicalism: A Physicalist Metaphysics for "Downward Causation". *Principia*, 6: 83–114.

Gillette, A. 2011. *Eugenics and the Nature-Nurture Debate in the Twentieth Century*, New York: Palgrave Macmillan.

Giubilini, A. & Minerva, F. 2013. After-Birth Abortion: Why Should the Baby Live? *Journal of Medical Ethics*, 39: 261–263.

Glover, V., O'Connor, T. G. & O'Donnell, K. 2010. Prenatal Stress and the Programming of the HPA Axis. *Neuroscience & Biobehavioral Reviews*, 35: 17–22.

Goddard, H. H. 1910. Four Hundred Feeble-Minded Children Classified by the Binet Method. *Journal of Psycho-Asthenics*, 15: 17–30.

1927. *The Kallikak Family: A Study in the Heredity of Feeble-Mindedness,* New York: Macmillan.

Goldberg, L. R. 1990. An Alternative "Description of Personality": The Big-Five Factor Structure. *J Pers Soc Psychol*, 59: 1216–29.

Goldhaber, D. 2012. *The Nature-Nurture Debate: Bridging the Gap*: Cambridge University Press.

Goodrich, J. K., Waters, J. L., Poole, A. C., et al. 2014. Human Genetics Shape the Gut Microbiome. *Cell*, 159: 789–99.

Gordon, L., Joo, J. E., Powell, J. E., et al. 2012. Neonatal DNA Methylation Profile in Human Twins Is Specified by a Complex Interplay between Intrauterine Environmental and Genetic Factors, Subject to Tissue-Specific Influence. *Genome Res*, 22: 1395–406.

Gottesman, I. I. & Wolfgram, D. L. 1991. *Schizophrenia Genesis: The Origins of Madness*, New York: Freeman.

Gottfredson, L. & Saklofske, D. H. 2009. Intelligence: Foundations and Issues in Assessment. *Can. Psychol.-Psychol. Can.*, 50: 183–195.

Gottlieb, G. 1992. *Individual Development and Evolution: The Genesis of Novel Behavior*: Oxford University Press.

1998. Normally Occurring Environmental and Behavioral Influences on Gene Activity: From Central Dogma to Probabilistic Epigenesis. *Psychol Rev*, 105: 792–802.

Gould, S. J. 1996. *The Mismeasure of Man*, New York: Norton.

Graff, J., Joseph, N. F., Horn, M. E., et al. 2014. Epigenetic Priming of Memory Updating During Reconsolidation to Attenuate Remote Fear Memories. *Cell*, 156: 261–76.

Grant, M. & Osborn, H. F. 1916. *The Passing of the Great Race: Or the Racial Basis of European History*, New York: C. Scribner's Sons.

Gravina, S., Ganapathi, S. & Vijg, J. 2015. Single-Cell, Locus-Specific Bisulfite Sequencing (SLBS) for Direct Detection of Epimutations in DNA Methylation Patterns. *Nucleic Acids Res.* 43: e93.

Gray, J. M., Karow, D. S., Lu, H., et al. 2004. Oxygen Sensation and Social Feeding Mediated by a C. Elegans Guanylate Cyclase Homologue. *Nature*, 430: 317–22.

Green, J. B. 2003. Science, Religion and the Mind-Brain Problem – the Case of Thomas Willis (1621–1675). *Science and Christian Belief*, 15: 165–185.

2004. *What About the Soul?: Neuroscience and Christian Anthropology*, Edinburgh: Alban.

2008. *Body, Soul, and Human Life: The Nature of Humanity in the Bible*, Grand Rapids, Mich.: Baker Academic.

Greene, E. & Cahill, B. S. 2012. Effects of Neuroimaging Evidence on Mock Juror Decision Making. *Behavioral Sciences & the Law*, 30: 280–296.

Greene, J. & Cohen, J. 2004. For the Law, Neuroscience Changes Nothing and Everything. *Philos Trans R Soc Lond B Biol Sci*, 359: 1775–85.

Greenfield, S. M. 2001. Nature/Nurture and the Anthropology of Franz Boas and Margaret Mead as an Agenda for Revolutionary Politics *Horizontes Antropológicos*, 7: 35–52.

Greenwood, J. D. 1999. Understanding the "Cognitive Revolution" in Psychology. *J. Hist. Behav. Sci.*, 35: 1–22.

Greer, E. L., Maures, T. J., Ucar, D., et al. 2011. Transgenerational Epigenetic Inheritance of Longevity in Caenorhabditis Elegans. *Nature*, 479: 365–71.

Gregory, C. 2009. Sent by the Scent of Death. *Nature*, 461: 181–182.

Grenz, S. J. 2001. *The Social God and the Relational Self: A Trinitarian Theology of the Imago Dei*, Louisville: Westminster John Knox Press.

Griffiths, P. & Stotz, K. 2013. *Genetics and Philosophy: An Introduction*: Cambridge University Press.

Griffiths, P. E. 2004. Instinct in the '50s: The British Reception of Konrad Lorenz's Theory of Instinctive Behavior. *Biol. Philos.*, 19: 609–631.

Griffiths, P. E., Lorenz, K. & Tinbergen, N. 2008. History of Ethology Comes of Age. *Biol. Philos.*, 23: 129–134.

Griffiths, P. E. & Tabery, J. 2008. Behavioral Genetics and Development: Historical and Conceptual Causes of Controversy. *New Ideas in Psychology*, 26: 332–352.

Groff, P. & Mcrae, L. 1998. The Nature-Nurture Debate in Thirteenth-Century France. *Paper presented at the annual meeting of the American Psychological Association, Chicago, August 1998.*

Guo, G., Ou, X. M., Roettger, M., et al. 2008. The VNTR 2 Repeat in MAOA and Delinquent Behavior in Adolescence and Young Adulthood: Associations and MAOA Promoter Activity. *Eur J Hum Genet*, 16: 626–34.

Gurian, M. 2009. *Nurture the Nature: Understanding and Supporting Your Child's Unique Core Personality*, San Francisco, Calif.: Jossey-Bass

Gushee, D. P. 2013. *The Sacredness of Human Life: Why an Ancient Biblical Vision Is Key to the World's Future*, Grand Rapids, Mich.: William B. Eerdmans Pub. Co.

Haberstick, B. C., Lessem, J. M., Hewitt, J. K., et al. 2014. MAOA Genotype, Childhood Maltreatment, and Their Interaction in the Etiology of Adult Antisocial Behaviors. *Biol Psychiatry*, 75: 25–30.

Haggard, P. 2015. "Free Will": Components and Processes. In: Mele, A. R. (ed.) *Surrounding Free Will*. Oxford University Press.

Hall, D. J. 1986. *Imaging God: Dominion as Stewardship*, Grand Rapids: W.B. Eerdmans Pub. Co;.

Hall, J., Philip, R. C., Marwick, K., et al. 2013. Social Cognition, the Male Brain and the Autism Spectrum. *PLoS One*, 7: e49033.

Hall, S. S. 2013a. A Gene of Rare Effect. *Nature*, 496: 152–155.

2013b. Neuroscience: As the Worm Turns. *Nature*, 494: 296–9.

Hallmayer, J., Cleveland, S., Torres, A., et al. 2011. Genetic Heritability and Shared Environmental Factors among Twin Pairs with Autism. *Arch Gen Psychiatry*, 68: 1095–102.

Hamer, D. H. 2004. *The God Gene: How Faith Is Hardwired into Our Genes*, New York: Doubleday.

Hamer, D. H. & Copeland, P. 1994. *The Science of Desire: The Search for the Gay Gene and the Biology of Behaviour*, New York: Simon and Schuster.

Hamer, D. H., Hu, S., Magnuson, V. L., et al. 1993. A Linkage Betweeen DNA Markers on the X-Chromosome and Male Sexual Orientation. *Science*, 261: 321–327.

Hammock, E. a. D. & Young, L. J. 2002. Variation in the Vasopressin V1a
    Receptor Promoter and Expression: Implications for Inter- and Intraspecific
    Variation in Social Behaviour. *European Journal of Neuroscience*, 16:
    399–402.
Hanage, W. P. 2014. Microbiology: Microbiome Science Needs a Healthy Dose of
    Scepticism. *Nature*, 512: 247–8.
Hart, D. B. 2009. *Atheist Delusions: The Christian Revolution and Its Fashionable
    Enemies*, New Haven Conn.: Yale University Press.
Hatemi, P. K. 2012. The Intersection of Behavioral Genetics and Political Science:
    Introduction to the Special Issue. *Twin Res Hum Genet*, 15: 1–5.
Hatemi, P. K., Dawes, C. T., Frost-Keller, A., et al. 2011. Integrating Social Science
    and Genetics: News from the Political Front. *Biodemography Soc Biol*,
    57: 67–87.
Hatemi, P. K., Funk, C. L., Medland, S. E., et al. 2009. Genetic and Environmental
    Transmission of Political Attitudes over a Life Time. *The Journal of Politics*,
    71: 1141–1156.
Hatemi, P. K., Hibbing, J. R., Medland, S. E., et al. 2010. Not by Twins Alone: Using
    the Extended Family Design to Investigate Genetic Influence on Political
    Beliefs. *American Journal of Political Science*, 54: 798–814.
Hatemi, P. K. & Mcdermott, R. 2012. The Genetics of Politics: Discovery,
    Challenges, and Progress. *Trends in Genetics*, 28: 525–533.
Hatemi, P. K., Medland, S. E., Klemmensen, R., et al. 2014. Genetic Influences
    on Political Ideologies: Twin Analyses of 19 Measures of Political Ideologies
    from Five Democracies and Genome-Wide Findings from Three Populations.
    Behav Genet, 44: 282–94.
Hatemi, P. K., Medland, S. E., Morley, K. I., et al. 2007. The Genetics of Voting: An
    Australian Twin Study. *Behav. Genet.*, 37: 435–448.
Hatemi, P. K. & Verhulst, B. 2015. Political Attitudes Develop Independently of
    Personality Traits. *PLoS One*, 10: e0118106.
Hayakawa, K., Shimizu, T., Ohba, Y., et al. 1992. Intrapair Differences of Physical
    Aging and Longevity in Identical Twins. *Acta Genet Med Gemellol (Roma)*,
    41: 177–85.
Heard, E. & Martienssen, R. A. 2014. Transgenerational Epigenetic Inheritance:
    Myths and Mechanisms. *Cell*, 157: 95–109.
Hedgecoe, A. 1998. Geneticization, Medicalisation and Polemics. *Medicine, health
    care, and philosophy*, 1: 235–43.
Heijmans, B. T., Tobi, E. W., Stein, A. D., et al. 2008. Persistent Epigenetic
    Differences Associated with Prenatal Exposure to Famine in Humans. *Proc
    Natl Acad Sci U S A*, 105: 17046–9.
Heldriss of Cornwall, R.-M., Sarah (Ed. And Trans.) c. 1250–1300. *Silence*, East
    Lansing, MI: Colleagues Press Ltd.
Herrmann, C. S., Pauen, M., Min, B. K., et al. 2008. Analysis of a Choice-Reaction
    Task Yields a New Interpretation of Libet's Experiments. *International jour-
    nal of psychophysiology: official journal of the International Organization of
    Psychophysiology*, 67: 151–7.
Herrnstein, R. J. & Murray, C. A. 1994. *The Bell Curve: Intelligence and Class
    Structure in American Life*, New York Free Press.

Heston, L. L. 1966. Psychiatric Disorders in Foster Home Reared Children of Schizophrenic Mothers. *Br J Psychiatry*, 112: 819–25.

Hibar, D. P., Stein, J. L., Renteria, M. E., et al. 2015. Common Genetic Variants Influence Human Subcortical Brain Structures. *Nature*, 520: 224–9.

Hines, M. 2011. Prenatal Endocrine Influences on Sexual Orientation and on Sexually Differentiated Childhood Behavior. *Front. Neuroendocrinol.*, 32: 170–182.

Hines, M., Ahmed, S. F. & Hughes, I. 2003. Psychological Outcomes and Gender-Related Development in Complete Androgen Insensitivity Syndrome. *Arch. Sex. Behav.*, 32: 93–101.

Hitler, A., Johnson, A. S., Chamberlain, J., et al. 1939. *Mein Kampf: Complete and Unabridged, Fully Annotated*, New York: Reynal & Hitchcock.

Hobert, O. 2010. Neurogenesis in the Nematode Caenorhabditis Elegans. *In*: Community, T. C. E. R. (ed.) *Wormbook: Doi/10.1895/Wormbook.1.12.1.*

Hodgkin, J. & Doniach, T. 1997. Natural Variation and Copulatory Plug Formation in Caenorhabditis Elegans. *Genetics*, 146: 149–164.

Hoekema, A. A. 1986. *Created in God's Image*, Grand Rapids: William B. Eerdmans.

Hoffmann, A. & Spengler, D. 2014. DNA Memories of Early Social Life. *Neuroscience*, 264: 64–75.

Hogben, L. 1932. *Genetic Principles in Medicine and Social Science*, New York: Alfred A. Knopf.

Hogben, L. T. 1933. *Nature and Nurture*, New York: W.W. Norton Company.

Holdcroft, B. 2006. What Is Religiosity? *Catholic Education*, 10: 89–103.

Holmes, S.J. 1948. *Life and Morals*. London: MacMillan.

Holmgren, M. & Rosenthal, J. J. 2015. Regulation of Ion Channel and Transporter Function through RNA Editing. *Curr Issues Mol Biol*, 17: 23–36.

Holton, R. 2009. *Willing, Wanting, Waiting*: Oxford University Press.

Hooker, E. 1957. The Adjustment of the Male Overt Homosexual. *Journal of Projective Techniques*, 21: 18–31.

Horgan, T. 1993. From Supervenience to Superdupervenience: Meeting the Demands of a Material World. *Mind & Language*, 102: 555–586.

Horvath, S. 2013. DNA Methylation Age of Human Tissues and Cell Types. *Genome Biol*, 14: R115.

Huberman, A. D., Feller, M. B. & Chapman, B. 2008. Mechanisms Underlying Development of Visual Maps and Receptive Fields. *Annu Rev Neurosci*, 31: 479–509.

Hudziak, J. J., Albaugh, M. D., Ducharme, S., et al. 2014. Cortical Thickness Maturation and Duration of Music Training: Health-Promoting Activities Shape Brain Development. *J Am Acad Child Adolesc Psychiatry*, 53: 1153–61.

Hughes, V. 2013. Can Research on Romanian Orphans Be Ethical? *Aeon*, July: 29th July.

Huizinga, D., Haberstick, B. C., Smolen, A., et al. 2006. Childhood Maltreatment, Subsequent Antisocial Behavior, and the Role of Monoamine Oxidase A Genotype. *Biol Psychiatry*, 60: 677–83.

Hume, D., Selby-Bigge, L. A. & Nidditch, P. H. 1975. *Enquiries Concerning Human Understanding and Concerning the Principles of Morals Sect X, Pt I* Oxford: Clarendon Press.

Hunt, E. B. 2011. *Human Intelligence*: Cambridge University Press.

Hyman, C. & Handal, P. J. 2006. Definitions and Evaluation of Religion and Spirituality Items by Religious Professionals: A Pilot Study. *J. Relig. Health*, 45: 264–282.

Insel, T. R. 2010. The Challenge of Translation in Social Neuroscience: A Review of Oxytocin, Vasopressin, and Affiliative Behavior. *Neuron*, 65: 768–779.

Ioannidis, J. P., Ntzani, E. E., Trikalinos, T. A., et al. 2001. Replication Validity of Genetic Association Studies. *Nat Genet*, 29: 306–9.

Iossifov, I., O'roak, B. J., Sanders, S. J., et al. 2014. The Contribution of De Novo Coding Mutations to Autism Spectrum Disorder. *Nature*, 515: 216–21.

Iourov, I. Y., Vorsanova, S. G., Liehr, T., et al. 2009. Aneuploidy in the Normal, Alzheimer's Disease and Ataxia-Telangiectasia Brain: Differential Expression and Pathological Meaning. *Neurobiol Dis*, 34: 212–20.

Ismael, J. T. 2015. On Being Someone. *In:* Mele, A. R. (ed.) *Surrounding Free Will.* Oxford University Press.

Jablonka, E. & Raz, G. 2009. Transgenerational Epigenetic Inheritance: Prevalence, Mechanisms, and Implications for the Study of Heredity and Evolution. *Q Rev Biol*, 84: 131–76.

Jacobs, P. A., Brunton, M., Melville, M. M., et al. 1965. Aggressive Behavior, Mental Sub-Normality and the XYY Male. *Nature*, 208: 1351–2.

James, W. 1899. *Talks to Teachers on Psychology*, London: Longmans, Green, and Co.

Jamieson, A. & Radick, G. 2013. Putting Mendel in His Place: How Curriculum Reform in Genetics and Counterfactual History of Science Can Work Together. *In:* Kampourakis, K. (ed.) *The Philosophy of Biology.* London: Springer.

Jang, K. L., Mccrae, R. R., Angleitner, A., et al. 1998. Heritability of Facet-Level Traits in a Cross-Cultural Twin Sample: Support for a Hierarchical Model of Personality. *J Pers Soc Psychol*, 74: 1556–65.

Jarrell, T. A., Wang, Y., Bloniarz, A. E., et al. 2012. The Connectome of a Decision-Making Neural Network. *Science*, 337: 437–44.

Jayaratne, T. E., Gelman, S. A., Feldbaum, M., et al. 2009. The Perennial Debate: Nature, Nurture, or Choice? Black and White Americans' Explanations for Individual Differences. *Rev. Gen. Psychol.*, 13: 24–33.

Jeeves, M. A. 2011. *Rethinking Human Nature: A Multidisciplinary Approach*, Grand Rapids, Mich.: William B. Eerdmans Pub. Co.

2015. *The Emergence of Personhood: A Quantum Leap?*, Grand Rapids, Michigan: William B. Eerdmans Publishing Company.

Jensen, A. R. 1969. How Much Can We Boost IQ and Scholastic Achievement? *Harvard Educational Review*, 39: 1–123.

1970. Race and the Genetics of Intelligence: A Reply to Lewontin. *Bulletin of the Atomic Scientists*, 26: 17–23.

2000. The G Factor: Psychometrics and Biology. *Novartis Found Symp*, 233: 37–47; discussion 47–57, 116–21.

Jewett, R. 1971. *Paul's Anthropological Terms; a Study of Their Use in Conflict Settings*, Leiden,: Brill.

Joel, D., Berman, Z., Tavor, I., et al. 2015. Sex Beyond the Genitalia: The Human Brain Mosaic. *Proc Natl Acad Sci U S A*, 112: 15468–73.

Johnson, M. B., Kawasawa, Y. I., Mason, C. E., et al. 2009. Functional and Evolutionary Insights into Human Brain Development through Global Transcriptome Analysis. *Neuron*, 62: 494–509.

Johnson, S. 1810. *The Works of the English Poets* Vol. 3. London: Whittingham

Johnson, W., Bouchard, T. J., Krueger, R. F., et al. 2004. Just One G: Consistent Results from Three Test Batteries. *Intelligence*, 32: 95–107.

Johnson, W., Penke, L. & Spinath, F. M. 2011. Heritability in the Era of Molecular Genetics: Some Thoughts for Understanding Genetic Influences on Behavioural Traits. *European Journal of Personality*, 25: 254–266.

Johnson, W., Te Nijenhuis, J. & Bouchard, T. J. 2008. Still Just 1 G: Consistent Results from Five Test Batteries. *Intelligence*, 36: 81–95.

Johnston, P. S. 2002. *Shades of Sheol – Death and Afterlife in the Old Testament*, Leicester: Apollos.

Jordan, D. M., Frangakis, S. G., Golzio, C., et al. 2015. Identification of Cis-Suppression of Human Disease Mutations by Comparative Genomics. *Nature*, 524: 225–9.

Jorde, L. B., Hasstedt, S. J., Ritvo, E. R., et al. 1991. Complex Segregation Analysis of Autism. *Am J Hum Genet*, 49: 932–8.

Joseph, J. 2004. *The Gene Illusion: Genetic Research in Psychiatry and Psychology under the Microscope*. New York: Algora Publishing.

2010. *Genetic Research in Psychiatry and Psychology*. Hood, K.E. Et Al Wiley-Blackwell.

2012. The Fruitless Search for Genes in Psychiatry and Psychology: Time to Re-Examine a Paradigm. *In:* Krimsky, S. & Gruber, J. (eds.) *Genetic Explanations: Sense and Nonsense*. Cambridge: Harvard University Press.

2013. The Use of the Classical Twin Method in the Social and Behavioral Sciences: The Fallacy Continues. *J. Mind Behav.*, 34: 1–39.

Joshi, P. K., Esko, T., Mattsson, H., et al. 2015. Directional Dominance on Stature and Cognition in Diverse Human Populations. *Nature*, 523: 459–62.

Joubert, B. R., Haberg, S. E., Nilsen, R. M., et al. 2012. 450K Epigenome-Wide Scan Identifies Differential DNA Methylation in Newborns Related to Maternal Smoking During Pregnancy. *Environ Health Perspect*, 120: 1425–31.

Jovelin, R., Ajie, B. C. & Phillips, P. C. 2003. Molecular Evolution and Quantitative Variation for Chemosensory Behaviour in the Nematode Genus Caenorhabditis. *Mol. Ecol.*, 12: 1325–1337.

Kaati, G., Bygren, L. O. & Edvinsson, S. 2002. Cardiovascular and Diabetes Mortality Determined by Nutrition During Parents' and Grandparents' Slow Growth Period. *Eur J Hum Genet*, 10: 682–8.

Kahneman, D. 2011. *Thinking, Fast and Slow*, London: Allen Lane.

Kamin, L. J. & Goldberger, A. S. 2002. Twin Studies in Behavioral Research: A Skeptical View. *Theor Popul Biol*, 61: 83–95.

Kammenga, J. E., Phillips, P. C., De Bono, M., et al. 2008. Beyond Induced Mutants: Using Worms to Study Natural Variation in Genetic Pathways. *Trends in Genetics*, 24: 178–185.

Kampourakis, K. 2015. That Gregor Mendel Was a Lonely Pioneer of Genetics, Being Ahead of His Time *In:* Numbers, R. L. & Kampourakis, K. (eds.) *Newton's Apple and Other Myths About Science*. Cambridge, MASS: Harvard University Press.

Kandler, C., Bleidorn, W. & Riemann, R. 2012. Left or Right? Sources of Political Orientation: The Roles of Genetic Factors, Cultural Transmission, Assortative Mating, and Personality. *J. Pers. Soc. Psychol.*, 102: 633–645.

Kandler, C. & Riemann, R. 2013. Genetic and Environmental Sources of Individual Religiousness: The Roles of Individual Personality Traits and Perceived Environmental Religiousness. *Behav Genet*, 43: 297–313.

Kane, R. 2005. *A Contemporary Introduction to Free Will*: Oxford University Press. 2011. *The Oxford Handbook of Free Will*: Oxford University Press.

Kang, D. H., Jo, H. J., Jung, W. H., et al. 2012. The Effect of Meditation on Brain Structure: Cortical Thickness Mapping and Diffusion Tensor Imaging. *Soc Cogn Affect Neurosci*, 8: 27–33.

Kang, D. W., Park, J. G., Ilhan, Z. E., et al. 2013. Reduced Incidence of Prevotella and Other Fermenters in Intestinal Microflora of Autistic Children. *PLoS One*, 8: e68322.

Kang, R., Wan, J., Arstikaitis, P., et al. 2008. Neural Palmitoyl-Proteomics Reveals Dynamic Synaptic Palmitoylation. *Nature*, 456: 904–9.

Kant, I., Guyer, P. & Wood, A. W. 1998. *Critique of Pure Reason*: Cambridge University Press.

Kaprio, J. 2012. Twins and the Mystery of Missing Heritability: The Contribution of Gene-Environment Interactions. *J Intern Med.* 272: 440–448.

Karg, K., Burmeister, M., Shedden, K., et al. 2011. The Serotonin Transporter Promoter Variant (5-HTTLPR), Stress, and Depression Meta-Analysis Revisited: Evidence of Genetic Modification. *Arch Gen Psychiatry*, 68: 444–54.

Kaun, K. R., Chakaborty-Chatterjee, M. & Sokolowski, M. B. 2008. Natural Variation in Plasticity of Glucose Homeostasis and Food Intake. *Journal of Experimental Biology*, 211: 3160–3166.

Kaun, K. R., Devineni, A. V. & Heberlein, U. 2012. Drosophila Melanogaster as a Model to Study Drug Addiction. *Hum. Genet.*, 131: 959–975.

Kaun, K. R., Hendel, T., Gerber, B., et al. 2007. Natural Variation in Drosophila Larval Reward Learning and Memory Due to a cGMP-Dependent Protein Kinase. *Learn. Mem.*, 14: 342–349.

Kawahara, Y., Ito, K., Sun, H., et al. 2004. Glutamate Receptors: RNA Editing and Death of Motor Neurons. *Nature*, 427: 801.

Keating, D. P. 2013. *Nature and Nurture in Early Child Development*, Cambridge University Press.

Keller, E. 1992. Nature, Nurture and the Human Genome Project. In: Kevles, D. a. H., L (ed.) *The Code of Codes: Scientific and Social Issues in the Human Genome Project*. Harvard University Press.

Keller, E. F. 2010a. Foreword. In: Hood, K. E., Halpern, C. T., Greenberg, G., et al. (eds.) *Handbook of Developmental Science, Behavior, and Genetics*. Chichester: Wiley-Blackwell.

2010b. *The Mirage of a Space between Nature and Nurture*, Durham, NC: Duke University Press.

Keller, J. 2005. In Genes We Trust: The Biological Component of Psychological Essentialism and Its Relationship to Mechanisms of Motivated Social Cognition. *J Pers Soc Psychol*, 88: 686–702.

Kelley, R. A., Castelli, F. R., Mabry, K. E., et al. 2013. Effects of Experience and Avpr1a Microsatellite Length on Parental Care in Male Prairie Voles (Microtus Ochrogaster). *Behavioral Ecology and Sociobiology*, 67: 985–992.

Kendler, K. S. 2005. A Gene for ...: The Nature of Gene Action in Psychiatric Disorders. *Am. J. Psychiat.*, 162: 1243–1252.

Kendler, K. S. & Gardner, C. O. 1998. Twin Studies of Adult Psychiatric and Substance Dependence Disorders: Are They Biased by Differences in the Environmental Experiences of Monozygotic and Dizygotic Twins in Childhood and Adolescence? *Psychol. Med.*, 28: 625–633.

Kendler, K. S., Gardner, C. O. & Prescott, C. A. 1997. Religion, Psychopathology, and Substance Use and Abuse: A Multimeasure, Genetic-Epidemiologic Study. *Am. J. Psychiat.*, 154: 322–329.

Kendler, K. S. & Myers, J. 2009. A Developmental Twin Study of Church Attendance and Alcohol and Nicotine Consumption: A Model for Analyzing the Changing Impact of Genes and Environment. *Am. J. Psychiat.*, 166: 1150–1155.

Kendler, K. S., Thornton, L. M., Gilman, S. E., et al. 2000. Sexual Orientation in a US National Sample of Twin and Nontwin Sibling Pairs. *Am. J. Psychiat.*, 157: 1843–1846.

Kent, C. F., Daskalchuk, T., Cook, L., et al. 2009. The *Drosophila Foraging* Gene Mediates Adult Plasticity and Gene-Environment Interactions in Behaviour, Metabolites, and Gene Expression in Response to Food Deprivation. *PLoS Genet*, 5: e1000609.

Kerr, A., Cunningham-Burley, S. & Amos, A. 1998. Eugenics and the New Genetics in Britain: Examining Contemporary Professionals' Accounts. *Science, Technology, & Human Values*, 23: 175–198.

Kevles, D. 1985. *In the Name of Eugenics*: Boston, Harvard University Press.

Khandaker, G. M. 2015. Inflammation and Immunity in Schizophrenia: Implications for Pathophysiology and Treatment. *The Lancet Psychiatry*, 2: 258–270.

Kheirbek, M. A. & Hen, R. 2014. Add Neurons, Subtract Anxiety. *Sci Am*, 311: 62–7.

Kiive, E., Laas, K., Akkermann, K., et al. 2014. Mitigating Aggressiveness through Education? The Monoamine Oxidase A Genotype and Mental Health in General Population. *Acta Neuropsychiatr*, 26: 19–28.

Kim, J. 1989. The Myth of Nonreductive Physicalism'. *Proceedings and Addresses of the Amerian Philosophical Association* 63: 31–47.

1994. Supervenience and Mind. In: Guttenplan, S. (ed.) *A Companion to the Philosophy of Mind.* Oxford: Blackwell.

Kim, Y. S., Leventhal, B. L., Koh, Y. J., et al. 2011. Prevalence of Autism Spectrum Disorders in a Total Population Sample. *Am J Psychiatry*, 168: 904–12.

Kim-Cohen, J., Caspi, A., Taylor, A., et al. 2006. MAOA, Maltreatment, and Gene-Environment Interaction Predicting Children's Mental Health: New Evidence and a Meta-Analysis. *Mol Psychiatry*, 11: 903–13.

Kimmelman, B. A. 1983. The American Breeders' Association: Genetics and Eugenics in an Agricultural Context, 1903–13. *Soc Stud Sci*, 13: 163–204.

King, J. S. 2012. Genetic Tests: Politics and Fetal Diagnostics Collide. *Nature*, 491: 33–4.

King, M. D. & Bearman, P. S. 2011. Socioeconomic Status and the Increased Prevalence of Autism in California. *Am Sociol Rev*, 76: 320–346.

Kiontke, K. & Sudhaus, W. 2006. Ecology of Caenorhabditis Species. *Wormbook*: Doi/10.1895/Wormbook.1.37.1.

Kirk, K. M., Eaves, L. J. & Martin, N. G. 1999. Self-Transcendence as a Measure of Spirituality in a Sample of Older Australian Twins. *Twin Research*, 2: 81–7.

Kirkpatrick, R. M., McGue, M., Iacono, W. G., et al. 2014. Results of a "GWAS Plus:" General Cognitive Ability Is Substantially Heritable and Massively Polygenic. *PLoS One*, 9: e112390.

Kitzinger, J. 2005. Constructing and Deconstructing the "Gay Gene": Media Reporting of Genetics, Sexual Diversity and "Deviance". *In:* Ellison, G. & Goodman, A. (eds.) *The Nature of Difference: Science, Society and Human Biology*. London: Taylor and Francis.

Kleiman, D. G. 1977. Monogamy in Mammals. *Q Rev Biol*, 52: 39–69.

Klemmensen, R., Hatemi, P. K., Hobolt, S. B., et al. 2012. Heritability in Political Interest and Efficacy across Cultures: Denmark and the United States. *Twin Res Hum Genet*, 15: 15–20.

Klingberg, T. 2013. *The Learning Brain: Memory and Brain Development in Children*: Oxford University Press.

Knopik, V. S., Maccani, M. A., Francazio, S., et al. 2012. The Epigenetics of Maternal Cigarette Smoking During Pregnancy and Effects on Child Development. *Development and Psychopathology*, 24: 1377–1390.

Knouse, K. A., Wu, J., Whittaker, C. A., et al. 2014. Single Cell Sequencing Reveals Low Levels of Aneuploidy across Mammalian Tissues. *Proc Natl Acad Sci U S A*, 111: 13409–14.

Koenig, H. G., King, D. E. & Carson, V. B. 2012. *Handbook of Religion and Health*: Oxford University Press.

Koenig, L. B., McGue, M. & Iacono, W. G. 2008. Stability and Change in Religiousness During Emerging Adulthood. *Dev. Psychol.*, 44: 532–543.

Koenig, L. B., McGue, M., Krueger, R. F., et al. 2005. Genetic and Environmental Influences on Religiousness: Findings for Retrospective and Current Religiousness Ratings. *J. Pers.*, 73: 471–488.

Kolber, A. J. 2014. Will There Be a Neurolaw Revolution? *Indiana Law Journal*, 89: 807–845.

Kondrashov, A. 2012. Genetics: The Rate of Human Mutation. *Nature*, 488: 467–8.

Kowal, E. & Frederic, G. 2012. Race, Genetic Determinism and the Media: An Exploratory Study of Media Coverage of Genetics and Indigenous Australians. *Genomics, Society and Policy*, 2?: 1–14.

Krackow, S. & König, B. 2008. Microsatellite Length Polymorphisms Associated with Dispersal-Related Agonistic Onset in Male Wild House Mice (Mus Musculus Domesticus). *Behavioral Ecology and Sociobiology*, 62: 813–820.

Krahn, T. M. & Fenton, A. 2012. The Extreme Male Brain Theory of Autism and the Potential Adverse Effects for Boys and Girls with Autism. *J Bioeth Inq*, 9: 93–103.

Krawczak, M., Trefilov, A., Berard, J., et al. 2005. Male Reproductive Timing in Rhesus Macaques Is Influenced by the 5HTTLPR Promoter Polymorphism of the Serotonin Transporter Gene. *Biology of Reproduction*, 72: 1109–1113.

Kriwaczek, P. 2010. *Babylon: Mesopotamia and the Birth of Civilization*, London: Atlantic.

Krueger, R. F., South, S., Johnson, W., et al. 2008. The Heritability of Personality Is Not Always 50%: Gene-Environment Interactions and Correlations between Personality and Parenting. *J Pers*, 76: 1485–522.

Kuepper, Y., Grant, P., Wielpuetz, C., et al. 2013. MAOA-UVNTR Genotype Predicts Interindividual Differences in Experimental Aggressiveness as a Function of the Degree of Provocation. *Behavioural Brain Research*, 247: 73–78.

Kumar, U., Guleria, A., Kishan, S. S., et al. 2013. Effect of Soham Meditation on Human Brain: A Voxel-Based Morphometry Study. *J Neuroimaging*, 24: 187–90.

Ladyman, J., Ross, D., Spurrett, D., et al. 2007. *Every Thing Must Go: Metaphysics Naturalized*: Clarendon Press.

Lander, E. S. & Schork, N. J. 1994. Genetic Dissection of Complex Traits. *Science*, 265: 2037–48.

Langstrom, N., Rahman, Q., Carlstrom, E., et al. 2010. Genetic and Environmental Effects on Same-Sex Sexual Behavior: A Population Study of Twins in Sweden. *Arch. Sex. Behav.*, 39: 75–80.

Lanphier, E., Urnov, F., Haecker, S. E., et al. 2015. Don't Edit the Human Germ Line. *Nature*, 519: 410–1.

Larson, E. 2010. Biology and the Emergence of the Anglo-American Eugenics Movement. *In:* Alexander, D. R. & Numbers, R. L. (eds.) *Biology and Ideology – from Descartes to Dawkins*. University of Chicago Press.

Lasco, M. S., Jordan, T. J., Edgar, M. A., et al. 2002. A Lack of Dimorphism of Sex or Sexual Orientation in the Human Anterior Commissure. *Brain Research*, 936: 95–98.

Lawrence, A. 2010. Societal Individualism Predicts Prevalence of Nonhomosexual Orientation in Male-to-Female Transsexualism. *Arch. Sex. Behav.*, 39: 573–583.

Lea, R. & Chambers, G. 2007. Monoamine Oxidase, Addiction, and the "Warrior" Gene Hypothesis. *N Z Med J*, 120: U2441.

Ledford, H. 2010. The Code within the Code. *Nature*, 465: 16–7.

2016. CRISPR, the Disruptor. *Nature*, 522: 20–24.

Lee, M. T., Bonneau, A. R. & Giraldez, A. J. 2014. Zygotic Genome Activation During the Maternal-to-Zygotic Transition. *Annu Rev Cell Dev Biol*, 30: 581–613.

Lennox, J. G. 2009. Aristotle. *In:* Ruse, M. & Travis, J. (eds.) *Evolution – the First Four Billion Years*. Cambridge: Harvard University Press.

Levay, S. 1991. A Difference in Hypothalamic Structure between Heterosexual and Homosexual Men. *Science*, 253: 1034–1037.

Levelt, C. N. & Hubener, M. 2012. Critical-Period Plasticity in the Visual Cortex. *Annu Rev Neurosci*, 35: 309–30.

Levitt, M. 2013. Genes, Environment and Responsibility for Violent Behavior: "Whatever Genes One Has It Is Preferable That You Are Prevented from Going around Stabbing People". *New Genet. Soc.*, 32: 4–17.

Levy, N. 2014. *Consciousness and Moral Responsibilty*: Oxford University Press.

Levy, S., Sutton, G., Ng, P. C., et al. 2007. The Diploid Genome Sequence of an Individual Human. *PLoS Biol*, 5: e254.

Lewis, G. B. 2009. Does Believing Homosexuality Is Innate Increase Support for Gay Rights? *Policy Studies Journal*, 37: 669–693.

Lewis, S. W. & Murray, R. M. 1987. Obstetric Complications, Neurodevelopmental Deviance, and Risk of Schizophrenia. *J Psychiatr Res*, 21: 413–21.

Lewis, T. L. & Maurer, D. 2005. Multiple Sensitive Periods in Human Visual Development: Evidence from Visually Deprived Children. *Dev Psychobiol*, 46: 163–83.

Lewontin, R. C. 1975. Genetic Aspects of Intelligence. *Annual Reviews of Genetics*, 9: 387–405.

Liang, J., Cai, W. & Sun, Z. 2014. Single-Cell Sequencing Technologies: Current and Future. *J Genet Genomics*, 41: 513–28.

Liang, P., Xu, Y., Zhang, X., et al. 2015. CRISPR/Cas9-Mediated Gene Editing in Human Tripronuclear Zygotes. *Protein Cell*, 6: 363–72.

Lichtenstein, P., Yip, B. H., Bjork, C., et al. 2009. Common Genetic Determinants of Schizophrenia and Bipolar Disorder in Swedish Families: A Population-Based Study. *Lancet*, 373: 234–9.

Lim, J. P. & Brunet, A. 2013. Bridging the Transgenerational Gap with Epigenetic Memory. *Trends Genet*, 29: 176–86.

Lim, M. M., Wang, Z. X., Olazabal, D. E., et al. 2004. Enhanced Partner Preference in a Promiscuous Species by Manipulating the Expression of a Single Gene. *Nature*, 429: 754–757.

Linden, D. J. 2007. *The Accidental Mind*, London: Belknap.

Lippa, R. 2008. Sex Differences and Sexual Orientation Differences in Personality: Findings from the BBC Internet Survey. *Arch. Sex. Behav.*, 37: 173–187.

Ljungquist, B., Berg, S., Lanke, J., et al. 1998. The Effect of Genetic Factors for Longevity: A Comparison of Identical and Fraternal Twins in the Swedish Twin Registry. *J Gerontol A Biol Sci Med Sci*, 53: M441–6.

Locke, J. 1979. *An Essay Concerning Human Understanding*, Oxford: Clarendon Press.

Loehlin, J. C. 1992. *Genes and Environment in Personality Development*, London: Sage.

Loehlin, J. C. & Nichols, R. C. 1976. *Heredity, Environment and Personality: A Study of 850 Sets of Twins.*: University of Texas Press.

Logan, C. A. & Johnston, T. D. 2007. Synthesis and Separation in the History of "Nature" and "Nurture". *Dev. Psychobiol.*, 49: 758–769.

Lu, R. B., Lee, J. F., Ko, H. C., et al. 2002. No Association of the MAOA Gene with Alcoholism among Han Chinese Males in Taiwan. *Prog Neuropsychopharmacol Biol Psychiatry*, 26: 457–61.

Ludeke, S., Johnson, W. & Bouchard Jr, T. J. 2013. "Obedience to Traditional Authority:" a Heritable Factor Underlying Authoritarianism, Conservatism and Religiousness. *Pers. Individ. Differ.*, 55: 375–380.

Ludmerer, K. M. 1972. *Genetics and American Society: A Historical Appraisal*, Baltimore,: Johns Hopkins University Press.

Lush, J. L. 1940. Intra-Sire Correlations or Regressions of Offspring on Dam as a Method of Estimating Heritability of Characteristics. *Proc. Am. Soc. Anim. Prod.*, 33: 293–301

Lutz, P. E. & Turecki, G. 2014. DNA Methylation and Childhood Maltreatment: From Animal Models to Human Studies. *Neuroscience*, 264: 142–56.

Lykken, D. T., McGue, M., Bouchard, T. J., Jr., et al. 1990. Does Contact Lead to Similarity or Similarity to Contact? *Behav Genet*, 20: 547–61.

Lynch, J., Bevan, J., Achter, P., et al. 2008. A Preliminary Study of How Multiple Exposures to Messages About Genetics Impact on Lay Attitudes Towards Racial and Genetic Discrimination. *New Genet. Soc.*, 27: 43–56.

Mabry, K. E., Streatfeild, C. A., Keane, B., et al. 2011. Avpr1a Length Polymorphism Is Not Associated with Either Social or Genetic Monogamy in Free-Living Prairie Voles. *Animal Behaviour*, 81: 11–18.

Macaulay, I. C., Haerty, W., Kumar, P., et al. 2015. G&T-Seq: Parallel Sequencing of Single-Cell Genomes and Transcriptomes. *Nat Methods*. 12: 519–522.

MacGregor, A. J., Snieder, H., Rigby, A. S., et al. 2000. Characterizing the Quantitative Genetic Contribution to Rheumatoid Arthritis Using Data from Twins. *Arthritis Rheum*, 43: 30–7.

Machin, G. 2009. Non-Identical Monozygotic Twins, Intermediate Twin Types, Zygosity Testing, and the Non-Random Nature of Monozygotic Twinning. *Am. J. Med. Genet. C*, 151C: 110–127.

Machin, G. A. 2004. Why Is It Important to Diagnose Chorionicity and How Do We Do It? *Best Pract Res Clin Obstet Gynaecol*, 18: 515–30.

Macke, J. P., Hu, N., Hu, S., et al. 1993. Sequence Variation in the Androgen Receptor Gene Is Not a Common Determinant of Male Sexual Orientation. *Am. J. Hum. Genet.*, 53: 844–852.

Mackie, J. L. 1980. *The Cement of the Universe: A Study of Causation*, Oxford: Clarendon Press.

Mackintosh, N. 2011. History of Theories and Measurement of Intelligence. *In:* Sternberg, R. & Kaufman, S. (eds.) *The Cambridge Handbook of Intelligence*. Cambridge University Press.

Madwin, G. 1999. *Myths About Queer by Choice People* [Online]. Available: http://www.queerbychoice.com/myths.html.

Maguire, E. A., Gadian, D. G., Johnsrude, I. S., et al. 2000. Navigation-Related Structural Change in the Hippocampi of Taxi Drivers. *Proc Natl Acad Sci U S A*, 97: 4398–403.

Maguire, E. A., Woollett, K. & Spiers, H. J. 2006. London Taxi Drivers and Bus Drivers: A Structural MRI and Neuropsychological Analysis. *Hippocampus*, 16: 1091–101.

Mahler, M. S., Furer, M. & Srttlage, S. F. 1959. Severe Emotiional Disturbances in Childhood: Psychosis. *In:* Arieti, S. (ed.) *American Handbook of Psychiatry*. New York: Basic Books.

Mahmoudzadeh, M., Dehaene-Lambertz, G., Fournier, M., et al. 2013. Syllabic Discrimination in Premature Human Infants Prior to Complete Formation of Cortical Layers. *Proc Natl Acad Sci U S A*, 110: 4846–51.

Mak, J. 2011. *Security – Its in Our DNA* [Online]. Oxygen Cloud. Available: http://blog.oxygencloud.com/2011/04/22/security-its-in-our-dna/ [Accessed March 22nd 2013].

Manuck, S. B., Flory, J. D., Ferrell, R. E., et al. 2000. A Regulatory Polymorphism of the Monoamine Oxidase-A Gene May Be Associated with Variability in Aggression, Impulsivity, and Central Nervous System Serotonergic Responsivity. *Psychiatry Res*, 95: 9–23.

Maoz, U., Mudrik, L., Rivlin, R., et al. 2015. On Reporting the Onset of the Intention to Move. *In:* Mele, A. R. (ed.) *Surrounding Free Will*. Oxford University Press.

Mapmygene. Available: http://www.mapmygene.com/genetic.htm [Accessed 20th March 2013].

Marlow, H. 2009. *Biblical Prophets and Contemporary Environmental Ethics: Re-Reading Amos, Hosea, and First Isaiah*: Oxford University Press.

Marsh, C. 2016. Human Dignity in Christian Theological Perspective. *In:* Gestrich, A. & King, S. B. (eds.) *Forthcoming*. Oxford University Press.

Maxmen, A. 2015. Three Technologies That Changed Genetics. *Nature*, 528: S2–3.
Mayer, E. A. 2011. Gut Feelings: The Emerging Biology of Gut-Brain Communication. *Nat Rev Neurosci*, 12: 453–66.
Mays, J. L. 2006. The Self in the Psalms and the Image of God' *In:* Soulen, R. K. & Woodhead, L. (eds.) *God and Human Dignity* Grand Rapids, Michigan: Eerdmans.
McGowan, P. O., Suderman, M., Sasaki, A., et al. 2011. Broad Epigenetic Signature of Maternal Care in the Brain of Adult Rats. *PLoS One*, 6: e14739.
McGue, M. 2010. The End of Behavioral Genetics?. *Behav Genet*, 40: 284–96.
McGue, M., Vaupel, J. W., Holm, N., et al. 1993. Longevity Is Moderately Heritable in a Sample of Danish Twins Born 1870–1880. *J Gerontol*, 48: B237–44.
McKaughan, D. J. 2012. Voles, Vasopressin, and Infidelity: A Molecular Basis for Monogamy, a Platform for Ethics, and More? *Biol. Philos.*, 27: 521–543.
McKaughan, D. J. & Elliott, K. C. 2012. Voles, Vasopressin, and the Ethics of Framing. *Science*, 338: 1285.
McMahon, E., Wintermark, P. & Lahav, A. 2012. Auditory Brain Development in Premature Infants: The Importance of Early Experience. *Annals of the New York Academy of Sciences*, 1252: 17–24.
Mead, M. 1928. *Coming of Age in Samoa: A Psychological Study of Primitive Youth for Western Civilization*, New York: Blue Ribbon Books.
    1963. *Sex and Temperament in Three Primitive Societies*, New York: Morrow.
Meaney, M. J. 2010. Epigenetics and the Biological Definition of Gene × Environment Interactions. *Child Development*, 81: 41–79.
Mehlman, P. T., Higley, J. D., Faucher, I., et al. 1995. Correlation of CSF 5-HIAA Concentration with Sociality and the Timing of Emigration in Free-Ranging Primates. *Am J Psychiatry*, 152: 907–13.
Mele, A. R. 2003. *Motivation and Agency*: Oxford University Press.
Mele, A. R. 2009a. Causation, Action, and Free Will. *In:* Beebee, H., Hitchcock, C. & Menzies, P. (eds.) *The Oxford Handbook of Causation*. Oxford University Press.
    2009b. *Effective Intentions – the Power of Conscious Will*. Oxford University Press.
    2010. Conscious Deciding and the Science of Free Will. *In:* Baumeister, R. F., Mele, A. R. & Vohs, K. D. (eds.) *Free Will and Consciousness*. Oxford University Press.
    2014. *Free: Why Science Hasn't Disproved Free-Will*: Oxford University Press.
    2015. *Surrounding Free Will: Philosophy, Psychology, Neuroscience*: Oxford University Press.
Merriman, T. & Cameron, V. 2007. Risk-Taking: Behind the Warrior Gene Story. *N Z Med J*, 120: U2440.
Mery, F., Belay, A. T., So, A. K. C., et al. 2007. Natural Polymorphism Affecting Learning and Memory in Drosophila. *Proc. Natl. Acad. Sci. U. S. A.*, 104: 13051–13055.
Metzinger, T. 2000. *Neural Correlates of Consciousness: Empirical and Conceptual Questions*, Cambridge, Mass.: MIT Press.
Meyer-Bahlburg, H. L., Dolezal, C., Baker, S., et al. 2008. Sexual Orientation in Women with Classical or Non-Classical Congenital Adrenal Hyperplasia as a Function of Degree of Prenatal Androgen Excess. *Arch. Sex. Behav.*, 37: 85–99.

Mick, E., Mcgough, J., Deutsch, C. K., et al. 2014. Genome-Wide Association Study of Proneness to Anger. *PLoS One*, 9: e87257.

Middleton, J. R. 2005. *The Liberating Image: The Imago Dei in Genesis 1*, Grand Rapids, MI: Brazos Press.

Mill, J. 1969. *Utility of Religion*, Farnborough: Gregg International Publishers.

Mill, J. S. 2004. *Principles of Political Economy*, Indianapolis: Hatchett Publishing Company.

Millard, A. R. & Bordreuil, P. 1982. A Statue from Syria with Assyrian and Aramaic Inscriptions. *The Biblical Archaeologist*, 45: 135–141.

Miller, G., Zhu, G., Wright, M. J., et al. 2012. The Heritability and Genetic Correlates of Mobile Phone Use: A Twin Study of Consumer Behavior. *Twin Research and Human Genetics*, 15: 97–106.

Mills, R. E., Bennett, E. A., Iskow, R. C., et al. 2007. Which Transposable Elements Are Active in the Human Genome? *Trends Genet*, 23: 183–91.

Minot, S., Bryson, A., Chehoud, C., et al. 2013. Rapid Evolution of the Human Gut Virome. *Proc Natl Acad Sci U S A*, 110: 12450–5.

Mitchell, K. S., Mazzeo, S. E., Bulik, C. M., et al. 2007. An Investigation of a Measure of Twins' Equal Environments. *Twin Research and Human Genetics*, 10: 840–847.

Mitman, G. 1992. *The State of Nature: Ecology, Community, and American Social Thought, 1900–1950*: University of Chicago Press.

Mock, S. E. & Eibach, R. P. 2012. Stability and Change in Sexual Orientation Identity over a 10-Year Period in Adulthood. *Arch. Sex. Behav.*, 41: 641–648.

Mokili, J. L., Rohwer, F. & Dutilh, B. E. 2012. Metagenomics and Future Perspectives in Virus Discovery. *Curr Opin Virol*, 2: 63–77.

Molenaar, P. C., Smit, D. J., Boomsma, D. I., et al. 2012. Estimation of Subject-Specific Heritabilities from Intra-Individual Variation: iFACE. *Twin Res Hum Genet*, 15: 393–400.

Moller-Levet, C. S., Archer, S. N., Bucca, G., et al. 2013. Effects of Insufficient Sleep on Circadian Rhythmicity and Expression Amplitude of the Human Blood Transcriptome. *Proc Natl Acad Sci U S A*, 110: E1132–41.

Moltmann, J. 1981. *The Trinity and the Kingdom of God: The Doctrine of God*: London: SCM Press.

Monk, C., Georgieff, M. K. & Osterholm, E. A. 2013. Research Review: Maternal Prenatal Distress and Poor Nutrition – Mutually Influencing Risk Factors Affecting Infant Neurocognitive Development. *J. Child Psychol. Psychiatry*, 54: 115–130.

Moore, D. S. 2008. Espousing Interactions and Fielding Reactions: Addressing Laypeople's Beliefs About Genetic Determinism. *Philosophical Psychology*, 21: 331–348.

2013a. Behavioural Genetics, Genetics and Epigenetics. *In:* Zelazo, P. D. (ed.) *Oxford Handbook of Developmental Psychology*. New York: Oxford University Press.

2013b. Current Thinking About Nature and Nurture. *In:* Kampourakis, K. (ed.) *The Philosophy of Biology*. London: Springer.

Morell, V. 1993. Evidence Found for a Possible 'Aggression Gene'. *Science*, 260: 1722–1723.

Moreno, J. D. 2003. Neuroethics: An Agenda for Neuroscience and Society. *Nat Rev Neurosci*, 4: 149–53.

Morrison, K. E., Rodgers, A. B., Morgan, C. P., et al. 2014. Epigenetic Mechanisms in Pubertal Brain Maturation. *Neuroscience*, 264: 17–24.

Morrone, M. C. 2010. Brain Development: Critical Periods for Cross-Sensory Plasticity. *Curr Biol*, 20: R934–6.

Morse, S. J. 2011. Genetics and Criminal Responsibility. *Trends Cogn. Sci.*, 15: 378–380.

Morsella, E. 2005. The Function of Phenomenal States: Supramodular Interaction Theory. *Psychol Rev*, 112: 1000–21.

Mueller, M., Martens, L. & Apweiler, R. 2007. Annotating the Human Proteome: Beyond Establishing a Parts List. *Biochim Biophys Acta*, 1774: 175–91.

Mulcaster, R. 1581. *Positions Vvherin Those Primitiue Circumstances Be Examined, Which Are Necessarie for the Training Vp of Children, Either for Skill in Their Booke, or Health in Their Bodie*. Written by Richard Mulcaster, Master of the Schoole Erected in London Anno. 1561. In the Parish of Sainct Laurence Povvntneie, by the Vvorshipfull Companie of the Merchaunt Tailers of the Said Citie. Printed at London: By Thomas Vautrollier for Thomas Chare [i.e. Chard].

1582. *Elementarie*: Oxford University Press.

Muller, G. B. 2007. Evo-Devo: Extending the Evolutionary Synthesis. *Nat Rev Genet*, 8: 943–9.

Muller, H. J. 1936. *Out of the Night: A Biologist's View of the Future*, London: V. Gollancz.

Müller-Wille, S. & Rheinberger, H.-J. 2012. *A Cultural History of Heredity*. University of Chicago Press.

Muller-Wille, S., Rheinberger, H. J., Gayon, J., et al. 2008. Race and Genomics. Old Wine in New Bottles? Documents from a Transdisciplinary Discussion. *NTM*, 16: 363–386.

Munafo, M. R., Durrant, C., Lewis, G., et al. 2009. Gene X Environment Interactions at the Serotonin Transporter Locus. *Biol Psychiatry*, 65: 211–9.

Murdoch, S. 2007. *IQ: A Smart History of a Failed Idea*, Hoboken, N.J.: J. Wiley and Sons.

Murphy, N. 1999. Supervenience and the Downward Efficacy of the Mental: A Nonreductive Physicalist Account of Human Action. *In*: Russell, R. J., Murphy, N., Meyering, T. C., et al. (eds.) *Neuroscience and the Person: Scientific Perspectives on Divine Action*. The Vatican Observatory and Center for Theology and Natural Science.

2006. Emergence and Mental Causation. *In*: Clayton, P. & Davies, P. (eds.) *The Re-Emergence of Emergence: The Emergentist Hypothesis from Science to Religion*. Oxford University Press.

Mustanski, B. S., Dupree, M. G., Nievergelt, C. M., et al. 2005. A Genomewide Scan of Male Sexual Orientation. *Hum. Genet.*, 116: 272–278.

Nahmias, E. 2015. Why We Have Free Will. *Sci Am*, 312: 76–9.

Nair, H. P. & Young, L. J. 2006. Vasopressin and Pair-Bond Formation: Genes to Brain to Behavior. *Physiology (Bethesda)*, 21: 146–52.

Naumova, O. Y., Lee, M., Koposov, R., et al. 2012. Differential Patterns of Whole-Genome DNA Methylation in Institutionalized Children and Children Raised by Their Biological Parents. *Dev Psychopathol*, 24: 143–55.

Needham, J. & Wang, L. 1956. *Science and Civilisation in China. Vol. 2, History of Scientific Thought*: Cambridge University Press.

Neisser, U. E. A. 1996. Intelligence: Knowns and Unknowns. *Am. Psychol.*, 51: 77–101.

Nelson, C. A., Fox, N. A. & Zeanah, C. H. 2013. Anguish of the Abandoned Child. *Sci Am*, 308: 62–7.

Nestler, E. J. 2011. Hidden Switches in the Mind. *Sci Am*, 305: 76–83.

Neyer, F. J. 2002. Twin Relationships in Old Age: A Developmental Perspective. *Journal of Social and Personal Relationships*, 19: 155–177.

Niebuhr, R. 1945. *The Nature and Destiny of Man. A Christian Interpretation … Gifford Lectures*, New York: Scribner.

Nikkels, P. G., Hack, K. E. & Van Gemert, M. J. 2008. Pathology of Twin Placentas with Special Attention to Monochorionic Twin Placentas. *J Clin Pathol*, 61: 1247–53.

Niu, W. & Brass, J. 2011. Intelligence in Worldwide Perspective. In: Sternberg, R. J. & Kaufman, S. B. (eds.) *The Cambridge Handbook of Intelligence*. Cambridge University Press.

Nowak, M. A. & Coakley, S. 2013. *Evolution, Games, and God: The Principle of Cooperation*, Cambridge, Massachusetts: Harvard University Press.

O'Connor, T. 2000. *Persons and Causes – the Metaphysics of Free Will*. New York: Oxford University Press,.

    2015. The Emergence of Personhood: Reflections on the Game of Life. In: Jeeves, M. (ed.) *The Emergence of Personhood – a Quantum Leap?* Grand Rapids: Eerdmans.

O'Connor, T. G., Heron, J., Golding, J., et al. 2003. Maternal Antenatal Anxiety and Behavioural/Emotional Problems in Children: A Test of a Programming Hypothesis. *J. Child Psychol. Psychiatry*, 44: 1025–1036.

O'Leary, D. D., Chou, S. J. & Sahara, S. 2007. Area Patterning of the Mammalian Cortex. *Neuron*, 56: 252–69.

Okasha, S. 2009. Causation in Biology. In: Beebee, H., Hitchcock, C. & Menzies, P. (eds.) *The Oxford Handbook of Causation*. Oxford University Press.

Olazabal, D. E. & Young, L. J. 2006. Oxytocin Receptors in the Nucleus Accumbens Facilitate "Spontaneous" Maternal Behavior in Adult Female Prairie Voles. *Neuroscience*, 141: 559–68.

Olby, R. 2013. Darwin and Heredity. In: Ruse, M. (ed.) *The Cambridge Encyclopedia of Darwin and Evolutionary Thought*. Cambridge University Press.

Ollikainen, M., Smith, K. R., Joo, E. J., et al. 2010. DNA Methylation Analysis of Multiple Tissues from Newborn Twins Reveals Both Genetic and Intrauterine Components to Variation in the Human Neonatal Epigenome. *Hum Mol Genet*, 19: 4176–88.

Ophir, A. G., Wolff, J. O. & Phelps, S. M. 2008. Variation in Neural V1aR Predicts Sexual Fidelity and Space Use among Male Prairie Voles in Semi-Natural Settings. *Proc. Natl. Acad. Sci. U. S. A.*, 105: 1249–1254.

Osborne, K. A., Robichon, A., Burgess, E., et al. 1997. Natural Behavior Polymorphism Due to a cGMP-Dependent Protein Kinase of Drosophila. *Science*, 277: 834–836.

Ott, M., Corliss, H., Wypij, D., et al. 2011. Stability and Change in Self-Reported Sexual Orientation Identity in Young People: Application of Mobility Metrics. *Arch. Sex. Behav.*, 40: 519–532.

Owen, R. & Sloan, P. R. 1992. *The Hunterian Lectures in Comparative Anatomy, May and June 1837,* University of Chicago Press.

Oyama, S. 2000. Causal Democracy and Causal Contributions in Developmental Systems. *Philosophy of Science (Proceedings),* 67: S332–347.

Painter, R. C., Osmond, C., Gluckman, P., et al. 2008. Transgenerational Effects of Prenatal Exposure to the Dutch Famine on Neonatal Adiposity and Health in Later Life. *BJOG,* 115: 1243–9.

Painter, R. C., Roseboom, T. J. & De Rooij, S. R. 2012. Long-Term Effects of Prenatal Stress and Glucocorticoid Exposure. *Birth Defects Research Part C: Embryo Today: Reviews,* 96: 315–324.

Pannenberg, W. 1970. *What Is Man?: Contemporary Anthropology in Theological Perspective,* Philadelphia: Fortress Press.

Panofsky, A. 2014. *Misbehaving Science: Controversy and the Development of Behavior Genetics:* The University of Chicago Press.

Partridge, T. 2011. Methodological Advances toward a Dynamic Developmental Behavioral Genetics: Bridging the Gap. *Research in Human Development,* 8: 242–257.

Pascalis, O., De Vivies, X. D., Anzures, G., et al. 2011. Development of Face Processing. *Wiley Interdiscip Rev Cogn Sci,* 2: 666–675.

Paul, D. B. 1995. *Controlling Human Heredity: 1865 to the Present,* Amherst, N.Y.: Humanity Books.

1998. *The Politics of Heredity: Essays on Eugenics, Biomedicine and the Nature-Nurture Debate:* State University of New York Press.

Pavlov, K. A., Chistiakov, D. A. & Chekhonin, V. P. 2012. Genetic Determinants of Aggression and Impulsivity in Humans. *Journal of Applied Genetics,* 53: 61–82.

Peacocke, A. 2006. Emergence, Mind, and Divine Action: The Hierarchy of the Sciences in Relation to the Human Mind-Brain-Body. In: Clayton, P. & Davies, P. (eds.) *The Re-Emergence of Emergence: The Emergentist Hypothesis from Science to Religion.* Oxford University Press.

Pearson, H. 2012. Children of the 90s: Coming of Age. *Nature,* 484: 155–8.

Pedersen, N. L., Plomin, R., Mcclearn, G. E., et al. 1988. Neuroticism, Extraversion, and Related Traits in Adult Twins Reared Apart and Reared Together. *J Pers Soc Psychol,* 55: 950–7.

Pembrey, M. E., Bygren, L. O., Kaati, G., et al. 2006. Sex-Specific, Male-Line Transgenerational Responses in Humans. *Eur J Hum Genet,* 14: 159–66.

Pennington, B. F., Mcgrath, L. M., Rosenberg, J., et al. 2009. Gene X Environment Interactions in Reading Disability and Attention-Deficit/Hyperactivity Disorder. *Dev Psychol,* 45: 77–89.

Perbal, L. 2013. The 'Warrior Gene' and the Maori People: The Responsibility of the Geneticists. *Bioethics,* 27: 382–387.

Petronis, A. 2010. Epigenetics as a Unifying Principle in the Aetiology of Complex Traits and Diseases. *Nature,* 465: 721–7.

Pfeiffer, R. H. 1935. *State Letters of Assyria [No 345],* New Haven American Oriental Society.

Phelps, S. M. 2010. From Endophenotypes to Evolution: Social Attachment, Sexual Fidelity and the Avpr1a Locus. *Current Opinion in Neurobiology*, 20: 795–802.

Philibert, R. A., Beach, S. R., Lei, M. K., et al. 2013. Changes in DNA Methylation at the Aryl Hydrocarbon Receptor Repressor May Be a New Biomarker for Smoking. *Clin Epigenetics*, 5: 19.

Phillips, B. A. 1846. Scrofula; Its Nature, Its Causes, Its Prevalence and the Principles of Treatment. *Journal of the Statistical Society of London*. 9:152–157.

Pink, T. 2004. *Free Will: A Very Short Introduction*: Oxford University Press.

Pinker, S. 2002. *The Blank Slate: The Modern Denial of Human Nature*: New York: The Penguin Press.

Piton, A., Poquet, H., Redin, C., et al. 2014. 20 Ans Apres: A Second Mutation in MAOA Identified by Targeted High-Throughput Sequencing in a Family with Altered Behavior and Cognition. *Eur J Hum Genet*, 22: 776–83.

Pitt, B. 2012. 'I Want Her Approval. Angie Is a Force ... I Want Her to Be Proud of Her Man': Brad Pitt on Guns, Violence and That Wedding[Online]. Available: http://www.dailymail.co.uk/home/moslive/article-2199295/Brad-Pitt-talks-Angelina-Jolie-I-want-approval-Angie-force–I-want-proud-man.html [Accessed March 22nd 2013].

Plato 2012. The Republic. Gutenberg eBook No 1497.

Plato & Bluck, R. S. 1955. *Plato's Phaedo*, London: Routledge & Kegan Paul.

Plato & Zeyl, D. J. 2000. *Timaeus*, Indianapolis: Hackett Pub. Co.

Plomin, R. 2011. Commentary: Why Are Children in the Same Family So Different? Non-Shared Environment Three Decades Later. *Int J Epidemiol*, 40: 582–92.

Plomin, R. & Bergeman, C. S. 1991. The Nature of Nurture: Genetic Influences on "Environmental" Measures. *Behavioral and Brain Sciences*, 14: 373–427.

Plomin, R., Coon, H., Carey, G., et al. 1991. Parent-Offspring and Sibling Adoption Analyses of Parental Ratings of Temperament in Infancy and Childhood. *J Pers*, 59: 705–32.

Plomin, R. & Deary, I. J. 2015. Genetics and Intelligence Differences: Five Special Findings. *Mol Psychiatry*, 20: 98–108.

Plomin, R., Defries, J. C., Knopik, V. S., et al. 2013a. *Behavioral Genetics: A Primer*, Duffield: Worth Publishers, 6th Edition.

Plomin, R., Defries, J. C., Mcclearn, G. E., et al. 2008. *Behavioral Genetics*: Duffield: Worth, 5th edition.

Plomin, R., Haworth, C. M., Meaburn, E. L., et al. 2013b. Common DNA Markers Can Account for More Than Half of the Genetic Influence on Cognitive Abilities. *Psychol Sci*, 24: 562–8.

Plomin, R. & Simpson, M. A. 2013. The Future of Genomics for Developmentalists. *Dev Psychopathol*, 25: 1263–78.

Poelmans, G., Franke, B., Pauls, D. L., et al. 2013. AKAPs Integrate Genetic Findings for Autism Spectrum Disorders. *Transl Psychiatry*, 3: e270.

Polderman, T. J., Benyamin, B., De Leeuw, C. A., et al. 2015. Meta-Analysis of the Heritability of Human Traits Based on Fifty Years of Twin Studies. *Nat Genet*, 47: 702–9.

Poldrack, R. A. & Farah, M. J. 2015. Progress and Challenges in Probing the Human Brain. *Nature*, 526: 371–9.

Pollock, S. 1999. *Ancient Mesopotamia: The Eden That Never Was*: Cambridge University Press.

Ponseti, J., Siebner, H. R., Kl√∂Ppel, S., et al. 2007. Homosexual Women Have Less Grey Matter in Perirhinal Cortex Than Heterosexual Women. *Plos One*, 2: e762.

Posner, S. F., Baker, L., Heath, A., et al. 1996. Social Contact, Social Attitudes, and Twin Similarity. *Behav Genet*, 26: 123–33.

Power, R. A. & Pluess, M. 2015. Heritability Estimates of the Big Five Personality Traits Based on Common Genetic Variants. *Transl Psychiatry*, 5: e604.

Prichard, Z., Mackinnon, A., Jorm, A. F., et al. 2008. No Evidence for Interaction between MAOA and Childhood Adversity for Antisocial Behavior. *Am J Med Genet B Neuropsychiatr Genet*, 147B: 228–32.

Proctor, R. 1988. *Racial Hygiene: Medicine under the Nazis*, Cambridge, Mass.: Harvard University Press.

Purcell, S. M., Moran, J. L., Fromer, M., et al. 2014. A Polygenic Burden of Rare Disruptive Mutations in Schizophrenia. *Nature*, 506: 185–90.

Rachels, J. 1990. *Created from Animals: The Moral Implications of Darwinism*: Oxford University Press.

Radford, E. J., Ito, M., Shi, H., et al. 2014. In Utero Effects. In Utero Undernourishment Perturbs the Adult Sperm Methylome and Intergenerational Metabolism. *Science*, 345: 1255903.

Rahm, A. K., Feigelson, H. S., Wagner, N., et al. 2012. Perception of Direct-to-Consumer Genetic Testing and Direct-to-Consumer Advertising of Genetic Tests among Members of a Large Managed Care Organization. *J. Genet. Couns.*, 21: 448–461.

Rajasethupathy, P., Sankaran, S., Marshel, J. H., et al. 2015. Projections from Neocortex Mediate Top-Down Control of Memory Retrieval. *Nature*, 526: 653–9.

Ramagopalan, S. V., Dyment, D. A., Handunnetthi, L., et al. 2010. A Genome-Wide Scan of Male Sexual Orientation. *J Hum Genet*, 55: 131–2.

Ramsey, E. M., Achter, P. J. & Condit, C. M. 2001. Genetics, Race, and Crime: An Audience Study Exploring the Bell Curve and Book Reviews. *Critical Studies in Media Communication*, 18: 1–22.

Rankin, C. H. 2002. From Gene to Identified Neuron to Behaviour in Caenorhabditis Elegans. *Nat. Rev. Genet.*, 3: 622–630.

Ratman, D., Vanden Berghe, W., Dejager, L., et al. 2013. How Glucocorticoid Receptors Modulate the Activity of Other Transcription Factors: A Scope Beyond Tethering. *Mol Cell Endocrinol*, 380: 41–54.

Reardon, S. 2014a. Gut-Brain Link Grabs Neuroscientists. *Nature*, 515: 175–7.

2014b. Science in Court: Smart Enough to Die? *Nature*, 506: 284–6.

2015. Global Summit Reveals Divergent Views on Human Gene Editing. *Nature*, 528: 173.

Reddy, K. C., Andersen, E. C., Kruglyak, L., et al. 2009. A Polymorphism in Npr-1 Is a Behavioral Determinant of Pathogen Susceptibility in C. Elegans. *Science*, 323: 382–4.

Relman, D. A. 2012. Microbiology: Learning About Who We Are. *Nature*, 486: 194–5.

Remy, J. J. 2010. Stable Inheritance of an Acquired Behavior in Caenorhabditis Elegans. *Curr Biol*, 20: R877–8.

Remy, J. J. & Hobert, O. 2005. An Interneuronal Chemoreceptor Required for Olfactory Imprinting in C. Elegans. *Science*, 309: 787–90.

Reyes, A., Haynes, M., Hanson, N., et al. 2010. Viruses in the Faecal Microbiota of Monozygotic Twins and Their Mothers. *Nature*, 466: 334–8.

Rice, F., Harold, G. T., Boivin, J., et al. 2010. The Links between Prenatal Stress and Offspring Development and Psychopathology: Disentangling Environmental and Inherited Influences. *Psychol. Med.*, 40: 335–345.

Rice, F., Jones, I. & Thapar, A. 2007. The Impact of Gestational Stress and Prenatal Growth on Emotional Problems in Offspring: A Review. *Acta Psychiatrica Scandinavica*, 115: 171–183.

Rice, G., Anderson, C., Risch, N., et al. 1999. Male Homosexuality: Absence of Linkage to Microsatellite Markers at Xq28. *Science*, 284: 665–667.

Richardson, K. & Norgate, S. 2005. The Equal Environments Assumption of Classical Twin Studies May Not Hold. *Br. J. Educ. Psychol.*, 75: 339–350.

Richardson, S. S., Daniels, C. R., Gillman, M. W., et al. 2014. Society: Don't Blame the Mothers. *Nature*, 512: 131–2.

Richmond, R. C., Simpkin, A. J., Woodward, G., et al. 2015. Prenatal Exposure to Maternal Smoking and Offspring DNA Methylation across the Lifecourse: Findings from the Avon Longitudinal Study of Parents and Children (ALSPAC). *Hum Mol Genet*, 24: 2201–17.

Ridley, M. 2003. *Nature Via Nurture: Genes, Experience and What Makes Us Human*, London: Fourth Estate.

Rieger, G., Linsenmeier, J. W., Gygax, L., et al. 2010. Dissecting 'Gaydar': Accuracy and the Role of Masculinity-Femininity. *Arch. Sex. Behav.*, 39: 124–140.

Rietveld, C. A., Medland, S. E., Derringer, J., et al. 2013. GWAS of 126,559 Individuals Identifies Genetic Variants Associated with Educational Attainment. *Science*, 340: 1467–1471.

Rietveld, M. J., Van Der Valk, J. C., Bongers, I. L., et al. 2000. Zygosity Diagnosis in Young Twins by Parental Report. *Twin Research*, 3: 134–41.

Rigoni, D., Kuhn, S., Gaudino, G., et al. 2012. Reducing Self-Control by Weakening Belief in Free Will. *Conscious Cogn*, 21: 1482–90.

Rigoni, D., KüHn, S., Giuseppe, S., et al. 2011. Inducing Disbelief in Free Will Alters Brain Correlates of Preconscious Motor Preparation: The Brain Minds Whether We Believe in Free Will or Not. *Psychol. Sci.*, 22: 613–618.

Rijsdijk, F. H. V. & Sham, P. C. 2002. Analytic Approaches to Twin Data Using Structural Equation Models. *Briefings in Bioinformatics*, 3: 119–133.

Risch, N., Herrell, R., Lehner, T., et al. 2009. Interaction between the Serotonin Transporter Gene (5-HTTLPR), Stressful Life Events, and Risk of Depression: A Meta-Analysis. *JAMA*, 301: 2462–71.

Roe, S. 2010. Biology, Atheism, and Politics in Eighteenth-Century France. In: Alexander, D. R. & Numbers, R. L. (eds.) *Biology and Ideology – from Descartes to Dawkins*. Chicago University Press.

Rogers, C., Reale, V., Kim, K., et al. 2003. Inhibition of Caenorhabditis Elegans Social Feeding by FMRFamide-Related Peptide Activation of NPR-1. *Nat. Neurosci.*, 6: 1178–1185.

Rose, R. J., Kaprio, J., Williams, C. J., et al. 1990. Social Contact and Sibling Similarity: Facts, Issues, and Red Herrings. *Behav Genet*, 20: 763–78.

Rose, S. P. R. 1997. *Lifelines: Biology, Freedom, Determinism*, London: Allen Lane.

Rose, S. P. R., Kamin, L. J. & Lewontin, R. C. 1984. *Not in Our Genes: Biology, Ideology and Human Nature*, Harmondsworth: Penguin.

Rothenberg, K. & Wang, A. 2006. The Scarlet Gene: Behavioral Genetics, Criminal Law, and Racial and Ethnic Stigma. *Law and Contemporary Problems*, 69: 343–365.

Rushton, J. P., Fulker, D. W., Neale, M. C., et al. 1986. Altruism and Aggression: The Heritability of Individual Differences. *J Pers Soc Psychol*, 50: 1192–8.

Russell, B. 1913. On the Notion of Cause. *Proceedings of the Aristotelian Society*, 13: 1–26.

Rutherford, A. 2004. 'A Visible Scientist': BF Skinner's Writings for the Popular Press. *European Journal of Behaviour Analysis*, 5: 109–120.

Rutter, M. 2006. *Genes and Behavior: Nature-Nurture Interplay Explained*, Malden, MA: Blackwell.

Sammut, M., Cook, S. J., Nguyen, K. C., et al. 2015. Glia-Derived Neurons Are Required for Sex-Specific Learning in C. Elegans. *Nature*, 526: 385–90.

Sanders, A. R., Martin, E. R., Beecham, G. W., et al. 2015. Genome-Wide Scan Demonstrates Significant Linkage for Male Sexual Orientation. *Psychol. Med.*, 45: 1379–1388.

Sartori, G., Pellegrini, S. & Mechelli, A. 2011. Forensic Neurosciences: From Basic Research to Applications and Pitfalls. *Current Opinion in Neurology*, 24: 371–377.

Saudino, K. J., Mcguire, S., Reiss, D., et al. 1995. Parent Ratings of EAS Temperaments in Twins, Full Siblings, Half Siblings, and Step Siblings. *J Pers Soc Psychol*, 68: 723–33.

Saudino, K. J., Wertz, A. E., Gagne, J. R., et al. 2004. Night and Day: Are Siblings as Different in Temperament as Parents Say They Are? *J Pers Soc Psychol*, 87: 698–706.

Savin-Williams, R. C. 2006. Who's Gay? Does It Matter? *Curr. Dir. Psychol.*, 15: 40–44.

2009. *How Many Gays Are There? It Depends – Contemporary Perspectives on Lesbian, Gay, and Bisexual Identities. In:* Hope, D. A. (ed.). New York: Springer

Savulescu, J., Hemsley, M., Newson, A., et al. 2006. Behavioural Genetics: Why Eugenic Selection Is Preferable to Enhancement. *J Appl Philos*, 23: 157–71.

Scarr, S. 1966. Genetic Factors in Activity Motivation. *Child Development*, 37: 663–673.

Scarr, S. & Weinberg, R. A. 1976. IQ Test Performance of Black Children Adopted by White Families. *Am. Psychol.*: 726–739.

Schafer, W. R. 2005. Deciphering the Neural and Molecular Mechanisms of C-Elegans Behavior. *Curr Biol*, 15: R723-R729.

Schaffner, K. F. 1998. Genes, Behavior, and Developmental Emergentism: One Process, Indivisible? *Philosophy of Science*, 65: 209–252.

2000. Behavior at the Organismal and Molecular Levels: The Case of C. Elegans. *Proceedings of the 1998 Biennial Meetings of the Philosophy of Science Association. Part II: Symposia Papers*: S273-S288.

2006. Behaviour: Its Nature and Nurture. *In:* Parens, E., Chapman, A. R. & Press, N. (eds.) *Wrestling with Behavioral Genetics: Science, Ethics, and Public Conversation.* Baltimore: Johns Hopkins University Press.

Scheiner, R., Sokolowski, M. B. & Erber, J. 2004. Activity of cGMP-Dependent Protein Kinase (PKG) Affects Sucrose Responsiveness and Habituation in Drosophila Melanogaster. *Learn. Mem.,* 11: 303–311.

Schifellite, J. 2011. *Biology after the Sociobiology Debate*: Peter Lang.

Schiff, M., Duyme, M., Dumaret, A., et al. 1982. How Much Could We Boost Scholastic Achievement and IQ Scores? A Direct Answer from a French Adoption Study. *Cognition,* 12: 165–96.

Schiff, M. & Lewontin, R. C. 1986. *Education and Class: The Irrelevance of IQ Genetic Studies,* Oxford: Clarendon.

Schlegel, A., Alexander, P., Sinnott-Armstrong, W., et al. 2015. Hypnotizing Libet: Readiness Potentials with Non-Conscious Volition. *Conscious Cogn,* 33: 196–203.

Schloissnig, S., Arumugam, M., Sunagawa, S., et al. 2013. Genomic Variation Landscape of the Human Gut Microbiome. *Nature,* 493: 45–50.

Schmidt, C. 2015. Mental Health: Thinking from the Gut. *Nature,* 518: S12–5.

Schmidt, W. H. 1983. *The Faith of the Old Testament: A History.* Oxford: Basil Blackwell.

Schmitz, S. 1994. Personality and Temperament. *In:* Defries, J. C., Plomin, R. & Fulker, D. W. (eds.) *Nature and Nurture During Middle Childhood.* Oxford: Blackwell.

Schubeler, D. 2015. Function and Information Content of DNA Methylation. *Nature,* 517: 321–6.

Schuurmans, C. & Kurrasch, D. M. 2013. Neurodevelopmental Consequences of Maternal Distress: What Do We Really Know? *Clin. Genet.,* 83: 108–117.

Schwartz, G., Kim, R. M., Kolundzija, A. B., et al. 2010. Biodemographic and Physical Correlates of Sexual Orientation in Men. *Arch. Sex. Behav.,* 39: 93–109.

Searle, J. R. 2001. *Rationality in Action,* Cambridge, Mass.: MIT Press.

2007. *Freedom and Neurobiology: Reflections on Free Will, Language, and Political Power,* New York: Columbia University Press.

2010. Consciousness and the Problem of Free Will. *In:* Baumeister, R. F., Mele, A. R. & Vohs, K. D. (eds.) *Free Will and Consciiousness – How Might They Work?*: Oxford University Press.

Segal, N. L. 2012. *Born Together – Reared Apart: The Landmark Minnesota Twin Study.* Cambridge, Mass.: Harvard University Press.

Segerstråle, U. C. O. 2000. *Defenders of the Truth: The Battle for Science in the Sociology Debate and Beyond*: Oxford University Press.

Sekar, A., Bialas, A. R., De Rivera, H., et al. 2016. Schizophrenia Risk from Complex Variation of Complement Component 4. *Nature,* 530: 177–83.

Selden, S. 1989. The Use of Biology to Legitimate Inequality: The Eugenics Movement within the High School Biology Txtbook, 1914–1949. *In:* Secada, W. (ed.) *Equity in Education.* New York: Falmer Press.

Sender, R., Fuchs, S. & Milo, R. 2016. Revised Estimates for the Number of Human and Bacteria Cells in the Body. *bioRxiv preprint first posted online Jan. 6, 2016; doi:*http://dx.doi.org/10.1101/036103.

Sesardic, N. 2005. *Making Sense of Heritability: How Not to Think About Behavior Genetics*: Cambridge University Press.

Shapiro, J. A. 2013. How Life Changes Itself: The Read-Write Genome. *Phys Life Rev*, 10: 287–323.

Sheldon, J. P., Pfeffer, C. A., Jayaratne, T. E., et al. 2007. Beliefs About the Etiology of Homosexuality and About the Ramifications of Discovering Its Possible Genetic Origin. *Journal of Homosexuality*, 52: 111–150.

Sheridan, M. A., Fox, N. A., Zeanah, C. H., et al. 2012. Variation in Neural Development as a Result of Exposure to Institutionalization Early in Childhood. *Proc Natl Acad Sci U S A*, 109: 12927–32.

Sherman, G. D., Lee, J. J., Cuddy, A. J., et al. 2012. Leadership Is Associated with Lower Levels of Stress. *Proc Natl Acad Sci U S A*, 109: 17903–7.

Shields, S. 2012. Aristotle. In: Zalta, E. N. (ed.) *The Stanford Encyclopedia of Philosophy*.

Shumay, E., Logan, J., Volkow, N. D., et al. 2012. Evidence That the Methylation State of the Monoamine Oxidase A Gene Predicts Brain Activity of MAO A Enzyme in Healthy Men. *Epigenetics*, 7: 1151–1160.

Shur, N. 2009. The Genetics of Twinning: From Splitting Eggs to Breaking Paradigms Introduction. *Am. J. Med. Genet.*, 151C: 105–109.

Sidanius, J. & Pratto, F. 1999. *Social Dominance: An Intergroup Theory of Social Hierarchy and Oppression*: Cambridge University Press.

Siegelman, M. 1974. Parental Background of Male Homosexuals and Heterosexuals. *Arch. Sex. Behav.*, 3: 3–18.

   1981. Parental Backgrounds of Homosexual and Heterosexual Men: A Cross National Replication. *Arch. Sex. Behav.*, 10: 505–513.

Siklenka, K., Erkek, S., Godmann, M., et al. 2015. Disruption of Histone Methylation in Developing Sperm Impairs Offspring Health Transgenerationally. *Science*, 350: aab2006.

Silberstein, M. & Mcgeever, J. 1999. The Search for Ontological Emergence. *The Philosophical Quarterly*, 49: 182–200.

Simicevic, J., Schmid, A. W., Gilardoni, P. A., et al. 2013. Absolute Quantification of Transcription Factors During Cellular Differentiation Using Multiplexed Targeted Proteomics. *Nat Methods*, 10: 570–6.

Simonson, I. & Sela, A. 2011. On the Heritability of Consumer Decision Making: An Exploratory Approach for Studying Genetic Effects on Judgment and Choice. *Journal of Consumer Research*, 37: 951–966.

Singer, P. 1993. *Practical Ethics, 2nd Edn.*: Cambridge University Press.

Singer, T., Mcconnell, M. J., Marchetto, M. C., et al. 2010. Line-1 Retrotransposons: Mediators of Somatic Variation in Neuronal Genomes? *Trends Neurosci*, 33: 345–54.

Smemo, S., Tena, J. J., Kim, K. H., et al. 2014. Obesity-Associated Variants within FTO Form Long-Range Functional Connections with IRX3. *Nature*, 507: 371–5.

Smerecnik, C. M., Mesters, I., De Vries, N. K., et al. 2009. Alerting the General Population to Genetic Risks: The Value of Health Messages Communicating the Existence of Genetic Risk Factors for Public Health Promotion. *Health Psychol*, 28: 734–45.

Smith, A. & Skinner, A. S. 1982. *The Wealth of Nations. Books I-Iii*, Harmondsworth: Penguin.

Smith, D. 2014. *How to Think Like Einstein*, London: Michael O'Mara Books.

Snyderman, M. & Rothman, S. 1988. *The IQ Controversy, the Media and Public Policy*, Piscataway, NJ: Transaction Publishers.

Sokolowski, M. B. 1980. Foraging Strategies of Drosophila Melanogaster – a Chromosomal Analysis. *Behav. Genet.*, 10: 291–302.

Solomon, N. G., Richmond, A. R., Harding, P. A., et al. 2009. Polymorphism at the Avpr1a Locus in Male Prairie Voles Correlated with Genetic but Not Social Monogamy in Field Populations. *Mol. Ecol.*, 18: 4680–4695.

Spalding, K. L., Bergmann, O., Alkass, K., et al. 2013. Dynamics of Hippocampal Neurogenesis in Adult Humans. *Cell*, 153: 1219–27.

Sparkman, A. M., Adams, J. R., Steury, T. D., et al. 2012. Evidence for a Genetic Basis for Delayed Dispersal in a Cooperatively Breeding Canid. *Animal Behaviour*, 83: 1091–1098.

Spector, T. D. 2012. *Identically Different: Why You Can Change Your Genes*, London: Weidenfeld & Nicolson.

Spitz, F. & Furlong, E. E. 2008. Transcription Factors: From Enhancer Binding to Developmental Control. *Nat Rev Genet*, 13: 613–26.

Steele, C. M. & Aronson, J. 1995. Stereotype Threat and the Intellectual Test Performance of African Americans. *J. Pers. Soc. Psychol.*, 69: 797–811.

Steensma, T. D., Van Der Ende, J., Verhulst, F. C., et al. 2013. Gender Variance in Childhood and Sexual Orientation in Adulthood: A Prospective Study. *Journal of Sexual Medicine*, 10: 2723–2733.

Stillman, T. F. & Baumeister, R. F. 2010. Guilty, Free, and Wise: Determinism and Psychopathy Diminish Learning from Negative Emotions. *Journal of Experimental Social Psychology*, 46: 951–960.

Stillman, T. F., Baumeister, R. F. & Mele, A. R. 2011. Free Will in Everyday Life: Autobiographical Accounts of Free and Unfree Actions. *Philosophical Psychology*, 24: 381–394.

Stochholm, K., Bojesen, A., Jensen, A. S., et al. 2012. Criminality in Men with Klinefelter's Syndrome and XYY Syndrome: A Cohort Study. *BMJ Open*, e000650 doi:10.1136/bmjopen-2011-000650.

Streatfeild, C. A., Mabry, K. E., Keane, B., et al. 2011. Intraspecific Variability in the Social and Genetic Mating Systems of Prairie Voles, Microtus Ochrogaster. *Animal Behaviour*, 82: 1387–1398.

Sturgis, P., Read, S., Hatemi, P., et al. 2010. A Genetic Basis for Social Trust? *Political Behavior*, 32: 205–230.

Sudjic, D. 2012. *The Urban DNA of London* [Online]. Available: http://www.domusweb.it/en/op-ed/the-urban-dna-of-london/ [Accessed March 22nd 2013].

Sugita, Y. 2008. Face Perception in Monkeys Reared with No Exposure to Faces. *Proc Natl Acad Sci U S A*, 105: 394–8.

Sullivan, J. 1963. *The Image of God: The Doctrine of St. Augustine and Its Influence*, Dubuque, Iowa: Priory Press.

Sullivan, P. F., Kendler, K. S. & Neale, M. C. 2003. Schizophrenia as a Complex Trait: Evidence from a Meta-Analysis of Twin Studies. *Arch Gen Psychiatry*, 60: 1187–92.

Swaab, D. F. & Hofman, M. A. 1990. An Enlarged Suprachiasmatic Nucleus in Homosexual Men. *Brain Research*, 537: 141–148.

Szyf, M. 2015. Nongenetic Inheritance and Transgenerational Epigenetics. *Trends Mol Med*, 21: 134–44.

Tabery, J. 2014. *Beyond Versus – the Struggle to Understand the Interaction of Nature and Nurture* Cambridge, Mass.: The MIT Press.

Talge, N. M., Neal, C., Glover, V., et al. 2007. Antenatal Maternal Stress and Long-Term Effects on Child Neurodevelopment: How and Why? *J. Child Psychol. Psychiatry*, 48: 245–261.

Tan, E. J. & Tang, B. L. 2006. Looking for Food: Molecular Neuroethology of Invertebrate Feeding Behavior. *Ethology*, 112: 826–832.

Taylor, A. & Kim-Cohen, J. 2007. Meta-Analysis of Gene-Environment Interactions in Developmental Psychopathology. *Dev Psychopathol*, 19: 1029–37.

Taylor, A. E. 1908. *Plato*, London: Constable.

Teh, A. L., Pan, H., Chen, L., et al. 2014. The Effect of Genotype and in Utero Environment on Interindividual Variation in Neonate DNA Methylomes. *Genome Res*, 24: 1064–74.

Teigen, K. H. 1984. A Note on the Origin of the Term Nature and Nurture – Not Shakespeare and Galton, but Mulcaster. *J. Hist. Behav. Sci.*, 20: 363–364.

Thaiss, C. A., Zeevi, D., Levy, M., et al. 2015. Transkingdom Control of Microbiota Diurnal Oscillations Promotes Metabolic Homeostasis. *Cell*, 159: 514–29.

Thiselton, A. 2015. The Image and Likeness of God. *In:* Jeeves, M. (ed.) *The Emergence of Personhood – a Quantum Leap?* Grand Rapids: Eerdmans.

Thomas, J. H. 2008. Genome Evolution in Caenorhabditis. *Brief Funct Genomic Proteomic*, 7: 211–6.

Thornhill, R. & Palmer, C. T. 2000. *A Natural History of Rape: Biological Bases of Sexual Coercion*. Cambridge, Mass., MIT Press.

Tieger, P. & Barbara, B.-T. 1997. *Nurture by Nature*, Boston, MA: Little.

Tiihonen, J., Rautiainen, M. R., Ollila, H. M., et al. 2015. Genetic Background of Extreme Violent Behavior. *Mol Psychiatry*, 20: 786–92.

Tooby, J. & Cosmides, L. 1989. The Innate Versus the Manifest: How Universal Does Universal Have to Be?. *Behavioral and Brain Sciences*, 12: 36–37.

1990. On the Universality of Human Nature and the Uniqueness of the Individual: The Role of Genetics and Adaptation. *J Pers*, 58: 17–67.

Trefilov, A., Berard, J., Krawczak, M., et al. 2000. Natal Dispersal in Rhesus Macaques Is Related to Serotonin Transporter Gene Promoter Variation. *Behav. Genet.*, 30: 295–301.

Trevena, J. & Miller, J. 2010. Brain Preparation before a Voluntary Action: Evidence against Unconscious Movement Initiation. *Conscious Cogn*, 19: 447–56.

Truett, K. R., Eaves, L. J., Meyer, J. M., et al. 1992. Religion and Education as Mediators of Attitudes – a Multivariate Analysis. *Behav. Genet.*, 22: 43–62.

Tse, P. 2013. *The Neural Basis of Free Will: Criterial Causation*, Cambridge, Mass.: MIT Press.

Tsuang, M. T., Williams, W. M., Simpson, J. C., et al. 2002. Pilot Study of Spirituality and Mental Health in Twins. *Am. J. Psychiat.*, 159: 486–488.

Turkheimer, E. 2000. Three Laws of Behavior Genetics and What They Mean. *Curr. Dir. Psychol.*, 9: 160–164.

2011. Still Missing. *Research in Human Development*, 8: 227–241.

Turkheimer, E., Haley, A., Waldron, M., et al. 2003. Socioeconomic Status Modifies Heritability of IQ in Young Children. *Psychol Sci*, 14: 623–8.

Turkheimer, E. & Harden, K. P. 2014. Behavior Genetic Research Methods. In: Reis, H. T. & Judd, C. M. (eds.) *Handbook of Research Methods in Social and Personality Psychology*. Second edition.: Cambridge University Press.

Turkheimer, E., Pettersson, E. & Horn, E. E. 2014. A Phenotypic Null Hypothesis for the Genetics of Personality. *Annu Rev Psychol*, 65: 515–40.

Turkheimer, E. & Waldron, M. 2000. Nonshared Environment: A Theoretical, Methodological, and Quantitative Review. *Psychol Bull*, 126: 78–108.

Ulbricht, R. J. & Emeson, R. B. 2014. One Hundred Million Adenosine-to-Inosine RNA Editing Sites: Hearing through the Noise. *Bioessays*, 36: 730–5.

Urbina, S. 2011. Tests of Intelligence. In: Sternberg, R. & Kaufman, S. (eds.) *The Cambridge Handbook of Intelligence*. Cambridge University Press.

Uslaner, E. M. & Brown, M. 2005. Inequality, Trust, and Civic Engagement. *American Politics Research*, 33: 868–894.

Van Den Bergh, B. R. H. & Marcoen, A. 2004. High Antenatal Maternal Anxiety Is Related to Adhd Symptoms, Externalizing Problems, and Anxiety in 8- and 9-Year-Olds. *Child Development*, 75: 1085–1097.

Van Den Bergh, B. R. H., Mulder, E. J. H., Mennes, M., et al. 2005. Antenatal Maternal Anxiety and Stress and the Neurobehavioural Development of the Fetus and Child: Links and Possible Mechanisms. A Review. *Neuroscience & Biobehavioral Reviews*, 29: 237–258.

Van Dongen, J., Slagboom, P. E., Draisma, H. H. M., et al. 2012. The Continuing Value of Twin Studies in the Omics Era. *Nature reviews. Genetics*, 13: 640–53.

Van Huyssteen, J. W. 2006. *Alone in the World?: Human Uniqueness in Science and Theology* Grand Rapids, Mich.: William B. Eerdmans Pub. Co.

Van Ijzendoorn, M. H., Belsky, J. & Bakermans-Kranenburg, M. J. 2012. Serotonin Transporter Genotype 5HTTLPR as a Marker of Differential Susceptibility? A Meta-Analysis of Child and Adolescent Gene-by-Environment Studies. *Transl Psychiatry*, 2: e147.

Van Praag, H. 2008. Neurogenesis and Exercise: Past and Future Directions. *Neuromolecular Med*, 10: 128–40.

Vance, T., Maes, H. H. & Kendler, K. S. 2010. Genetic and Environmental Influences on Multiple Dimensions of Religiosity a Twin Study. *Journal of Nervous and Mental Disease*, 198: 755–761.

Vanderlaan, D. P., Forrester, D. L., Petterson, L. J., et al. 2013. The Prevalence of Fa'afafine Relatives among Samoan Gynephilic Men and Fa'afafine. *Arch. Sex. Behav.*, 42: 353–359.

Vassos, E., Collier, D. A. & Fazel, S. 2014. Systematic Meta-Analyses and Field Synopsis of Genetic Association Studies of Violence and Aggression. *Mol Psychiatry*. 19: 471–477.

Veenendaal, M. V., Painter, R. C., De Rooij, S. R., et al. 2013. Transgenerational Effects of Prenatal Exposure to the 1944–45 Dutch Famine. *BJOG*, 120: 548–53.

Velasquez-Manoff, M. 2015a. Genetics: Relative Risk. *Nature*, 527: S116–7.

2015b. Gut Microbiome: The Peacekeepers. *Nature*, 518: S3–11.

Verhulst, B., Eaves, L. J. & Hatemi, P. K. 2012. Correlation Not Causation: The Relationship between Personality Traits and Political Ideologies. *Am J Pol Sci*, 56: 34–51.

Virgin, H. W. 2014. The Virome in Mammalian Physiology and Disease. *Cell*, 157: 142–50.

Visscher, P. M., Hill, W. G. & Wray, N. R. 2008. Heritability in the Genomics Era – Concepts and Misconceptions. *Nat. Rev. Genet.*, 9: 255–266.

Visscher, P. M., Medland, S. E., Ferreira, M. A., et al. 2006. Assumption-Free Estimation of Heritability from Genome-Wide Identity-by-Descent Sharing between Full Siblings. *PLoS Genet*, 2: e41.

Vohs, K. D. & Schooler, J. W. 2008. The Value of Believing in Free Will: Encouraging a Belief in Determinism Increases Cheating. *Psychol. Sci.*, 19: 49–54.

Voland, E. 2009. Evaluating the Evolutionary Status of Religiosity and Religiousness. In: Voland, E. S., W (ed.) *The Biological Evolution of Religious Mind and Behavior*. Springer.

von Rad, G. 1972 . Genesis: *A Commentary*, London: SCM.

Vonasch, A. J. & Baumeister, R. F. 2013. Implications of Free Will Beliefs for Basic Theory and Societal Benefit: Critique and Implications for Social Psychology. *British Journal of Social Psychology*, 52: 219–227.

Wachs, T. D. 1992. *The Nature of Nurture*, Newbury Park; London: Sage.

Waddington, C. 1952. *Organisers and Genes*: Cambridge University Press.

Wade, N. J. 2014. *A Troublesome Inheritance: Genes, Race, and Human History*, New York: The Penguin Press.

Waldron, J. 2010. The Image of God: Rights, Reason, and Order. *University Public Law and Legal Theory Working Papers*, Paper 246.

Walker, B. 2013. When the Facts and the Law Are against You, Argue the Genes?: A Pragmatic Analysis of Genotyping Mitigation Defenses for Psychopathic Defendants in Death Penalty Cases. *Washington University Law Review*, 90: 1779–1817.

Wall, S. 2009. *The Nurture of Nature: Childhood, Antimodernism, and Ontario Summer Camps, 1920–55*, Vancouver, BC: UBC Press.

Waller, J. 2001. Ideas of Heredity, Reproduction and Eugenics in Britain, 1800–1875. *Stud. Hist. Phil. Biol. & Biomed. Sci.*, 32: 457–489.

Waller, N. G., Kojetin, B. A., Bouchard, T. J., et al. 1990. Genetic and Environmental Influences on Religious Interests, Attitudes, and Values – a Study of Twins Reared Apart and Together. *Psychol. Sci.*, 1: 138–142.

Wallis, C. 2014. Gut Reactions. *Sci Am*, 310: 30–33.

Walton, J. H. 2009. *The Lost World of Genesis One: Ancient Cosmology and the Origins Debate*, Downers Grove, IL: IVP Academic.

Wang, B., Zhou, S., Hong, F., et al. 2012. Association Analysis between the Tag SNP for Sonic Hedgehog Rs9333613 Polymorphism and Male Sexual Orientation. *Journal of Andrology*, 33: 951–954.

Wang, H., Duclot, F., Liu, Y., et al. 2013. Histone Deacetylase Inhibitors Facilitate Partner Preference Formation in Female Prairie Voles. *Nat Neurosci*, 16: 919–24.

Wang, L., Liu, Q., Shen, H., et al. 2015. Large-Scale Functional Brain Network Changes in Taxi Drivers: Evidence from Resting-State fMRI. *Hum Brain Mapp*, 36: 862–71.

Wang, X. & He, C. 2014. Dynamic RNA Modifications in Posttranscriptional Regulation. *Mol Cell*, 56: 5–12.

Warden, C. 2007. *Nurture through Nature*, Auchtertarder, Perthshire: Mindstretchers.

Warner, J. H. 1997. *The Therapeutic Perspective: Medical Practice, Knowledge, and Identity in America, 1820–1885*, Princeton, N.J.: Princeton University Press.

Wasserman, D. 2004. Is There Value in Identifying Individual Genetic Predispositions to Violence? *J Law Med Ethics*, 32: 24–33.

Watrin, J. P. & Darwich, R. 2012. On Behaviorism in the Cognitive Revolution: Myth and Reactions. *Rev. Gen. Psychol.*, 16: 269–282.

Watson, J. B. 1925. *Behaviorism*, London: Kegan Paul, Trench, Trubner & Company.

Watson, J. D. & Crick, F. H. 1953. Molecular Structure of Nucleic Acids; a Structure for Deoxyribose Nucleic Acid. *Nature*, 171: 737–8.

Weindling, P. 2010. Genetics, Eugenics, and the Holocaust. *In*: Alexander, D. R. & Numbers, R. L. *Biology and Ideology – from Descartes to Dawkins*. University of Chicago Press.

Weingart, P., Maasen, S., Segerstrale, U. 1997. Shifting Boundaries between the Biological and the Social: The Social and Political Contexts. *In*: Weingart, P., Mitchell, S.D., Richerson, P.J. and Maasen, S (ed.) *Human by Nature: Between Biology and the Social Sciences*. Lawrence Erlbaum.

Weinstock, M. 2008. The Long-Term Behavioural Consequences of Prenatal Stress. *Neuroscience & Biobehavioral Reviews*, 32: 1073–1086.

Weintraub, K. 2011. The Prevalence Puzzle: Autism Counts. *Nature*, 479: 22–4.

Welz, C. 2011. Imago Dei – References to the Invisible. *Studia Theologica*, 65: 74–91.

West-Eberhard, M. J. 2003. *Developmental Plasticity and Evolution*: Oxford University Press.

Westermann, C. 1984. *Genesis 1–11: A Commentary*, Minneapolis: Augsburg Pub. House.

White, J. N. 2006. *Intelligence, Destiny, and Education: The Ideological Roots of Intelligence Testing*, London: Routledge.

White, L. J. 1967. The Historical Roots of Our Ecologic Crisis. *Science*, 155: 1203–1207.

Whiteway, E. & Alexander, D. R. 2015. Understanding the Causes of Same Sex Attraction. *Science and Christian Belief*, 27: 17–40.

Wiggam, A. E. 1925. *The Fruit of the Family Tree*, London: T. Werner Laurie.

Wilhelm, M., Schlegl, J., Hahne, H., et al. 2014. Mass-Spectrometry-Based Draft of the Human Proteome. *Nature*, 509: 582–7.

Wilkerson, W. S. 2009. Is It a Choice? Sexual Orientation as Interpretation. *Journal of Social Philosophy*, 40: 97–116.

Wilson, E. O. 1975a. *Sociobiology – the New Synthesis* Cambridge: Harvard University Press.

Oct. 12th 1975. *The New York Times Magazine*.

Wilson, H. & Widom, C. 2010. Does Physical Abuse, Sexual Abuse, or Neglect in Childhood Increase the Likelihood of Same-Sex Sexual Relationships and Cohabitation? A Prospective 30-Year Follow-Up. *Arch. Sex. Behav.*, 39: 63–74.

Winter, T., Kaprio, J., Viken, R. J., et al. 1999. Individual Differences in Adolescent Religiosity in Finland: Familial Effects Are Modified by Sex and Region of Residence. *Twin Research*, 2: 108–14.

Witelson, S. F., Kigar, D. L., Scamvougeras, A., et al. 2008. Corpus Callosum Anatomy in Right-Handed Homosexual and Heterosexual Men. *Arch. Sex. Behav.*, 37: 857–863.

Witte, J. & Alexander, F. S. 2010. *Christianity and Human Rights: An Introduction*, Cambridge University Press.

Wolterstorff, N. 2008. *Justice: Rights and Wrongs*: Princeton University Press.

Wong, C. C., Meaburn, E. L., Ronald, A., et al. 2013. Methylomic Analysis of Monozygotic Twins Discordant for Autism Spectrum Disorder and Related Behavioural Traits. *Mol Psychiatry*, 19: 495–503.

Wood, A. R., Esko, T., Yang, J., et al. 2014. Defining the Role of Common Variation in the Genomic and Biological Architecture of Adult Human Height. *Nat Genet*, 46: 1173–86.

Woodhouse, J. 1982. Eugenics and the Feeble-Minded: The Parliamentary Debates of 1912–14. *History of Education*, 11: 127–37.

Woodward, J. 2003. *Making Things Happen: A Theory of Causal Explanation*: Oxford University Press.

  2010. Causation in Biology: Stability, Specificity, and the Choice of Levels of Explanation. *Biology and Philosophy*, 25: 287–318.

Wright, S. 1921. Correlation and Causation. *Journal of Agricultural Research*, 20: 557–585.

Wu, Q., Wen, T. Q., Lee, G., et al. 2003. Developmental Control of Foraging and Social Behavior by the Drosophila Neuropeptide Y-Like System. *Neuron*, 39: 147–161.

Xiao, R. & Boehnke, M. 2009. Quantifying and Correcting for the Winner's Curse in Genetic Association Studies. *Genet Epidemiol*, 33: 453–62.

Xu, S., Wilf, R., Menon, T., et al. 2014. Epigenetic Control of Learning and Memory in Drosophila by Tip60 HAT Action. *Genetics*, 198: 1571–86.

Xu, X., Mar, R. A. & Peterson, J. B. 2013. Does Cultural Exposure Partially Explain the Association between Personality and Political Orientation? *Pers Soc Psychol Bull*, 39: 1497–517.

Xue, Z., Huang, K., Cai, C., et al. 2013. Genetic Programs in Human and Mouse Early Embryos Revealed by Single-Cell RNA Sequencing. *Nature*, 500: 593–7.

Yamashita, T. & Kwak, S. 2013. The Molecular Link between Inefficient Glua2 Q/R Site-RNA Editing and TDP-43 Pathology in Motor Neurons of Sporadic Amyotrophic Lateral Sclerosis Patients. *Brain Res*, 1584: 28–38.

Yampolsky, L. Y., Glazko, G. V. & Fry, J. D. 2012. Evolution of Gene Expression and Expression Plasticity in Long-Term Experimental Populations of Drosophila Melanogaster Maintained under Constant and Variable Ethanol Stress. *Mol. Ecol.*, 21: 4287–4299.

Yang, B. Z., Zhang, H., Ge, W., et al. 2013. Child Abuse and Epigenetic Mechanisms of Disease Risk. *Am J Prev Med*, 44: 101–7.

Yau, S. Y., Gil-Mohapel, J., Christie, B. R., et al. 2014. Physical Exercise-Induced Adult Neurogenesis: A Good Strategy to Prevent Cognitive Decline in Neurodegenerative Diseases? *Biomed Res Int*, 2014: 403120.

Yong, E. 2013. Chinese Project Probes the Genetics of Genius. *Nature*, 497: 297–9.

Young, L. J. & Hammock, E. A. 2007. On Switches and Knobs, Microsatellites and Monogamy. *Trends Genet*, 23: 209–12.

Young, L. J., Nilsen, R., Waymire, K. G., et al. 1999. Increased Affiliative Response to Vasopressin in Mice Expressing the V-1a Receptor from a Monogamous Vole. *Nature*, 400: 766–768.

Zeman, A. 2002. *Consciousness: A User's Guide*: Yale University Press.

Zinnbauer, B. J., Pargament, K. I., Cole, B., et al. 1997. Religion and Spirituality: Unfuzzying the Fuzzy. *Journal for the Scientific Study of Religion*, 36: 549–564.

Zoghbi, H. Y. & Bear, M. F. 2012. Synaptic Dysfunction in Neurodevelopmental Disorders Associated with Autism and Intellectual Disabilities. *Cold Spring Harb Perspect Biol*, 4.

Zucker, K., Mitchell, J., Bradley, S., et al. 2006. The Recalled Childhood Gender Identity/Gender Role Questionnaire: Psychometric Properties. *Sex Roles*, 54: 469–483.

Zuk, O., Hechter, E., Sunyaev, S. R., et al. 2012. The Mystery of Missing Heritability: Genetic Interactions Create Phantom Heritability. *Proc Natl Acad Sci U S A*, 109: 1193–8.

Zwarts, L., Magwire, M. M., Carbone, M. A., et al. 2011. Complex Genetic Architecture of Drosophila Aggressive Behavior. *Proc. Natl. Acad. Sci. U. S. A.*, 108: 17070–17075.

Zwarts, L., Versteven, M. & Callaerts, P. 2012. Genetics and Neurobiology of Aggression in Drosophila. *Fly*, 6: 35–48.

Zwijnenburg, P. J. G., Meijers-Heijboer, H. & Boomsma, D. I. 2010. Identical but Not the Same: The Value of Discordant Monozygotic Twins in Genetic Research. *Am J Med Genet B*, 153B: 1134–1149.

Index

364 Index

free will, 254–78; genetic, 1–61; genetic, in complex biological organisms, 62–86; genetic, in contemporary discourse, 2–5; genetic, direct-linear-causal model, 8, 10, 53–4, 273; genetic, and medical conditions. See medical genetics; genetic, in public discourse, 2–11. See also genes, in the media and popular culture; hard, definition, 1, 2, 276; and nature-nurture discourse, 15–61; philosophical, 1, 3, 8, 17–22; and social attitudes, 8–13; soft, definition, 1, 277
development; adult, as complex multi-causal process, 103–8; biological, and understanding of human personhood, 64–6; birth order, significance of, 225–6; in C. elegans, 111–20; chicks, Aristotle on, 64; developmental integrated complementary interactionism. See DICI; dispersal of macaques, as development process, 128–30; of Drosophila, 120–3; epigenetics, 64, 74–5; evolutionary developmental biology, 65; fetal, as multi-causal process, 87–97; and heredity. See heredity; human, as a complex multi-causal process, 66–108; human, and DICI, 87–108; human, and direct-to-consumer genetic testing, 7–8; and human identity. See identity, human; and inheritance, as single process, 27; postnatal, as complex multi-causal process, 97–103; preformationism, 64; in voles, 123–8
Developmental Dual-Aspect Monistic Emergentism (DAME), 258, 259–68, 270, 275–6, 289, 302
Developmental Integrated Complementary Interactionism. See DICI
diabetes, 6, 71, 80
dichotomy; dichotomous ideas. See nature-nurture discourse
DICI (Developmental Integrated Complementary Interactionism); and adult development, 103–8; and aggressive behaviour, 272; and animal behaviour, 109–33, 110, 111, 112, 119, 128; and animal choice, possibility of, 132–3; and C. elegans, 111, 111–20, 112, 119; concept of, 87, 268, 302; and democracy, causal, 268–9; and dispersal of mammals, 128–32; and Drosophila, 120–3; and emergentism, 260–8; and environment, 272, 273, 274,

289; and fetal development, 87–97, 89, 93, 95; and free will, 274–8, 277; and human behaviour, 134–187, 135, 139, 143, 147, 149, 157, 160, 186–7; and human development, 87–108, 89, 93, 95, 108; and human personhood, 268, 269, 272, 274, 276–302, 277, 289, 296, 299; and image of God, 289–90, 295, 299; and intelligence, 194–201, 195, 200, 210, 210–2; and multi-causality, 268, 269, 272, 274; and molecular behavioural genetics, 187; and neuronal development, 93, 95; and personality traits, 276–7; and post-natal development, 87–103; and religiosity, 201–06; and stress, 186–7; and voles, 123–8, 128
diet; as environmental factor, 6, 80, 94–5, 139, 151, 155; and genomic modification, 103; sociological explanation, 269–70
differentiation; cells, 76, 112; sexual, 224, 231; undifferentiated cells, 97
dihydrotestosterone, 228
dimorphism, sexual, 226, 228, 230
diploid cells, 165
direct-to-consumer (DTC) genetic testing, 7–8
discordance, in twins, 95, 150–1, 173, 223
disease. See medical genetics
dispersal, 128–32; macaque monkeys, 128–31; in mammals, 128–32; mice, 131; wolves, 131–2
DNA (deoxyribonucleic acid); amino acid sequence, coding of, 66, 67, 68, 117; base pairs, 66, 75, 111, 120, 151, 163, 233; chromosomes, 74, 120, 141, 151, 163–5, 170, 173, 175–6, 184, 223–5, 236, 295; codons, 67, 69, 70; CRISPR-Cas technique, 296–7; in culture, popular, 4, 10–11, 60; cutting, 296; cytosine, 74–6; DNA-binding proteins, 72; double-helical structure, discovery of, 55; duplication, 76, 163, 169; ENCODE project, 4, 67, 72; enhancers, 71–2, 73, 74–5, 78, 81; epigenetic marks, 66, 74–81, 82, 93–8, 102–3, 104–7, 118–9, 121, 144, 151–3, 173, 175, 200, 204, 211, 223, 230, 300; exon, 66–8, 111; genetic code, breaking of, 55; as genetic material, identification of, 55; genome, definition of, 1; genomic sequencing, 3, 60, 72, 82, 83, 111–2, 120, 153, 162–4, 167, 172, 200, 295–6; identification of, as genetic